Fractured Rock Hydrogeology

Selected papers on hydrogeology

20

Series Editor: Dr. Nick S. Robins
Editor-in-Chief IAH Book Series, British Geological Survey, Wallingford, UK

INTERNATIONAL ASSOCIATION OF HYDROGEOLOGISTS

Fractured Rock Hydrogeology

Editor

John M. Sharp, Jr.
The University of Texas, Austin, USA

CRC Press
Taylor & Francis Group
Boca Raton London New York

CRC Press is an imprint of the
Taylor & Francis Group, an **informa** business

A BALKEMA BOOK

Published by:
CRC Press/Balkema
P.O. Box 447, 2300 AK Leiden, The Netherlands
e-mail: Pub.NL@taylorandfrancis.com
www.crcpress.com – www.taylorandfrancis.com

First issued in paperback 2020

ISBN 13: 978-0-367-57614-1 (pbk)
ISBN 13: 978-1-138-00159-6 (hbk)

Except: Chapter 6 'From geological complexity to hydrogeological understanding
using an integrated 3D conceptual modelling approach – insights from the
Cotswolds, UK' © British Geological Survey

Typeset by V Publishing Solutions Pvt Ltd., Chennai, India

Library of Congress Cataloging-in-Publication Data

Fractured rock hydrogeology/editor: John M. Sharp. – Edition 1.
    pages cm – (Selected papers on hydrogeology)
    Includes bibliographical references and index.
    ISBN 978-1-138-00159-6 (hardback: alk. paper) 1. Hydrogeology.
    2. Rocks–Fracture. I. Sharp, John Malcolm, Jr., 1944- II. Series: Hydrogeology
    (International Association of Hydrogeologists)
    GB1001.7.F73 2014
    553.7'9–dc23
                                                                2014006966

Table of contents

Dedication

We dedicate this volume to Jiri Krásný who has been at the forefront of fractured rock hydrogeology and the leader of the hardrock hydrogeology efforts of the International Association of Hydrogeologists (IAH) for nearly three decades. He convened the 2003 Conference in Prague and has been instrumental in inspiring colleagues, particularly the younger scientists, to study fractured rock hydrogeology. He is the founder of the IAH Commission on Hardrock Hydrogeology and collaborates with the working groups throughout Europe. An important change from the original philosophy of the IAH Commission on Hardrock Hydrogeology to exchange knowledge about hardrock (crystalline rock) hydrogeology was, after a long discussions in the Commission then chaired by Jiri, expanded topics that include all fractured rocks, including karstic carbonates. This resulted in larger, more diverse, and innovative scientific community as well as an IAH Green Book (Krásný and Sharp, 2007) with important papers from the 2003 conference. He has also served the IAH in a number of roles, including as its scientific vice president.

Jiri Krásný in the field

Jiri received his PhD from Charles University and worked at the Geological Survey Prague before taking several academic positions at his alma mater. In addition to his services to IAH, he was the principal author of *Groundwater in the Czech Republic* and has received numerous honors, including the Ota Hynie Prize (named after the founder of Czechoslovak hydrogeology) and the IAH's Honorary Membership in 2009.

Jiri Krásný's good nature, friendliness, encouragement, scientific acumen, and dedication to the study of fractured rocks, the hydrogeology of massifs, and hardrock hydogeology in Europe, Africa, and Central America are unparalleled.

Thank you, Jiri Krásný, from the IAH and past, present, and future hardrock hydrogeologists everywhere.

Foreword

Approximately every decade the International Association of Hydrogeologists (IAH) has sponsored or co-sponsored a conference on the hydrogeology of fractured rocks. The first was the 1993 IAH Congress on the Hydrogeology of Hard Rocks, in Norway convened by Erik Rohr-Torp and Jens Olaf-Englund, amongst others. The two-volume set of memoires of this, the 24th Congress, were edited by David and Sheila Banks (1993). The second was the 2003 IAH Conference in Prague, convened by Jiri Krásný, which led to an IAH Selected Papers (Green Book), *Groundwater in Fractured Rocks*, edited by Krásný and Sharp (2007). The third, co-sponsored with the International Association of Hydrological Sciences, the IAH National Chapter of the Czech Republic, Charles University, and the Masaryk Water Research Institute, was the 2012 International Conference on Groundwater in Fractured Rocks, convened in Prague by Zbyněk Hrkal and Karel Kovar (GwFR 2012). This Green Book (IAH Selected Papers), *Fractured Rock Hydrogeology*, is a direct result of that conference.

A selection of papers from the conference were invited for publication based upon the presentations made at the conference and the range of topics covered. Papers selected include both theoretical and practical analyses that use numerical modelling, geochemistry and isotope chemistry, aquifer tests, lab tests, field mapping, geophysics, geologic analyses, and unique combinations of these types of investigation. Studies are included from the consulting community to document the current problems and currently used techniques being used to address those problems. In addition, there are studies from the academic and governmental agency scientific communities.

The papers presented at this 3rd conference emphasized the need for sound geologic data in the analysis of fractured rock systems and the importance of the zone of weathering. There were other papers presented in the conference that are not included in this compendium because of length limitations. See the conference programme (GwFR 2012) for their titles and there were some that would have fit nicely, but were already in press or had been previously submitted elsewhere. Those seeking these researches should search appropriate journals and governmental reports. Many of the papers selected include geological interpretation in analysing the hydrogeology of fractured, mostly crystalline rocks. This is indeed hydrogeology in its broadest sense. Also clear is both the difficulty and the importance of extending or scaling fracture rock hydraulic properties. This remains a great challenge.

The 20 chapters in this volume include studies from Asia, Australia, Europe, the Middle East, and North America. The review chapter by Krásný *et al.* notes the importance of hardrock hydrogeology and future possibilities and presents the history of the IAH Commission on Hardrock Hydrogeology. The lead chapter by Lassachsagne *et al.* offers a conceptual model of weathered hard rock aquifers. Weathered crystalline rocks are often the best source for groundwater supplies in these environments. The authors demonstrate how geological and hydrogeological observations can be combined to assess these aquifer systems. Davies *et al.* use hydrogeological analyses and borehole data to understand aquifers in fractured crystalline rocks and their weathered zones. Chambel's chapter encapsulates the practical aspects of finding and producing groundwater in hard rocks, specifically in semi-arid regions and in the developing world. Baiocchi *et al.* examine sustainable yield in fractured crystalline rocks and also stress the importance of the upper weathered layer in these systems.

Bricker *et al.* develop 3D conceptual hydrogeological models in the United Kingdom to assist in modelling and predictions of hydraulic responses of fractured aquifer systems, particularly at the local catchment scale. Chou *et al.* characterise the spatial distribution of transmissivity of fractured aquifers in Taiwan using borehole logs and packer tests. The shallow weathering zone is here shown to be important. Continuing the consideration of sustainable yields, Gargini *et al.* couple pumping tests and numerical models to ascertain sustainability for crystalline rocks and the overlying weathered zone in Italy. Frampton evaluates fracture transmissivity with natural gradient flow measurements and compares them with those derived from pumping tests. The laboratory studies by Sharp *et al.* demonstrate the impracticality in upscaling fracture properties and show that while fracture roughness may be scalable, fracture permeability can not be adequately determined by point or scanline estimates of aperture. Winkler and Reichl use scale dependent field and laboratory tests to deduce permeability anisotropy in faulted crystalline rocks. Normani *et al.* develop a methodology to generate fracture network models that simulate flow and transport at a study site in the Canadian Shield.

Several chapters deal with springs and tunnel observations. Bauer and Draser evaluate the Barada Spring catchment in Syria using structural geology mapping and geophysics to infer flow directions and zones of high permeability. Mayo and Bruthans use heat flow and ^{14}C data to infer recharge areas for thermal springs in crystalline rocks. Hokr *et al.* use tunnel inflow data, geology, and geophysics to constrain an inverse model that estimates hydraulic conductivity zones in the Bohemian massif.

Frengstad and Banks discuss uranium distribution in fractured crystalline rocks while Gaut and Storro relate groundwater vulnerability to well construction and siting in crystalline rocks. Both of these focus on data from Norway but are applicable elsewhere as many fractured crystalline rocks have high levels of naturally-occurring radioactivity.

The articles by Blandford *et al.*, David *et al.*, and Krusic-Hrustanpasic and Cosme are based upon current consultancy reports for groundwater resource development. Blandford *et al.* use a variety of methods, including sophisticated pump tests, to evaluate sustainability of deep groundwaters in fractured rocks in New Mexico, USA. David *et al.* combine a variety of data sets with a numerical model in fractured metamorphic rocks. Krusic-Hrustanpasic and Cosme evaluate the hydrogeology of a fractured basalt aquifer. The latter two case histories focus on sites in Australia. These three case histories demonstrate the importance of fractured rock hydrogeology.

I thank Uwe Troeger for his help in selecting the invited papers and in preparing the dedication; the many manuscript reviewers for their in reviewing, criticisms, and helpful suggestions; the authors of the chapters for their patience and scientific inquisitiveness; and Jeff Horowitz of The University of Texas for emergency drafting support. All the chapters of *Fractured Rock Hydrogeology* have had several reviews, including at least one internal (another chapter author) and at least one external reviewer. I also thank Nick Robins, the IAH Books Editor-in-Chief, for his assistance and copy editing and JanJaap Blom of Taylor and Francis for his encouragement.

Finally, I can do no better than to cite Gurrieri (2013) – "Read on and take time to listen to the lessons these...systems have to tell us. These authors are pioneers and... [we] thank them for their insight, careful observation, and inspiration."

John M. Sharp, Jr.
Carlton Professor of Geology
Department of Geological Sciences
The University of Texas, Austin, Texas

REFERENCES

Banks, D., and Banks, S. (eds.), 1993, Hydrogeology of Hard Rocks. Memoires of the 24th Congress, International Association of Hydrogeologists, Oslo, Norway, v. 24, 2 volumes.

Gurrieri, J.T., 2013, Foreword: in Groundwater and Ecosystems (Ribeiro, L., Stigter T.Y., Chambel A., Condesso de Melo, M.T., Monteiro J.P., and Medeiros A., eds.), International Association of Hydrogeologists Selected Papers on Hydrogeology, 18.

GwFR 2012, International Conference on Groundwater in Fractured Rocks, Abstracts with conference programme [http://web.natur.cuni.cz/gwfr2012]

Krásný, J., and Sharp, J.M., Jr. (eds.), 2007, Groundwater in Fractured Rocks. International Association of Hydrogeologists Selected Papers on Hydrogeology, 9.

Krásný, J., Sharp, J.M., Jr., and Troeger, U., 2014, IAH Commission on Hardrock Hydrogeology (HyRoC): Past and present activities, future possibilities. This volume.

About the editor

John M. (Jack) Sharp, Jr. is the Carlton Professor of Geology in Department of Geological Sciences at The University of Texas. He has a Bachelor of Geological Engineering from the University of Minnesota and MS and PhD degrees in hydrogeology from the University of Illinois. He has served as the President of the Geological Society of America (GSA) and as an officer in both the American Institute of Hydrology (AIH) and the International Association of Hydrogeologists (IAH). His honours include the GSA's Meinzer, AIH's Theis, IAH's Presidents', and the Association of Engineering Geologists' Publication Awards as well as Phi Kappa Phi and Tau Beta Pi. He is a Fellow of the GSA and the Alexander von Humboldt Stiftung. His research covers flow in fractured and carbonate rocks, thermohaline free convection, sedimentary basin hydrogeology, subsidence and coastal land loss, groundwater management, and the effects of urbanization. His hobbies include gardening, genealogy, fishing, duck hunting, Australia, opera, UT football, and (before bad knees) handball.

List of contributors

AL-JOHAR Mishal M.   Arcadis, Portland, Oregon, USA

BAIOCCHI Antonella   Dipartimento di Scienze Ecologiche e Biologiche, Università degli Studi della Tuscia, Viterbo, Italy

BALVÍN Aleš   Faculty of Mechatronics, Informatics and Interdisciplinary Studies, Technical University of Liberec, Studentska 2, 461 17 Liberec, Czech Republic

BANKS David   Holymoor Consultancy Ltd., 8 Heaton Street, Chesterfield, Derbyshire, S40 3AQ, UK

BARRON A.J.M.   British Geological Survey, Kingsley Dunham Centre, Keyworth, Nottingham, NG12 5GG, UK

BAUER Florian   Berlin Institute of Technology, Department of Applied Geosciences, Berlin, Germany

BLANDFORD T. Neil   Daniel B. Stephens & Associates, Inc., 6020 Academy NE, Suite 100, Albuquerque, New Mexico, USA

BRICKER Stephanie H.   British Geological Survey, Kingsley Dunham Centre, Keyworth, Nottingham, NG12 5GG, UK

BRUTHANS Jiri   Faculty of Science, Charles University in Prague, Albertov 6, 128 43 Praha 2, Czech Republic

BUSHNER Greg L.   Vidler Water Company, 3480 GS Richards Blvd., Suite 101, Carson City, Nevada, USA

CHAMBEL António   Geophysics Centre of Évora, University of Évora, Rua Romão Ramalho 59, 7000-671 Évora, Portugal

CHEN Nai-Chin   Geotechnical Engineering Research Center, Sinotech Engineering Consultants, Inc., 280 Xinhu 2nd Rd., Neihu Dist., Taipei City 11494, Taiwan

CHEN Po-Jui   Geotechnical Engineering Research Center, Sinotech Engineering Consultants, Inc., 280 Xinhu 2nd Rd., Neihu Dist., Taipei City 11494, Taiwan

CHENEY Colin	States of Jersey, Planning and Environment Department, Howard Davis Farm, La Route de la Trinité, Trinity, Jersey JE3 5JP, Channel Islands
CHOU Po-Yi	Geotechnical Engineering Research Center, Sinotech Engineering Consultants, Inc., 280 Xinhu 2nd Rd., Neihu Dist., Taipei City 11494, Taiwan
COSME Frederick	Golder Associates Pty Ltd, Building 7, Botanicca Corporate Park, 570-588 Swan Street, Richmond, Vic 3121, Australia
DAVID Katarina	4/29 Portia Road, Toongabbie, NSW 2146, Australia
DAVID Vladimir	Argent Minerals, Suite 6, Level 6, 50 Clarence Street, NSW 2000, Australia
DAVIES Jeffrey	British Geological Survey, Maclean Building, Wallingford, OX10 8BB, UK
DE NARDO Maria Teresa	Geologic, Seismic and Soil Survey, Emilia Romagna Region, Bologna, Italy
DEWANDEL Benoît	BRGM, Water Division, New Water Resources Unit, 1039 rue de Pinville, 34000 Montpellier, France
DRAGONI Walter	Dipartimento di Scienze della Terra, Università degli Studi di Perugia, Perugia, Italy
DRASER Jens Harold	Berlin Institute of Technology, Department of Applied Geosciences, Berlin, Germany
FRAMPTON Andrew	Department of Physical Geography and Quaternary Geology, Stockholm University, 10691 Stockholm, Sweden
FRENGSTAD Bjorn S.	Geological Survey of Norway, P.O. Box 6315, Sluppen, NO 7491 Trondheim, Norway
GARGINI Alessandro	Biological, Geological and Environmental Sciences Department, Alma Mater Studiorum – University of Bologna, Italy
GAUT Sylvi	Sweco, Professor Brochsgt. 2, 7030 Trondheim, Norway
HOKR Milan	Centre for Nanomaterials, Advanced Technologies, and Innovations, Technical University of Liberec, Studentska 2, 461 17 Liberec, Czech Republic
HSU Shih-Meng	Geotechnical Engineering Research Center, Sinotech Engineering Consultants, Inc., 280 Xinhu 2nd Rd., Neihu Dist., Taipei City 11494, Taiwan
HUANG Chun-Chieh	Geotechnical Engineering Research Center, Sinotech Engineering Consultants, Inc., 280 Xinhu 2nd Rd., Neihu Dist., Taipei City 11494, Taiwan

HUGHES Andrew G.   British Geological Survey, Kingsley Dunham Centre,
                                Keyworth, Nottingham, NG12 5GG, UK

JACKSON Chris   British Geological Survey, Kingsley Dunham Centre,
                                Keyworth, Nottingham, NG12 5GG, UK

KE Chien-Chung   Geotechnical Engineering Research Center,
                                Sinotech Engineering Consultants, Inc.,
                                280 Xinhu 2nd Rd., Neihu Dist., Taipei City 11494,
                                Taiwan

KETCHAM Richard A.   Department of Geological Sciences, Jackson School
                                of Geosciences, The University of Texas,
                                Austin, Texas 78712, USA

KRÁSNÝ Jirí   Faculty of Science, Charles University in Prague,
                                Albertov 6, 128 43 Praha 2, Czech Republic

KRUSIC-HRUSTANPASIC   Golder Associates Pty Ltd., Building 7,
Irena   Botanicca Corporate Park, 570-588 Swan Street,
                                Richmond, Vic 3121, Australia

LACHASSAGNE Patrick   Danone Waters, Evian Volvic World Sources,
                                11 Avenue du Général Dupas,
                                BP 87-74503, Evian-les-Bains Cedex, France

LEE Feng-Mei   Geotechnical Engineering Research Center,
                                Sinotech Engineering Consultants, Inc.,
                                280 Xinhu 2nd Rd., Neihu Dist., Taipei City 11494,
                                Taiwan

LEE Wong-Ru   Geotechnical Engineering Research Center,
                                Sinotech Engineering Consultants, Inc.,
                                280 Xinhu 2nd Rd., Neihu Dist., Taipei City 11494,
                                Taiwan

LIN Jung-Jun   Geotechnical Engineering Research Center,
                                Sinotech Engineering Consultants, Inc.,
                                280 Xinhu 2nd Rd., Neihu Dist., Taipei City 11494,
                                Taiwan

LIU Tingting   Heritage Computing, 12 King Road, Hornsby,
                                NSW 2077, Australia

LO Hung-Chieh   Geotechnical Engineering Research Center,
                                Sinotech Engineering Consultants, Inc.,
                                280 Xinhu 2nd Rd., Neihu Dist., Taipei City 11494,
                                Taiwan

LOTTI Francesca   Dipartimento di Scienze Ecologiche e Biologiche,
                                Università degli Studi della Tuscia, Viterbo, Italy

MARLEY Robert   Daniel B. Stephens & Associates, Inc.,
                                6020 Academy NE, Suite 100, Albuquerque,
                                New Mexico, USA

MAYO Alan. L.   Department of Geosciences, Brigham Young University,
                                Provo, UT 84602, USA

NORMANI Stefano D.	Department of Civil and Environmental Engineering, University of Waterloo, Waterloo, Ontario, Canada
PEACH Denis	British Geological Survey, Kingsley Dunham Centre, Keyworth, Nottingham, NG12 5GG, UK
PICCININI Leonardo	Department of Geosciences, University of Padua, Italy
PISCOPO Vincenzo	Dipartimento di Scienze Ecologiche e Biologiche, Università degli Studi della Tuscia, Viterbo, Italy
RÁLEK Petr	Centre for Nanomaterials, Advanced Technologies, and Innovations, Technical University of Liberec, Studentska 2, 461 17 Liberec, Czech Republic
REICHL Peter	Resources – Institute for Water, Energy and Sustainability, Joanneum Research Forschungsges mbH, Elisabethstrasse 18/II, 8010 Graz, Austria
ROBINS Nick	British Geological Survey, Maclean Building, Wallingford, OX10 8BB, UK
SEGADELLI Stefano	Earth Sciences and Physics Department "Macedonio Melloni", University of Parma, Italy
SHARP John M., Jr.	Department of Geological Sciences, Jackson School of Geosciences, The University of Texas, Austin, Texas 78712, USA
ŠKARYDOVÁ Ilona	Faculty of Mechatronics, Informatics and Interdisciplinary Studies, Technical University of Liberec, Studentska 2, 461 17 Liberec, Czech Republic
SLOTTKE Donald T.	Schlumberger, Inc., Houston, Texas 77077, USA
STORRØ Gaute	Groundwater and Urban Geology Section, Geological Survey of Norway, PO Box 6315, Sluppen, 7491 Trondheim, Norway
SYKES Jonathan F.	Department of Civil and Environmental Engineering, University of Waterloo, Waterloo, Ontario, Canada
TROEGER Uwe	Lehrstuhl für Hydrogeologie, Institut für Angewandte Geowissenschaften, Sekr. BH 3-2 Ernst-Reuter-Platz 1, Technical University of Berlin, D-10587, Berlin, Germany & Dean of Water Engineering, Campus El Gouna, Mohamed Ibrahim Kamel Str., El Gouna, Red Sea, Egypt
UMSTOT Todd	Daniel B. Stephens & Associates, Inc., 6020 Academy NE, Suite 100, Albuquerque, New Mexico, USA

VINCENZI Valentina	Geotema s.r.l., Ferrara, Italy
WEN Hui-Yu	Geotechnical Engineering Research Center, Sinotech Engineering Consultants, Inc., 280 Xinhu 2nd Rd., Neihu Dist., Taipei City 11494, Taiwan
WINKLER Gerfried	Institute for Earth Sciences, University of Graz, Heinrichstrasse 26, 8010 Graz, Austria
WOLF Christopher	Daniel B. Stephens & Associates, Inc., 6020 Academy NE, Suite 100, Albuquerque, New Mexico, USA
WYNS Robert	BRGM, Geology Division, BP 36009, 45060 Orléans Cedex, France
YIN Yong	Department of Civil and Environmental Engineering, University of Waterloo, Waterloo, Ontario, Canada

Chapter 1

IAH Commission on Hardrock Hydrogeology (HyRoC): Past and present activities, future possibilities

Jiří Krásný[1]*, John M. Sharp, Jr.*[2] *& Uwe Troeger*[3]
[1]*Faculty of Science, Charles University in Prague, Czech Republic*
[2]*Department of Geological Sciences, Jackson School of Geosciences, The University of Texas, Texas, USA*
[3]*Lehrstuhl für Hydrogeologie, Institut für Angewandte Geowissenschaften, Technical University of Berlin, Berlin*

ABSTRACT

The IAH Commission on Hardrock Hydrogeology (HyRoC) became a significant feature of the International Association of Hyrogeologists since the early 1990s. It has focussed research on crystalline and consolidated sedimentary rocks. We review the importance of the field (groundwater production, environmental protection, geotechnical issues, and geothermal energy); the hisory of the Commission, and future significant issues.

1.1 INTRODUCTION

Crystalline (igneous and metamorphic) and consolidated sedimentary rocks – so called hard or fractured rocks – extend over much of the world. They crop out at the land surface commonly in large areas – shields and massifs and also in cores of major mountain ranges. They cover more than 20 per cent (approximately 30 million square km^2) of the present land surface.

In addition, these mostly old rocks form the basement of all younger sedimentary rocks that are often concentrated into large basins. Thus, hard rocks represent in the depth a continuous environment enabling extended and deep regional or even continental/global groundwater flow.

During past decades inadequate attention has been paid to groundwater in this specific hydrogeologic environment except in arid and semi-arid regions. There, because of the typical nonavailability of surface water, groundwater has traditionally represented the only source of water. In temperate climatic zones, water managers have paid attention mostly to sedimentary basins that offered water-supply solutions by concentrated groundwater withdrawals, which are economically and technically more profitable than those by scattered small-yield intake objects in hard rocks.

In the last decades of the 20th century and at the beginning of the 21st century, however, increasing interest in hardrock hydrogeology has been noted due to many important theoretical and applied issues concerning groundwater in hard rocks in both tropical and temperate climatic zones that can be summed up as follows:

1.1.1 Groundwater systems

Groundwater systems in hardrock regions in temperate zones, mainly those originating in mountains, have often been proved strong enough to maintain flow of surface water courses in adjacent piedmont zones during dry periods. As natural recharge everywhere determines regional possibilities of long-term groundwater abstraction, studies on groundwater flow conditions and on water balance are indispensable for sustainable integrated water management.

1.1.2 Groundwater abstraction

With respect to groundwater abstraction, under recent trends when water demand increases in many regions, adequately sited water wells or other water intake systems (e.g., horizontal drainage facilities) in hardrock areas typically can cover requirements on water supply for small communities, industries, or farms and for domestic water consumption. In some areas, groundwater abstraction possibilities are high enough to supply small towns where values of transmissivity sufficiently high and can be compared with prevailing values of several other groundwater environments (Fig. 1.1). Siting of water wells in open fracture systems or thick permeable overburden, depth related-changes in rock permeability, and assessment of economically adequate depth of wells are some important issues of applied hardrock hydrogeology. Social and economic considerations of groundwater development help to decide and justify whether scattered local groundwater withdrawals in hardrock areas should be preferred ahead of other water-supply possibilities, such as extended regional groundwater or surface water-supply systems.

1.1.3 Groundwater pollution and environmental protection

Complex issues are connected with groundwater pollution and environmental protection. Impacts of industrialisation, urbanisation, landfills, and deep hazardous waste repositories and of fertilisers and pesticides used in agriculture on soil, rock, and water have been studied and monitored in hardrock environments. Knowledge of complicated groundwater flow and contaminant transport in weathered and fractured zones of hard rocks is decisive for siting both surficial and subsurface waste repositories and assessment of unfavourable environmental impact of a contamination source.

1.1.4 Geotechnical and engineering-geological activities

Clear hydrogeologic understanding and quantitative hydrogeologic assessments are important for geotechnical and engineering-geological activities such as construction of tunnels and underground cavities as well as in mining. Studies of deep repositories for radioactive, toxic and other dangerous wastes, carried out in the recent years and often concentrate on crystalline rocks. These studies have extended our hydrogeologic knowledge of this environment to depths of many hundreds or even thousands meters. On the other hand, all these data represent an important and useful feedback to results provided by other hydrogeologic methodologies and techniques.

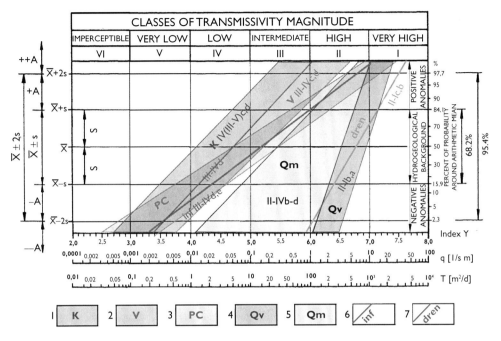

Figure 1.1 Distribution of transmissivity values of selected hydrogeological environments in the Czech Republic expressed as cumulative relative frequencies (after Krásný *et al.* 2012). (*See colour plate section, Plate 1*).

Index Y = index of transmissivity Y = log (10^6 q), q = specific capacity [l/s m], T = coefficient of transmissivity [m^2/d]; $\overline{x}$ = arithmetic mean of statistical samples, s = standard deviation, ++A, +A, −A, −−A = fields of positive and negative anomalies (+A, −A) and extreme anomalies (++A, −−A) outside the interval $\overline{x} \pm$ s of prevailing transmissivity (= hydrogeological background).

Classes of transmissivity magnitude and variation after Krásný (1993a).

Prevailing transmissivity values in different hydrogeological environments expressed as fields or lines of cumulative relative frequencies determined by aquifer tests in hydrogeological boreholes. Represented ranges of transmissivity encompass the greater part of hydrogeological environments in Czech between the lowest values less than 1 m^2/d (field K) and the highest values of more than 1,000 m^2/d (field Qv):

- 1 − near-surface aquifer of most of "hard rocks" (hydrogeological massif − K) except of environments represented in the field V (2);
- 2 − crystalline limestones and other rocks of hydrogeological massifs with relatively higher prevailing transmissivity − V;
- 3 − Permo-Carboniferous basins PC − data from boreholes up to the depths of several tens of meters;
- 4 − most of Quaternary fluvial deposits along the main water courses − Qv (rivers Labe and Morava etc.); adjacent field Qm (5) represents Quaternary deposits along smaller water courses and of accumulations at higher terrace benches;
- 6, 7 − lines reflect possible considerable differences in transmissivity of sandstones of the Bohemian Cretaceous basin as determined in boreholes located at landform elevations and slopes (hydrogeologically groundwater recharge zones − inf) and in valleys (zones of groundwater discharge − dren).

1.1.5   Geothermal studies

Results of recently drilled deep boreholes in crystalline rocks, connected with geothermal studies, have re-opened the issue of deep-seated groundwater flow and brine occurrences and discussion on mineral and thermal water origin that are often connected with hard rocks.

1.2   DATA

Increasing amounts of available data have stimulated efforts to regionalise and generalise results from different hydrogeologic environments. Knowledge of the hierarchy of inhomogeneity elements and of a scale effect influencing spatial distribution of hydraulic properties enables reasonably simplify real natural conditions when defining conceptual and numerical models of groundwater flow and solute transport.

Data on hardrock permeability and transmissivity, groundwater resources and quality, obtained by different methodological approaches in different regions all over the world, offer excellent possibilities of correlative hydrogeologic studies. Thus new important results and conclusions can be drawn in order to understand hydrogeologic properties of fractured rocks both in local and regional scales.

1.3   HYROC HISTORY

Many publications and international meetings have reflected increasing attention paid to groundwater in hard rocks. Probably the first hydrogeological congresses focused explicitly or preferentially on hardrocks issues were those in Porto Alegre, Brasil (Hanke *et al.* 1975) and in Tucson, USA (Neuman *et al.* 1985). The first real benchmark in hardrock hydrogeology was 24th Congress of the International Association of Hydrogeologists held in 1993 in Oslo (Banks & Banks, 1993). Afterwards several other professional meetings covering the same topic were the IAH International Conferences held in Prague in 2003 [Krásný *et al.* eds (2003), Krásný & Sharp, eds (2007)] and in 2012 (this SP volume) and the 37th IAH International Congress in Hyderabad (2009) in India. Also IAH Congresses held in Cape Town, South Africa (2000), Munich, Germany (2001), Mar del Plata, Argentina (2002), Zacatecas, México (2004), and Lisbon (2007) where special symposia on thermal and mineral waters in hardrock terrains was organised (Marques *et al.* 2007)] and many other international conferences contributed significantly to hardrock hydrogeology.

Being aware of the importance of groundwater in the hardrock environment, Jens-Olaf Englund (Norway, Fig. 1.2), who unfortunately passed away in 1996, and Jiří Krásný (Czech Republic) discussed in 1993 during the Oslo Congress possibility of establishing the IAH **Commission on Hardrock Hydrogeology.** In the next years the Commission stimulated international co-operation and exchange of information between hardrock hydrogeologists. Specialists from about 50 countries all over the world expressed their willingness to support Commission activities. Among the most active professionals could be mentioned Giovanni Barrocu, Tatiana Bocheńska, Jiří Bruthans, Antonio Chambel, Zbyněk Hrkal, Gert Knutsson, José Manuel Marques,

Figure 1.2 Jens-Olaf Englund (Norway) at the field-trip during the Oslo IAH International Congress in 1993.

Henryk Marszałek, S.N. Rai, Luís Ribeiro, Erik Rohr-Torp, Esa Rönkä, Ahmed Shakeel, Jack Sharp, Stanisław Staśko, Georgios Stournaras, Uwe Troeger, Fermin Villaroya, Stefan Wohnlich, and Javier Yélamos.

In Europe, four Regional Working Groups (RWG) of the Bohemian massif, Fennoscandinavia, Iberia and Middle and East Mediterranian (MEM) convened workshops on different topics of hardrock hydrogeology starting in 1994. Until 2008, twelve workshops had been organised, all with published proceedings: 1994 – Rohanov, Czechia (Krásný & Mls, 1996), 1996 – Borowice, Poland (Bocheńska & Staśko, 1997), 1997 – Miraflores, Spain (Yélamos & Villarroya, 1997), 1998 – Äspö, Sweden (Knutsson, 2000), 1998 – Windischeschenbach, Germany (Annau *et al.*, 1998), 2001 – Oslo, Norway (Rohr-Torp & Roberts, 2002), 2002 – Tinos, Greece (Stournaras, 2003), 2004 – Helsinki, Finland (Rönkä *et al.*, 2005), 2005 – Evora, Portugal (Chambel, 2006), 2005 – Athens, Greece (Stournaras *et al.*, 2005), 2006 – Jugowice (Marszałek & Chudy, 2007), 2008 – Athens (Migiros *et al.*, 2008).

In 2009, during the meeting of the IAH Commission on Hardrock Hydrogeology in Hyderabad, Jiří Krásný resigned from the chairmanship of the Commission and was changed in this position by Georgios Stournaras from Greece. For several reasons, there was a retardation in HyRoC activities followed in the next several years.

A revival started during the International Conference on Hardrock Hydrogeology, convened to Prague in 2012 by Zbyněk Hrkal and Karel Kovar. At this conference, Uwe Troeger chaired a meeting of HyRoC that gathered participants from different hardrock regions all over the world willing to support next Commmission activities.

1.4   FUTURE POSSIBILITIES

Hard rocks (hydrogeologic massifs and platforms) are an important hydrogeologic environment (even those sometimes of smaller extension, such as karst and volcanic complexes) that differs from hydrogeologic basins in changes of vertical distribution in permeability/transmissivity/storativity (Figs. 1.1 and 1.3), groundwater flow, natural and artifical resources (Fig. 1.4), and chemical and physical changes of groundwater properties (Fig. 1.5). Hard rock porosity is generally dominated by fracture porosity; porosity in hydrogeologic basins is intergranular or double porosity.

Another important difference in extension of the study area (upscaling) where local results differ considerably from regional approches: laboratory tests, results of well loggings, data from aquifer tests to regional extensions. This can lead to extraordinary differences of hydraulic parameters by five or more orders of magnitude. This

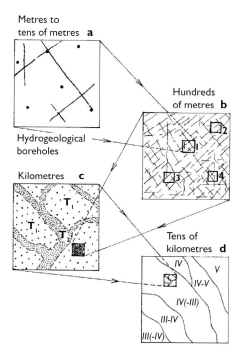

Figure 1.3 Relation of size of inhomogeneity elements in a hydrogeological massif to the extension of a study area (after Krásný & Sharp 2007).

     a – fractures and fracture zones in a local scale as determined by drilling (can be also seen at outcrops of rocks or determined by well logging);

     b – sub-regional more or less regular fissuring representing hydrogeological background; squares 1–4 represent different statistical samples characteristic of usually similar mean transmissivity magnitude and variation;

     c – sub-regional inhomogeneities often following valleys, mostly with water courses: T_1 – lower prevailing transmissivity, T_2 – higher prevailing transmissivity;

     d – regional changes in transmissivity caused by different neo-tectonic activity; roman numbers express the class of transmissivity magnitude after Krásný (1993).

is the result of laboratory to regional treatment as proved by so called scale effect and Representative Elementary Volume (REV). Fractured rock environment is an intricate hierarchic system consisting of inhomogeneties on local to regional scales as represented in Fig. 1.3. Also these features can change with time. Permeability of fractures and of fault zones may increase in discharge areas, all this may decrease in recharge

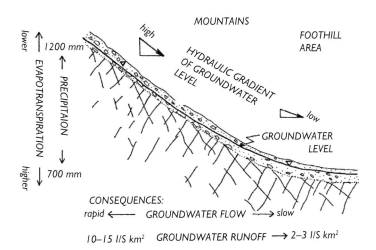

Figure 1.4 Groundwater flow and natural resources formation in the near-surface aquifer of mountainous hardrock terrains under temperate climatic conditions (after Krásný 1996).

Depth of zones can reach up to		Hydrodynamical zone (Groundwater flow)	Hydrochemical zone - main groundwater constituents	TDS	General increase with the depth in
in basins	in massifs				
hundreds of m	several tens of m	**upper / local** (intensive, shallow)	**Ca(-Mg) -HCO$_3$(-SO$_4$)**	0.0x – 0.x g/l	temperature and
few thousands of m	hundreds of m	**middle / regional** (intermediate)	**Na-HCO$_3$(-SO$_4$)**	up to several g/l	gas content ↓ ↓
thousands of m	↑ many thousands of m	**lower / retarded** (slow, deep, negligible up to stagnant)	**Na-Cl** ↓ ↓ ↓	up to several hundreds g/l	↓ ↓ ↓ ↓
global flow often insignificant	↓ ↓	**global** (continental, planetary)	**Na(-Ca)-Cl** ↓		↓ ↓

Figure 1.5 Global scheme combining vertical hydrodynamic and hydrochemical zonality (after Krásný & Sharp, 2007, slightly modified).

zones because of processes as clogging, hydrothermal alteration, mineral precipitation and mechanical clogging. These features are described in detail in Krásný & Sharp (2007).

Hard rocks (hydrogeologic massif, platform) in contrast to hydrogeologic basins as another important hydrogeologic environment (even those sometimes of smaller extent are karst and volcanic complexes) are different mostly in changes of vertical distribution in permeability, transmissivity, and storativity (Figs. 1.1 and 1.3), groundwater flow (Fig. 1.4), natural and artifical resources, and chemical and physical changes of groundwater properties (Fig. 1.5), etc. Most present-day hydrogeologic studies are performed in detailed extensions that are considerably different than regional hydrogeologic studies. Based on these results, hardrock environment can be generally divided into three vertical zones:

a. *Upper or weathered zone* formed by regolith, colluvium, talus, etc. often juxtaposed with mostly Quaternary alluvial, fluvial, glacial, or lacustrine deposits. The usual thickness is several meters but under special conditions may be much thicker.
b. *Middle or fractured zone* usually represented by fractured bedrock to depths of some tens or hundreds of meters. Fracture openings depend mostly on exogeneous geologic processes so that permeability in this zone generally decreases with depth.
c. *Deep or massive zone* in massive bedrock where fractures, faults, or fracture-fault zones are relatively scarce and fracture apertures are commonly less than in the middle zone. Deep fractures may sometimes act as isolated, more or less individual hydraulic bodies. In a regional scale these inhomogeneities may form interconnected networks enabling extended and deep, regional and continental/global groundwater flow reaching depths of many hundreds or even thousands of meters (Figs. 1.6 and 1.7). Under suitable structural conditions or in a specific hard rocks (granites, quarzites, etc.), mineral and thermal waters ascend there along deep faults.

Vertical position of boundaries of the three zones depends on natural conditions, but human effects can also be important. There is a possibility for comparing hydrogeological conditions in fractured rocks globally and differentiating/comparing hydrogeologic backgrounds with particular extreme features. In addition to geological-hydrogeological conditions, important changes include the global climatic zones, toopgraphy, and other regionally prevailing features. Even the comparison of particular areas with distinct climatic and other features (e.g., high mountains, areas of extreme climatic features, zones along the banks of rivers and seas, etc.) would be important. These could offer a regional review of global hydrogeologic conditions and their changes as hard rocks are important and widespread even below the extended hydrogeologic basins.

For these reasons,we suggest that specific attention for HyRoC could focus on the stable group of members who represent all the areas where hard rocks occur (South and North America, Africa, Asia, and Australia in addition to previous several European Working groups) to provide mutual cooperation, congresses, conferences and workshops, and publications.

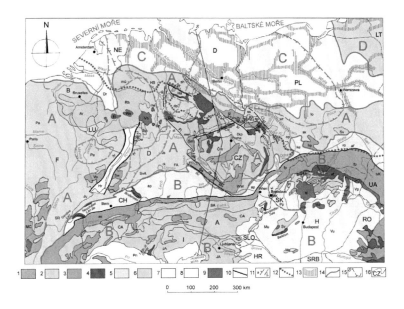

Figure 1.6 Geological and hydrogeological position of the Czech Republic in the Central Europe (after Krásný *et al.* 2012, simplified). (*See colour plate section, Plate 2*).

Defined hydrogeological megaprovinces: A – Megaprovince of European Variscan units and their platform cover; B – Alpine-Carpathian Megaprovince; C – Megaprovince of Central European Lowland; D – Megaprovince of East-European and North-European (Fennosarmatian) Platform.

1–9 – Main types of hydrogeological environment:

1–3 – *hydrogeological massif:*
    1 – igneous rocks of different age except for young Tertiary and Quaternary volcanic rocks,
    2 – metamorphic rocks and intensively folded preorogenic sediments, mosty of Precambrian and Paleozoic age,
    3 – Mesozoic and Tertiary intensively folded, usually non-carbonate sediments of Alpides (e.g. Outer Carpathians Flysch belt that mostly has a character of hydrogeological massif).

4–8 – *hydrogeological basins:*
    4 – mostly less intensively folded post-orogenic sediments of different age: Upper Paleozoic of the Variscid platform cover, Intracarpathian Paleogene;
    5 – Mesozoic and Tertiary unmetamorphosed and slightly metamorphosed sediments of Variscid platform cover,
    6 – Mesozoic carbonate rocks of Alpides, often forming hydrogeologic basins;
    7 – Neogene sediments of Alpine and Carpathian Foredeeps („Molasse") and intramountain basins of Alpides;
    8 – Quaternary sediments of the Middleeuropean Lowland, of the Upper-Rhein graben and of the Po-Lowland with their Tertiary and Mesozoic basements;

    9 – *young volcanic rocks;*
10 – *important structural elements, faults and fault zones;*
11 – *regions with frequent occurrences of salt deposits and domes* in hydrogeological megaprovinces A and C: a) mostly of Permian age, b) mostly of Triassic, partly also of Jurassic age;
12 – *southernmost limit of the Pleistocene continental ice-sheet;*
13 – *approximate extension of the main overdeepened valleys of Quaternary glaciofluvial origin* („Urstromtäller") in the Middle-Europian Lowland, partially extended to the south into the Megaprovince of European Variscan units;
14 – *boundaries of hydrogeological megaprovinces;*
15 – *line of the section represented in Fig. 1.7.*
16 – *state boundaries and symbols of states.*

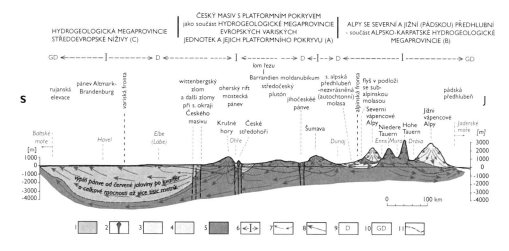

Figure 1.7 Schematic geological-hydrogeological section through the Central Europe between the Baltic sea and the Adriatic sea (after Krásný *et al.* 2012). Section is twenty times exaggerated, the line of the section is represented in Fig. 1.6. (*See colour plate section, Plate 3*).

The section in the north-south direction represents the main geological units and defined hydrogeologic megaprovinces A–C, character of hydrogeological environments, main features of vertical hydrodynamic and hydrochemical zonalities and position of the Bohemian massif in the framework of regional and global groundwater flow between the Baltic and Adriatic seas with the most important zones of groundwater recharge and discharge. Some units are projected into the section. Relatively small thickness of shallow groundwaters with small TDS contents compared with deeper saline groundwaters (brines) can be observed.

A–C: Hydrogeological megaprovinces: A – Megaprovince of European Variscan units and their platform cover, B – Alpine-Carpathian Megaprovince, C – Megaprovince of Central European Lowland.

1 – Cenozoic, Mesozoic and Upper Paleozoic deposits of Variscan platform cover and of other platform units occurring more to the north,
2 – Tertiary and Quaternary volcanic rocks at the Ohře (Eger) rift,
3 – sediments of Alpine Foredeep (Molasse),
4 – Alpine carbonate rocks,
5 – igneous, metamorphic and diagenetically lithificated and/or intensively folded Variscan rocks, rocks of other platform units, of Alpine core and of Flysch Belt ("hard rocks")
6 – zones of prevailing groundwater recharge, with only local groundwater discharge,
7 – main inferred directions of local to regional groundwater flow,
8 – main inferred directions of regional to continental/global groundwater flow, somewhere directed out of the line of the section or even with stagnant groundwater,
9 – important zones of groundwater discharge,
10 – zones of discharge of continental groundwater flow (Baltic and Adriatic seas),
11 – inferrred boundary between groundwaters with small TDS contents and deep saline groundwaters (brines).

SELECTED REFERENCES ON HARDROCK HYDROGEOLOGY

Annau, R., Bender, S. &Wohnlich, S. (eds.) (1998) Hardrock Hydrogeology of the Bohemian Massif. Proc. 3rd Internat. Workshop 1998, Windischeschenbach. Münchner Geol. Hefte, B8. München.

Banks, D. & Banks, S. (eds.) (1993) Hydrogeology of hard rocks. Memoires, 24th Congress, Int. Association Hydrogeologists 24. 2 volumes. – Ås, Oslo.

Bocheńska, T. & Staśko, S. (eds.) (1997) Hydrogeology, 2nd Workshop on hardrock hydrogeology of the Bohemian Massif. – Acta Univ. Wratislav. 2052.

Chambel, A. (ed.) (2006) Proceedings of the 2nd Workshop of the IAH Iberian Regional Working Group on Hard rock hydrogeology, May 18–21, 2005. Évora.

Cook, P.G. (2003) A guide to regional groundwater flow in fractured rock aquifers: CSIRO, Australia.

Hanke, A.K. et al. (1975) Memoires, International Association of Hydrogeologists 11. – Congress of Porto Alegre.

Karrenberg, H. (1981) Hydrogeologie der nichtverkarstungsfähigen Festgesteine. – Springer. Wien, New York.

Knutsson, G. (ed.) (2000) Hardrock hydrogeology of the Fennoscandian shield: Proc. Workshop on Hardrock Hydrogeologie, Äspö, Sweden. May 1998, Nordic Hydrol. Programme Report 45. Stockholm.

Krásný, J. (1993) Classification of transmissivity magnitude and variation: Ground Water 31, 2, 230–236. Dublin, Ohio.

Krásný, J. & Mls, J. (eds.) (1996) Proceedings of the First Workshop on "Hardrock hydrogeology of the Bohemian Massif", Oct. 3–5, 1994. – Acta Univ. Carol., Geol. 40, 2, 81–292.

Krásný, J., Hrkal, Z. & Bruthans, J. (eds.) (2003) Proceedings – IAH International Conference on "Groundwater in fractured rocks", Sept. 15–19, 2003. Prague. UNESCO IHP-VI, Series on Groundwater 7, Prague, 15–19 September 2003.

Krásný, J. & Sharp, J.M. (eds.) (2007) Groundwater in fractured rocks. – IAH Selected Papers 9, Taylor and Francis.

Krásný, J., Císlerová, M., Čurda S., Datel, J., Dvořák J., Grmela A., Hrkal Z., Kříž, H., Marszałek, H., Šantrůček J. & Šilar J. (2012): Podzemní vody České republiky: Regionální hydrogeologie prostých a minerálních vod. (In Czech: Groundwater in the Czech Republic. Regional hydrogeology of groundwaters and mineral waters with English summary). – Česká geologická služba. Praha.

Larsson, I. et al. (1987): Les eaux souterraines des roches dures du socle. – UNESCO Études et rapports d'hydrologie 33.

Lloyd, J.M. et al. (1999): Water resources of hard rock aquifers in arid and semi-arid zones. – UNESCO Studies and Reports in Hydrology 58.

Marszałek, H. & Chudy, K. (eds.) (2007): Selected hydrogeologic problems of the Bohemian Massif and of other hard rock terrains in Europe. – Acta Univ. wratislav. 3041, Hydrogeologia.

Marques, J.M., Chambel, A. & Ribeiro, L., (eds.) (2007): Proceedings of the Symposium of thermal and mineral waters in hard rock terrains. – Lisbon.

Migiros, G., Stamatis, G., & Stournaras, G., (eds.) (2008): 8th Hellenic Hydrogeological Conference – 3rd MEM Workshop on Fissured rocks hydrology. Athens.

Neuman, S.P. et al. (eds.) (1985): Hydrogeology of rocks of low permeability. – Mem. IAH 17 (2 Volumes). – Tucson.

Robins, N.S. & Misstear, B.D.R. (eds.) (2000): Groundwater in the Celtic regions: Studies in Hard Rock and Quaternary Hydrogeology. – Geol. Society, Spec. Publication 182. London.

Rohr-Torp, E & Roberts, D. (eds.) (2002): Hardrock hydrogeology. – Proc. Nordic Workshop Oslo 2001. Trondheim.

Rönkä, E., Niini, H. & Suokko, T. (eds.) (2005) Proceedings of the Fennoscandian 3rd Regional Workshop on hardrock Hydrogeology, Finnish Environment Institute. Helsinki.

Shakeel, A. Jayakumar, R. & Abdin, S. (2007): Groundwater dynamics in hard rock aquifers. – Capital Publishng Co. New Delhi, Kolkata, Bangalore.

Singhal, B.B.S. & Gupta, R.P. (1999): Applied hydrogeology of fractured rocks. – Kluwer Academic Publishers. Dordrecht.

Stober I. & Bucher K. (eds.) (2000): Hydrogeology of crystalline rocks. – Kluwer Acad. Publishers. Dordrecht.

Stournaras G. (ed.) (2003) Proc. 1st Workshop on Fissured rocks hydrogeology. – Tinos Island 2002.

Stournaras G., Pavlopoulos K. & Bellos, T. (eds.) (2005): 7th Hellenic Hydrogeological Conference – 2nd MEM Workshop on Fissured rocks hydrology. Athens.

Yélamos J.G. & Villarroya F. (eds.) (1997): Hydrogeology of hard rocks. Some experiences from Iberian Peninsula and Bohemian Massif. – Proc. Workshop of Iberian Subgroup on Hard Rock Hydrogeology. – Miraflores de la Sierra, Madrid.

Chapter 2

The conceptual model of weathered hard rock aquifers and its practical applications

*Patrick Lachassagne[1], Benoît Dewandel[2] &
Robert Wyns[3]*

[1]*Danone Waters, Evian Volvic World Sources, Evian-les-Bains Cedex, France*
[2]*BRGM, Water Division, New Water Resources Unit, Montpellier, France*
[3]*BRGM, Geology Division, Orléans Cedex, France*

ABSTRACT

Most hard rocks are or were exposed to deep weathering processes, as in large shields of Africa, India, North and South America, Australia and Europe. It turns out that the hydraulic conductivity of hard rocks is inherited from these weathering processes, within their Stratiform Fissured Layer located immediately below the unconsolidated weathered layer (saprolite) and, to a much lesser extent, within the vertical fissured layer at the periphery of or within pre-existing geological discontinuities (e.g. joints, dykes or veins). This concept unifies the geological and hydrogeological observations and data about hard rocks. Its recognition opens up large perspectives in terms of applied hydrogeology and geology: mapping hydrogeological potential, water well siting, quantitative management and modelling of the groundwater resource, computing the drainage discharge of tunnels and even quarrying in hard rocks.

2.1 INTRODUCTION

2.1.1 Hydrogeological definition of hard rocks

Hard rocks, or crystalline rocks, are plutonic and metamorphic rocks (Michel & Fairbridge, 1992). These rocks are originally of very low primary porosity, thus of low hydraulic conductivity (granites *s.l.*), or have lost their original hydrodynamic characteristics because of metamorphic processes (metamorphic rocks). Marble is excluded because it can become a karst aquifer. Limestones and non-metamorphic volcanic rocks (see Charlier *et al.*, 2011), even if they are often mechanically hard, particularly to drill, are not considered as hard rocks as their hydrogeological properties are very different from those of other hard rocks. Some authors also consider as hard rocks, or fractured rocks, some of the very hard sedimentary rocks, like, for example, the quartzites from the Table Mountain Group in South Africa (see Roets *et al.*, 2008). These have lost their primary porosity during geological time due to diagenetic processes, and the main permeability results from fractures that are not related to the weathering processes. The origin of fracturing, and thus the origin of the permeability of such rocks, is often not evaluated. It is of interest to address this issue.

Hard rocks make up the basement of the continents and outcrop over large areas throughout the world, mostly in tectonically stable regions such as ancient cratons.

They constitute about 20% of the Earth's surface; namely a large part of Africa, South and North America, India, and Australia. It has been shown (Lachassagne *et al.*, 2011) that, even if plutonic (various kind of granites) and metamorphic rocks (meta-volcanites, -sediments and -granites resulting in schists, micaschists, quartzites and gneisses) are different one from another particularly in terms of mineralogy, petrofabrics or texture, they exhibit similar hydrogeological properties (*e.g.* Dewandel *et al.*, 2006). These hydrodynamic similarities are related to the weathering processes.

2.1.2   Importance of Hard Rock Aquifers in the world

The groundwater resource in hard rock aquifers is small in terms of sustainable discharge per productive well, from several 100 l/h to a few tens of m³/h (Courtois *et al.*, 2009) compared to that from porous, karstic and (recent) volcanic aquifers. Moreover, dry wells are common in hard rock aquifers. Hard rock aquifers are, however, well-suited for water supply to scattered rural populations and small-to-medium-size cities and periurban areas. The groundwater resource contributes to the well-being of the population and to economic development, especially in areas exposed to arid and semi-arid climatic conditions where the surface water resource is limited in time during the year and in space. Africa, South America, Australia rely on this groundwater resource. In India, hard rock aquifers contributed to the 'Green Revolution' that promoted food self sufficiency in the country, but not without any drawbacks such as overexploitation and groundwater quality degradation (Maréchal & Ahmed, 2003; Dewandel *et al.*, 2007, 2010; Perrin *et al.*, 2012).

2.2   STRUCTURE AND HYDRODYNAMIC PROPERTIES OF HARD ROCK AQUIFERS

2.2.1   The fissure/fracture permeability of hard rock aquifers

Hard rock aquifers intrinsically present a very low (below 10^{-8} m/s) matrix hydraulic conductivity. These rock masses are good candidates for nuclear waste storage at least at depth (several hundreds of meters). Their hydraulic conductivity exclusively relies on secondary fissure/fracture permeability. A well providing a significant yield from hard rocks, i.e. a few m³/h (Maréchal *et al.*, 2004; Dewandel *et al.*, 2006; Dewandel *et al.*, 2011), crosscuts the low permeability matrix alternating from a few millimeters to decimeter thick permeable fractures/fissures. Thus, a pumping well must tap this secondary fracture/fissure permeability.

2.2.2   Origin of hard rock aquifer hydrodynamic properties – previous concepts

The groundwater resource in hard rocks was mainly discovered in the 1960s in tropical countries (mostly Africa) by hydrogeologists. The development of a new drilling method, the down-hole-hammer, very suitable for hard rocks, helped to reveal their permeability, their exploitability by small diameter (a few decimeters) deep (a few tens of meters) wells and thus made it possible to describe their geological structure. Before the development of that drilling technique, only the superficial meter to decameter thick unconsolidated (weathered) fringe was exploited, by large diameter wells (Lelong & Lemoine, 1968). Since that period, many deep wells have been drilled in hard rock aquifers throughout the world, particularly in the Sahel in Africa, with the main aim of providing water, and in some cases safer water, for drinking (Lenck, 1977; Wright & Burgess, 1992) and particularly in India, mostly for irrigation (Foster, 2012).

The work revealed the existence of a permeable fissured/fractured zone located in the first meters or few tens of meters of the unweathered hard rock underlying the unconsolidated weathered layers – regolith or saprolite – (Acworth, 1987; Wright & Burgess, 1992; Taylor & Howard, 1999, 2000). In isotropic granites, this permeable zone is composed from (sub)horizontal joints, or sheet fractures, with a depth-related increasing spacing (Jahns, 1943; Twidale, 1982; Ollier, 1988; Wright & Burgess, 1992; Chilton & Foster, 1995; Hill *et al.*, 1995; Shaw, 1997; Taylor & Eggleton, 2001; Mabee *et al.*, 2002; Mandl, 2005; Lachassagne *et al.*, 2011) (Figure 2.1). This permeable horizon is the fissured layer or the stratiform fissured layer (Lachassagne *et al.*, 2011). Descriptions of this fissured layer are mostly provided by hydrogeologists because of its remarkable hydrodynamic properties (cf. all the previous references and also Hsieh and Shapiro, 1996). The shallow unconsolidated weathered profile was described in details by Acworth (1987), Mac Farlane (1992) and Tardy & Roquin (1998).

Tectonics, unloading processes, and emplacement and cooling of plutonic rocks have been the three most common genetic hypotheses formulated to explain the development of such fissures/fractures, but without scientific demonstration of the proesses; for unloading processes see Farmin (1937), Twidale (1982), Houston & Lewis (1988), Holzhausen (1989), Wright & Burgess (1992), Chilton & Foster (1995), Taylor & Howard (1999, 2000), Mabee *et al.* (2002), and for tectonic processes: Wright & Burgess (1992), Key (1992), Razack & Lasm (2006), Neves & Morales (2007).

Jahns (1943) demonstrated that the joints of plutonic rocks have no genetic relationships with the emplacement and cooling of these rocks.

More recently, Lachassagne *et al.* (2011), synthesized the previous work to show that:

i   The fissured layer cannot be explained by natural unloading processes. With the exception of some specific conditions such as: artificial unloading – deep wells, mine galleries or deep sedimentary basins, there is no existing natural phenomenon able to induce a strong compression parallel to the surface. This is the only strain state, with a rapid (a question of hours or days) decrease of the vertical stress component, that may produce sub-horizontal fractures. In addition, gravity structures can only be invoked in very specific contexts such as high reliefs, for instance in Alpine valleys. Such processes do not exist in the near subsurface. Moreover, such a process, if it exists, would also be observed in other types of rocks than hard rocks.

ii With the exception of tectonically active areas the fissured layer and their associated hydrodynamic properties cannot be explained by tectonic fracturing. Tectonically active areas are mostly plate boundaries i.e. actively spreading ridges and transform faults, rifts, subduction zones, and convergent or collisional plate

Figure 2.1 The fissured layer in granites, Lozère, France, outcropping (top: near the village of Serverette; the ~5 m high phone pile gives the scale) and in a former quarry (bottom: near the village of Rieutord de Randon).

boundaries. They only occupy a small area of the earth's surface. The concentration of strain makes them anomalous with regard to the average properties of the crust. The tectonic fracturing theory faces several inconsistencies. In tectonically stable areas such as most hard rock areas in the world it requires:

- a tectonic process to create the fracture. It is clear that the occurrence of such fractures is very rare both in time and space;
- that the resulting fracture be permeable. It is obvious that a tectonic fracture is a complex structure that is far from being systematically permeable (see Caine *et al.*, 1996);
- that the resulting tectonic fracture reaches the subsurface, i.e. within the depth of a hard rock water well;
- that a permeable fracture is not fast sealed at the geological timescale, when in fact rejuvenation is overbalanced by sealing. Outcropping tectonic fractures are very often old and thus sealed;
- the ubiquity of such fractures. Tectonic fractures are spatially unevenly distributed and thus cannot account for the tens of millions of evenly distributed productive water wells in HR areas.

In addition other arguments can be put forward: the absence of thermal springs in such tectonically stable areas whereas the main active faults quite systematically exhibit several of these thermal springs; the fact that most tectonic fractures have a strong dip, are difficult to tap by vertical drilling, as opposed to the subhorizontal jointing of granitoids and the variously dipping fractures of the other hard rock fissured layers which are systematically penetrated by vertical wells.

2.2.3 Origin of hard rock aquifer hydrodynamic properties – fracture permeability as a consequence of weathering processes

The origin of hard rock aquifer fissured layer permeability has been demonstrated from observations performed around the world: France (Lachassagne *et al.*, 2001b; Wyns *et al.*, 2004; Roques *et al.*, 2012), South Korea (Cho *et al.*, 2003), Burkina Faso (Courtois *et al.*, 2009), India (Dewandel *et al.*, 2007, 2010, 2012; Maréchal *et al.*, 2004, 2006), and also from several other observations by the authors in various regions of France (Brittany, Massif Central, Vosges, Corsica), other African countries and French Guiana, mostly reported within BRGM reports, see www.brgm.fr. The basic concept is that the stratiform fissured layer belongs to the weathering profile as well as the overlying unconsolidated layers (saprolite) (Figure 2.2).

The development of a weathering profile thick enough to be hydrogeologically effective is first of all a matter of time: millions to tens of millions years (My) are required to create a few tens of meters thick weathering profile (Wyns, 2002; Wyns *et al.*, 2003). In addition, such a development requires the emersion of the rocks from the sea, and liquid water; hot or cold deserts are not favorable to weathering, nor areas with permafrost (Acworth, 1987; Wyns *et al.*, 2003). The weathering profiles are of the lateritic type, characterised by leaching, with the evacuation of dissolved chemical elements from the rock. These conditions require a favorable geodynamical environment; regional uplift

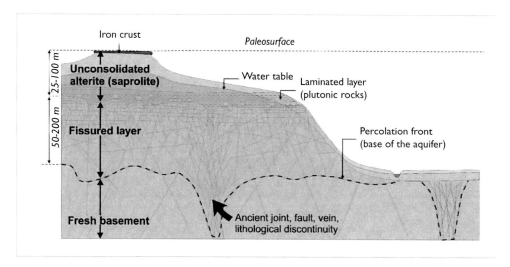

Figure 2.2 Conceptual model of a partly eroded paleo-weathering profile on hard rocks. As stated below (see also Figures 2.4, 2.5 and 2.6), the thickness of the various horizons constituting the weathering profile can be much lower, as a result of erosion, of the shorter duration of the weathering process. (*See colour plate section, Plate 4*).

associated with a long wavelength lithospheric deformation leads to the formation of peneplains (Wyns, 2002; Wyns *et al.*, 2003) with gentle slopes. More precisely it requires a topographical surface verging on the equilibrium, i.e. with an erosion rate not exceeding the weathering rate (Wyns *et al.*, 2003), which allows the weathering profile to thicken. The temperature, linked to the climatic conditions, is a secondary order factor acting only on the kinetics of the weathering process, with a factor of about 1.7 between a tropical (28°C) and a temperate climate (15°C) (Oliva *et al.*, 2003). Moreover, the characteristic period of the climate, 10^2 to 10^3 years, cannot explain the functioning of the weathering profiles which is much longer (Wyns *et al.*, 2003).

When the geodynamic context changes, i.e. passing from an absence of tectonic activity to an active one, the weathering profile can be partially or totally eroded. Conditions again favorable for the development of a new weathering profile lead to polyphased structures with superimposed and/or totally or partly eroded weathering profiles (Dewandel *et al.*, 2006). Thus most HR areas of the world exhibit relics of one or several generations of weathering profiles (Jahns, 1943; Chilton & Smith-Carington, 1984; Theveniaut & Freyssinet, 1999, 2002; Thiry *et al.*, 2005; Dewandel *et al.*, 2006). In Europe, since the end of the Primary era, during which at least Carboniferous, infra Permian, infra Triassic and infra Liassic weathering profiles are recognised, the Early Cretaceous, and the Early and Middle Eocene, having lasted 45 and 25 My, respectively, have constituted the most recent periods during which such processes lasted long enough to develop 70 to 100 m thick weathering profiles of 50 to 70 m thick fissured layer underlying a 20 to 30 m layer of unconsolidated saprolite (Wyns *et al.*, 2003).

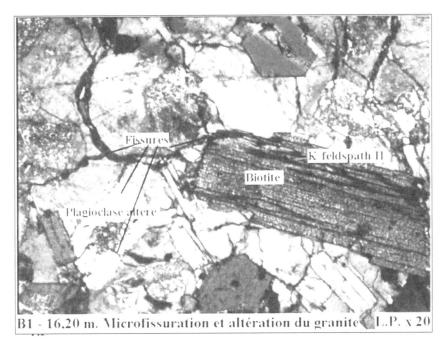

B1 - 16.20 m. Microfissuration et altération du granite L.P. x 20

Figure 2.3 Development of microcracks around a biotite crystal in the fissured layer of a weathering profile in granite, Brittany, France. The biotite cleavages are bent along two spindle-shaped zones of chloritisation, illustrating the swelling of biotite during weathering.

2.2.3.1 *The stratiform weathered-fissured layer*

The result of the weathering process is a stratiform weathered layer parallel to the paleotopography contemporaneous with the weathering process (Figure 2.2). The development of its fracture network is the response to the stress induced by the swelling of some minerals, particularly biotite (Figure 2.3), which is progressively hydrated and turned into hydro-biotite, then vermiculite, then mixed clay layers. (Hill, 1996; Mac Farlane, 1992; Wyns *et al.*, 2003).

The weathering of biotite is an early weathering process particularly in granitic type formations (Tieh *et al.*, 1980). The increase in volume of the weathered mineral can be as much as 30% (Banfield & Eggleton, 1990) and that of the total rock 50% (Folk & Patton, 1982). In granular rocks like granites with a quasi-random orientation of swelling minerals, the potential expansion tensor is isotropic, but expansion is more difficult in the horizontal plane as the medium is infinite in this direction. Consequently, the horizontal stress component accumulates during weathering. In the vertical axis however, the stress increases until the lithostatic component is offset, then allows vertical expansion, while horizontal stress continues to increase. Consequently, the resulting stress tensor is characterised by a minor vertical component (σ_3), and two major ones (σ_1 and σ_2) that are horizontal. When the stress deviator reaches the elastic limit of the rock, tension cracks appear (Wyns *et al.*, 2004). For granitic rocks (Pollard & Aydin, 1988; Mandl,

2005), the resulting fractures are perpendicular to the minor stress (sub-vertical) and consequently are subhorizontal, parallel to the gentle topography contemporaneous with the weathering, and lead to the formation of the subhorizontal jointing of granites (see section 2.2.2). In foliated and folded rocks, the variability of the orientation of the minerals able to swell as well as the ones of the weaker surfaces of the rock (foliation, schistosity) induce an anarchic fracturing, without any preferential orientation. It is, however, the early-stage of biotite weathering that induces fracturing (Figure 2.3). In fact, the later-stages produce clays which exhibit a tendency to fill newly formed pores around the fractured and weathered minerals (Bisdom *et al.*, 1982). Nevertheless, even if the early weathering of biotites leads to a local clogging of the biotite mineral and its surroundings at a millimeter scale, the density of biotites in the parent rocks is not high enough to clog the aperture of all the fractures of the fissured layer at a meter to decameter scale.

The thickness of the fissured layer is approximately twice that of the saprolite (Dewandel *et al.*, 2006) (Figure 2.2). The density of the fractures decreases with depth towards the base of the weathering profile; thus the decrease of the hydraulic conductivity of a hard rock aquifer is not a consequence of a lower permeability of the fractures or their closure but a consequence of their disappearance in depth (Maréchal *et al.*, 2004; Dewandel *et al.*, 2006). Geological observations make it possible to identify several tens of such fractures within the fissured layer; however, due to various processes, further healing of the fracture, its weathering and channeling within the fracture plane, only a maximum of 4 to 5 of these fractures are permeable enough to be detected as water strikes in a water well. This explains, in turn, the large variability in well discharges even for dry well close to a productive one (Maréchal *et al.*, 2004; Dewandel *et al.*, 2006). Ancient profiles may have been completely sealed particularly as the result of diagenesis due to burial below marine sediments.

In granitoids, the dimension of each subhorizontal fracture constituting the stratiform fissured layer is a few tens of meters (Maréchal *et al.*, 2004). Such horizontal permeable fracture sets are on average ten times more permeable, and more numerous, than the subvertical joint sets. The hydraulic conductivity of these permeable fractures does not show a very large variability (Maréchal *et al.*, 2004, Dewandel *et al.*, 2006). This constitutes another argument for an unique origin: the weathering.

2.2.3.2   Structure and hydrodynamic properties of the stratiform weathered-fissured layer

A typical weathering profile (Figure 2.2) comprises the following layers that have specific hydrodynamic properties. Altogether, where and when saturated with groundwater, these various layers form a composite aquifer. From the top to bottom, the layers are (Dewandel *et al.*, 2006):

–   The iron or bauxitic crust, that can be absent, due to erosion or rehydratation of hematite in a latosol (for iron crusts), or resilicification of gibbsite/boehmite into kaolinite (for bauxitic crusts). Where preserved from later erosion and recharged by heavy rainfall the iron or bauxitic crust can give rise to small perched aquifers with some springs locally that exhibit an epikarst-type function. This is particularly observed in tropical humid regions, for instance in French Guiana, or during the wet season in western Africa.

- The saprolite or alterite, or regolith, a clay-rich material, derived from prolonged in situ decomposition of bedrock, a few tens of metres thick where this layer has not been eroded. The saprolite layer can be divided into two sub-units (Figure 2.2; Wyns *et al.*, 1999): the alloterite and the isalterite. The alloterite is mostly a clayey horizon where, and due to the volume reduction related to mineralogical weathering processes, the structure of the mother rock is lost. In the underlying isalterite, the weathering processes only induce slight or no change in volume and preserve the original rock structure; in most of the cases this layer takes up half to two thirds of the entire saprolite layer. In plutonic rocks, such as granites, the base of the isalterite is frequently laminated, and is called the laminated layer. This layer has a relatively consolidated highly-weathered parent rock with coarse sand-size clasts and a millimeter-scale dense horizontal lamination crosscutting the biggest minerals (e.g., porphyritic feldspars), but still preserving the original structure of the rock. This clayey saprolite layer is generally of quite low hydraulic conductivity, about 10^{-6} m/s. Nevertheless, because of its clayey-sandy composition, in coarse granites, the saprolite layer can reach a high porosity, which depends on the lithology of the parent rock; its bulk porosity is mainly between 5% and 30%. Where this layer is saturated, it mainly constitutes the capacitive function of the composite aquifer. In fine-grained or low quartz content rocks (e.g. schists), this layer is mostly clayey and of very low hydraulic conductivity.
- The fissured layer is characterised by dense fissuring in the first few meters and a decreasing density of the fractures with depth, mostly sub-horizontal in granite-type rocks or in horizontally foliated rocks, anarchic in orientation in metamorphic rocks. In granites, the overlying laminated layer results from an increased fissuring at the top of the stratiform fissured layer. This mainly assumes the transmissive function of the composite aquifer and is drawn from most of the wells drilled in hard-rock areas. However, the covering saprolite layer may have been partially or totally eroded, or may be unsaturated. In these cases, the fissured layer assumes also the capacitive function of the composite aquifer; e.g., in French Brittany 80–90% of the stored groundwater in the composite aquifer is in the fissured layer (Wyns *et al.*, 2004).
- The fresh basement is permeable only locally, where deep fractures are present (see section 2.2.3.3). Even if these fractures are as permeable as the fissures belonging to the stratiform fissured layer, their density both with depth and laterally is much lower. At catchment scale, and for water resources applications, the fresh basement can be considered to be impermeable and of very low storativity (Maréchal *et al.*, 2004).

More complex weathering profiles can result from multiphase weathering and erosion processes. In South Korea, for instance, at least two main phases of weathering have been identified (Cho *et al.*, 2003; Figure 2.4):

- An "ancient" one which deeply affects hard rocks, up to >50 m deep, and results in a typical weathering profile, similar to those observed in other regions of the world (Africa, South America, Europe, etc.).
- A more recent one which only affects the rocks, and also the ancient weathering profile, within their first 5–8 m below the topographic surface. It is later than the

Figure 2.4 Red recent weathering, related to the present day topography, intersecting a much older weathering profile, Namwon area, South Korea.

previous weathering phase(s) as it crosscuts all the horizons of that weathering profile. It is probably quite recent as it clearly follows the actual topographic surface. This recent weathering phase does not develop a thick fissured layer. It is thus of little hydrogeological interest.

In southern India (Figure 2.5), the non-laminated saprolite is very thin (1–3 m) and is almost constant in thickness while it should be thicker in the plateau areas than in the valleys if the weathering had occurred earlier than the erosion that shaped the present topography (Dewandel *et al.*, 2006). The thickness of the laminated layer, 15–20 m, is disproportionate to that of the non-laminated saprolite, while in the classical weathering profile (Figure 2.2) the laminated layer occupies only one third to one half of the entire saprolite layer. The thickness of the fissured layer is small compared to that of the saprolite, with a ratio of ~1 instead of ~2. Moreover, the laminated layer presents preserved fissures, which are usually not observed in the classical model.

The weathered zone appears to be composed of an old, probably Mesozoic, weathering profile (Figure 2.5a), where only a part of the fissured layer has been preserved. An erosion phase, due to regional uplift, caused the erosion of the entire saprolite layer and a part of the fissured layer (Figure 2.5b). Then at least one more recent weathering phase, the latest being probably still active, is responsible for the saprolitisation of this truncated profile and explains the development of 1–3 m of non-laminated saprolite locally capped by an iron crust, and the lamination of a large part of the ancient fissured layer (Figure 2.5c). Consequently,

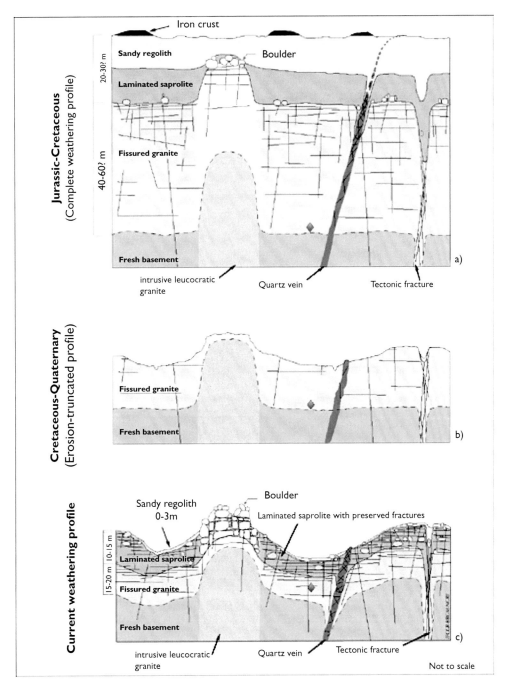

Figure 2.5 Idealised multiphase weathering conceptual model, Maheswaram, Andhra Pradesh, India; vertical scale is deliberately exaggerated. (a) Complete Mesozoic Era (Jurassic-Cretaceous) in age weathering profile, (b) truncated by erosion profile due to Cretaceous to Quaternary uplifts, and (c) current weathering profile. Diamond = benchmark.

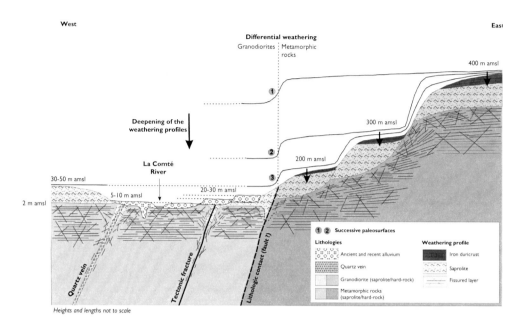

Figure 2.6 Differential and multiphase uplift, erosion and weathering in the Cacao area, French Guiana (after Courtois *et al.* 2003).

the profile structure is controlled by a multiphase weathering process that was induced by the geodynamic history of the Indian Peninsula. This (or these) more recent weathering phase(s) were efficient enough to hydrogeologically rejuvenate the old weathering profile through the development of new permeable weathering induced fractures.

In the Cacao village area, in French Guiana (Figure 2.6) differential weathering between metamorphic rocks and a granodiorite, and also a polyphased weathering, uplift and stripping history explain the complex structure of the weathering profile. It has at least four successive palaeo-surfaces, including the present day fluviatile erosion surface, and four "cut-and-filled" weathering profiles (Courtois *et al.*, 2003).

The weathering profiles or weathering structures not only rely on the present day climate, but also and mainly on ancient ones. This is why, for example, Norway, Scotland, France, Central Europe and Korea all exhibit such weathering profiles. The main deterministic feature of the weathering profiles is not the climate, but the geodynamical context (morphology, erosion rate and duration of the favorable period at the geological timescale), and also, of course, the presence of water. Consequently, present day arid or semi-arid regions, or areas nowadays with permafrost, may exhibit relicts of ancient weathering profiles as seen in the Arabian and Scandinavian shields and in Greenland. In addition, several areas where hard rock outcrop exhibit a significant relief where the ancient weathering profiles, if any, may have been partly or totally eroded. As a consequence, the present day weathering profiles may only have been introduced in the late Tertiary or Quaternary and are thin and much less developed.

Iron caps are nowadays mostly observed in present day tropical countries. However, relicts of old weathering profiles are locally present in Europe (for instance in France; see for instance Thiry *et al.*, 2005). Several examples of lateritic and bauxitic paleoprofiles can also be cited in the Scandinavian Shield or in Greenland. The absence of iron duricrust at the top of lateritic profiles is generally due either to erosion of the top of the profile before it is buried below sediments, or to rehydratation of hematite of the duricrust into iron hydroxides (goethite and limonite) within a latosoil. This absence is thus not a consequence of the present day temperate climate. The reason is that the actual climatic belts are not older than 10 My, due to the polar ice caps: palynology and stable isotope data show that lateritic weathering was possible (T°~18°C, rainfall > 1.5 m) at least up to 60° of latitude. Conversely, not all the iron or bauxitic caps existing in present day tropical countries are actual. Most of them are also related to ancient weathering profiles (see for instance Theveniaut & Freyssinet, 1999; 2002).

2.2.3.3 *Deep vertical discontinuities*

Pre-existing heterogeneities within the hard rocks such as veins (quartz veins, pegmatite or aplite), dykes, ancient faults or joints, or even contacts between different geological units (see for instance Neves & Morales, 2007; Le Borgne *et al.*, 2004, 2006; Durand *et al.*, 2006), locally favour the weathering process (see for instance Chilton & Smith-Carington, 1984; Acworth, 1987; Sander, 1997; Owen *et al.*, 2007; Dewandel *et al.*, 2011). It results in a local deepening of the weathering profile which can reach – in the vicinity of such heterogeneities – several hundreds of meters below the surface topography (Roques *et al.*, 2012), with good hydraulic relationships locally between the subsurface aquifers and these deep structures. In this case, the weathering process turns out to be at the origin of the fissures development and of an enhanced local hydraulic conductivity in the surrounding hard rock at the periphery of these heterogeneities and also within decameter wide discontinuities or veins of quartz or pegmatite (Dewandel *et al.*, 2011) (Figure 2.2 and Figure 2.7).

Near these discontinuities, the hard rock weathering profile (saprolite and fissured layer) is characterised by a sharp deepening of the weathered layers in the hard rok close to the contact. Because, in this case the stress tensor is deviated, σ_3 becomes sub-orthogonal to the heterogeneity and the fissured layer becomes sub-vertical as it develops parallel to the vein or the former fault/joint (Figure 2.7). Consequently the fissured layer and the saprolite reach much greater depths than through the classic expansion of biotite (up to several hundred meters) (Dewandel *et al.*, 2011; Roques *et al.*, 2012). Such geological structures are of primary hydrogeological interest in areas where the stratiform fissured layer has been eroded or is unsaturated and they constitute the only available targets for water-supply wells. However, intrusive bodies or veins whose weathering products are of low permeability, such as those of dolerite dykes, do not offer better hydrodynamic properties, neither in the fissure, nor around it. These discontinuities neither locally enhance the thickness of weathered layers (saprolite, stratiform fissured layer) in granite nor its hydraulic conductivity, making such structures unfavorable hydrogeological targets (Dewandel *et al.*, 2011; Perrin *et al.* 2011).

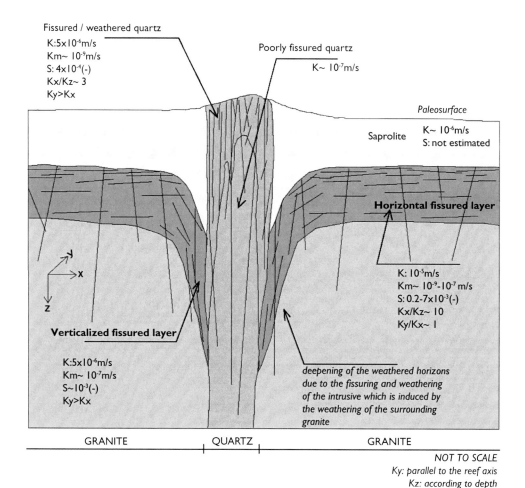

Figure 2.7 Conceptual hydrodynamic model of a vertical discontinuity in hard rock (after Dewandel *et al.* 2011). (*See colour plate section, Plate 5*).

2.2.4   Influence of mineralogy, texture and structure on the development of the fissured layer

2.2.4.1   *Mineralogy*

Biotite is the most common mineral that is responsible for creating fissure networks in hard rocks during weathering. It can be found in many acidic plutonic rocks (granitoids) and in metamorphic rocks such as micaschists, paragneisses and orthogneisses. White micas are also prone to swelling during weathering, but their transformation into clay minerals starts later than biotite because they need more activation energy in two-micas granites, the weathering of muscovite generally begins when the rock has been transformed into sandy saprolite. In this case, muscovite plays no part in the fissuring of hard rocks, and most of the fissures are caused by the weathering of biotites. However, in some white mica-bearing rocks such as sericitoschists, the weathering of sericite can contribute to creating a fissure network, but these cases are poorly documented. Feldspars are not known to swell during weathering even if the transformation of plagioclase to clay minerals begins as early as for biotite (Figure 2.3).

In basic and ultrabasic rocks, the existence of a fissured layer is dependent on the presence of up to only two minerals, pyroxene and/or olivine. Basalts containing phenocrysts of pyroxene and/or olivine generally suffer fissuring during early stages of weathering. Other highly weatherable minerals in basic rocks, such as plagioclase and amphibole, do not induce fissuring, so that when basic lavas undergo slight metamorphism or propylitisation, they cannot be fissured later by the weathering process. That is particularly the case for palaeovolcanics or some doleritic dykes where pyroxenes and olivines have been completely retromorphosed.

In peridotites, weathering induces swelling of olivine and intense fissuring of the rock (Dewandel *et al.*, 2005). Karstification may develop in such hard rocks, below the saprolite cover mainly made up of limonite (Genna *et al.*, 2005).

2.2.4.2 *Texture*

During weathering, the pressure of swelling minerals is dependent upon the mineral size. Therefore, hard rock fissuring is easier in granular rocks than in microgranular ones: for instance, microgranites, even containing biotite, are less fissured than granites.

2.2.4.3 *Structure*

During weathering, the swelling of biotite mineral occurs perpendicular to the cleavages. The layer spacing is changing from 10 Å to 14 Å as biotite turns into chlorite or vermiculite. This is a consequence of the exchange of ions (K+) initially present between the layers by larger, hydrated cations (Mg, Na, Mn, Ca) and water molecules. Consequently, the swelling pressure is exerted perpendicular to the cleavage and the resulting stress ellipsoid shape depends on the orientation of the minerals. Thus the rock permeability tensor is dependent on mineral orientation. Figures 2.8 to 11 show respectively, from base to top and from left to right, the mineral (mica) orientation, the potential deformation ellipsoid, the resulting stress and permeability ellipsoid, and the geometry of the induced fissure network.

When swelling minerals are randomly oriented (Figure 2.8) (e.g., in plutonic rocks), the theoretical deformation ellipsoid is isotropic. Vertical stress increases until the lithostatic charge is offset, whereas horizontal stresses continue to increase. Horizontal tension joints are created when the difference between minor and major stress components reaches the elastic limit. The resulting reservoir properties (permeability, porosity) are good. The major permeability component is horizontal.

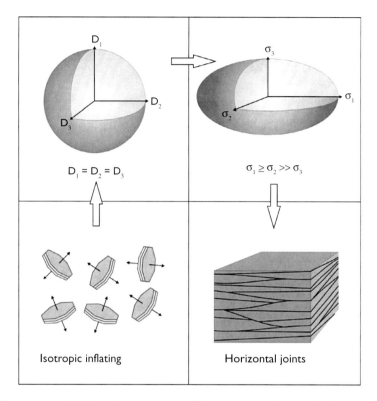

Figure 2.8 Parameters of deformation, stress and fissure creation for randomly oriented swelling minerals (plutonic rocks). From base to top and from left to right: the mica orientation, the potential deformation ellipsoid, the resulting stress and permeability ellipsoid, and the geometry of induced fissure network.

When the swelling minerals are vertically oriented (Figure 2.9), the deformation ellipsoid is anisotropic, with the major axis orthogonal to mineral orientation (i.e., horizontal). The maximum stress component has the same orientation, and the resulting tension joints are horizontal and generally largely open. The resulting reservoir properties (permeability, porosity) are good, the major permeability component being horizontal.

When the swelling minerals are horizontally oriented (Figure 2.10), the whole swelling potential is vertical, the horizontal stress being very slight. In this case there are no open tension joints, but the rock acquires an horizontal structure. The resulting reservoir properties are poor.

In case of metasedimentary folded rocks, such as schists, micaschists, gneisses, (Figure 2.11), the orientation of swelling minerals varies laterally, and the bedding creates lithological brittle planes. Breaking joints are created in random directions.

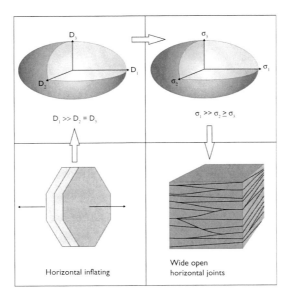

Figure 2.9 Parameters of deformation, stress and fissure creation for vertically oriented swelling minerals (gneisses, orthogneisses, micaschists). From base to top and from left to right: the mica orientation, the potential deformation ellipsoid, the resulting stress and permeability ellipsoid, and the geometry of induced fissure network.

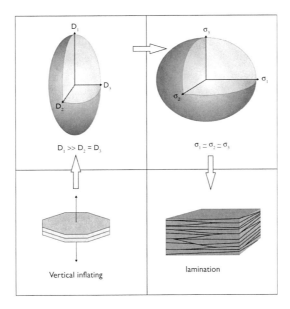

Figure 2.10 Parameters of deformation, stress and fissure creation for horizontally oriented swelling minerals (gneisses, orthogneisses, micaschists). From base to top and from left to right: the mica orientation, the potential deformation ellipsoid, the resulting stress and permeability ellipsoid, and the geometry of induced fissure network.

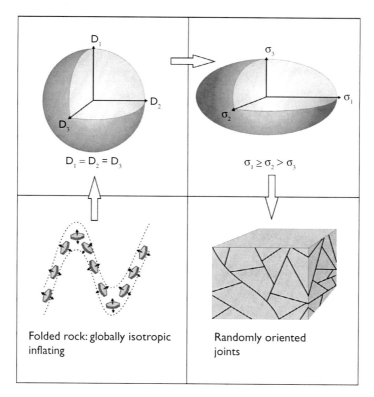

Figure 2.11 Parameters of deformation, stress and fissure creation for folded rocks (gneisses, micaschists, schists). From base to top and from left to right: the mica orientation, the potential deformation ellipsoid, the resulting stress and permeability ellipsoid, and the geometry of induced fissure network.

2.2.4.4 Conclusion: Most favorable criteria for good aquifer properties

For plutonic rocks, granular, biotite-bearing granites, granodiorites and diorites along with unmetamorphosed gabbros are generally good aquifers. For metamorphic rocks, orthogneisses, paragneisses and micaschists can be good aquifers if their foliation is near-vertical, and are poor ones when foliation is near-horizontal. Consequently, biotite-rich rocks (e.g. biotite-granites or -gneisses, or micaschists) exhibit a well-developed stratiform fissured layer, with a high hydraulic conductivity as well as a proportionally thick saprolite, whereas biotite-poor rocks, such as leucogranites, are poorly weathered/fissured. This often results in a granitic landscape where leucogranite intrusions in the biotite-granite remaining unweathered and forming hills gently emerging from the weathered, often partly eroded, biotite-granite flat weathering cover. A similar landscape results from differential weathering in metamorphic-rocks where unweathered sandstone ridges emerge from flat weathered schists or micaschists.

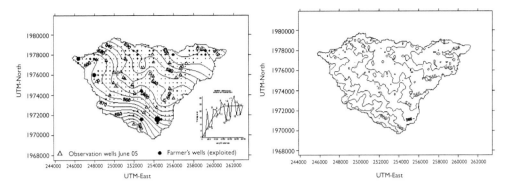

Figure 2.12 Piezometric map (left) and topographic map (right) of the 83 km² Gajwel watershed, Andhra Pradesh, India, characterised by a total groundwater abstraction of about 20 10⁶ m³/y.

In granite-type rocks, a high hydraulic conductivity stratiform fissured layer develops in coarse-grained rocks, such as porphyritic granite, whereas in fine-grained rocks, like aplitic granites, the stratiform fissured layer has a lower hydraulic conductivity.

In foliated rocks, the fracturing intensity is maximum where the foliation is sub-vertical whereas it decreases and reaches a minimum where the foliation is subhorizontal.

Basic volcanic rocks can form a medium quality aquifer if they contain pyroxene and/or olivine. Paleovolcanics and amphibolites are generally poor aquifers, as are acidic volcanic rocks.

Peridotites can be good aquifers, with dual aquifer behaviour, some also karstic and fissured.

2.2.4.5   *Hydrogeological functioning of hard rock aquifers*

Due to their low hydraulic conductivity, hard rock aquifers generally exhibit quite high hydraulic gradients and piezometric heads that mimic the topographic surface (Figure 2.12). At the exceptions of areas of low recharge and overexploited areas, the piezometric surface is shallow. Consequently, the streams drain the aquifer. In wet regions, each stream is flowing and most of the topographic depressions give rise to a spring, albeit of low discharge.

In natural conditions, most of the groundwater flow resulting from the recharge occurs in the upper layers of the aquifer, from the topographic surface to the narrowest perennial stream or spring, mostly through the saprolite. Only a few flow lines and a small proportion of the recharge reaches the deepest levels. The result is an aquifer in which groundwater age and mineral content increase with depth.

Pumping from the stratiform fissured layer modifies the groundwater flow lines in the aquifer as groundwater from the upper horizon is forced to flow vertically downwards towards the pumped stratiform fissured layer. Consequently, groundwater chemistry is also largely modified.

2.3   PRACTICAL HYDROGEOLOGICAL APPLICATIONS OF THIS CONCEPTUAL MODEL

This new genetic conceptual model of hard rock aquifers leads to a revision of the methodologies used for characterising the structure and hydraulic processes of such aquifers. It also allows improvement of the analytical methodologies and enables a wide range of practical applications in hydrogeology as well as in other domains of applied geosciences to be applied.

The strategy for siting a water well is different in the general case, looking for sub-horizontal fractures of the stratiform fissured layer that will efficiently be tapped by vertical boreholes (see Figure 2.2), to the case where the stratiform fissured layer has been completely removed by erosion, searching for a vertical fissured layer located along the walls of a sub-vertical discontinuity (see Figure 2.7). In the former case, the survey focusses on the stratiform fissured layer preserved from erosion, not clogged by further weathering or diagenesis, and saturated with water, i.e. below the piezometric level. In the latter case, techniques for well siting such as lineament analysis or radon surveys (Lachassagne & Pinault, 2001b) can be applied. In that case inclined boreholes are more effective than vertical ones to increase the probability of crosscutting a vertical fracture.

2.3.1   Mapping the layers (saprolite, stratiform fissured layer) constituting the hard rock aquifers: From the local to the country-size scale

The hard rock aquifer can be considered as continuous aquifers. It is now possible to map, or model in 3-D, the geometry (elevation, thickness) and the physical properties of the various layers: saprolite, with its storage hydrodynamic properties, and the stratiform fissured layer, with mostly transmissive but also storage properties (Lachassagne *et al.*, 2001b; Wyns *et al.*, 2004; Dewandel *et al.*, 2006). This mapping methodology is based on geomorphological (topographic maps, DEM – digital elevation models), geological (outcrop observations, well logs), geophysical (both field and aerial), and hydrodynamic approaches (Figure 2.13). It comprises a twofold methodology:

– Characterising the local weathering history, at the regional scale, and consequently the vertical structure of the weathering profile, for the main rock types of the area (see Figures 2.1, 2.2, 2.4, 2.5 and 2.6).
– Mapping the elevation of the interfaces separating the main horizons, namely the saprolite, the stratiform fissured layer, and the fresh substratum (Figure 2.13). The methodology may vary according to the available data, and the presence or absence of outcrops. The resulting maps can either show the intersection of these horizons with the topography, as a classical geological map (Figure 2.14a), or isolines for these interfaces (Figure 2.14b), or thickness of these horizons (Figure 2.14c).

In most cases, the geometry of the base of the saprolite is determined at first, through the combined use of well data, geophysical data if any, and field survey data.

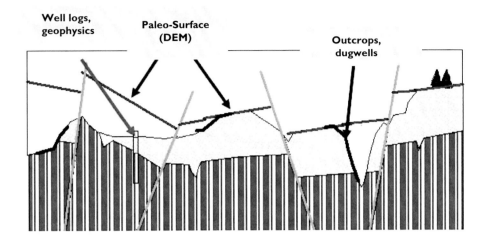

Figure 2.13 Methodology and data used for mapping hard rock aquifer layers. In grey: weathering profile (saprolite + stratiform fissure layer). Vertical stripes: unweathered rock.

Where the surface of the base of the saprolite is cut by the present-day ground surface,
for instance near valleys, it is relatively simple in the field to determine the position
of the saprolite/weathered-fissured layer interface. On the basis of observations of the
thickness of the stratiform fissured layer (observations on outcrops, statistical treatment of existing well data), the altitude of its base can also be directly computed from
the alterite base. The altitudes of (i) the saprolite/stratiform fissured layer interface
and (ii) the elevation of the stratiform fissured layer/fresh rock interface can, therefore, be subtracted from that of the present ground surface, for instance as inferred
from DEM data, in order to compute the residual thickness of respectively the saprolite and the SFL. In regions where the paleosurface(s) have partly been preserved from
erosion, slope analysis from DEM data can also be used for mapping in order to
reconstruct paleaosurfaces.

2.3.2 Water well siting in hard rock aquifers and
mapping hydrogeological potential at various
scales

Water well siting, either for water supply, or for agricultural or industrial purposes,
provides for an important need. A complete methodological toolkit, especially devoted
to hard rock aquifers, has been developed on the basis of these concepts.

2.3.2.1 *Regional to country scale approach*

The 'blow discharge' of a well measured with an air lift pump once drilling is completed is a very robust indicator of the local hydraulic conductivity/transmissivity. It
is also a good indicator of the future exploitable discharge of the well, particularly
for villages water supplies where quite low discharges, often only > 0.5 m³/h, are

required (Lenck, 1977). The depth to the base of the saprolite is also a parameter measured with a good accuracy during drilling, either by a geologist or a hydrogeologist, but also and more often by the driller himself. These data are very often reported and stored in data bases (Courtois *et al.*, 2009). Statistics computed from these data, particularly those gathered in Africa during large villages water supply projects, are used to infer, at the regional scale, the hydrodynamic properties of the various lithologies in a study area (Lenck, 1977). These statistical methodologies have been refined on the basis of the above conceptual model (Courtois *et al.*, 2009). The hydraulic conductivity of a well, the blow discharge, appear to be strongly correlated to the depth of the well below the base of the saprolite, i.e. to the linear discharge, the blow discharge/depth of the well below the base of the saprolite. The shallowest wells, cross-cutting the more densely fissured part of the stratiform fissured layer, provide a higher linear discharge than the deeper ones. The statistical processing of this parameter allows computing, for each lithotype, two important parameters (Courtois *et al.*, 2009): (i) the thickness of the most productive part of the stratiform fissured layer (in Burkina Faso this thickness is between 35 and 40 m according to the lithotype), (ii) the hydrodynamic parameters, namely basic statistics about the potential discharge of a well: mean, median, standard deviation, percentage of dry wells, i.e. below a discharge threshold.

In association with the thickness of the saprolite (see Figure 2.14c), these data provide an estimation of optimal well length, i.e. the length beyond which it is not necessary to drill for there is little gain in discharge that does not justify the increased drilling cost. Consequently, they are of particular interest for the planning (duration) and for assessing cost of drilling campaigns (cumulative length amount and depth of the wells to be drilled, including the number of dry wells). They provide an estimated minimum depth for water wells necessary to cross the saprolite and reach the stratiform fissured layer, and as such can be used to evaluate the drilling constraints, i.e. the technical characteristics of the drilling machine. Other applications can also be inferred from the results, such as mapping the mean discharge of wells at regional/country scale for different depths and coupling these results with economic parameters such as drilling costs. They can also be used in other fields such as town and country planning, soils engineering, the search for areas prone to quarrying.

2.3.2.2   *Local scale*

The results are also of interest at the local scale. However, the available data from existing wells might not be numerous enough to compute robust statistics, but high resolution mapping (see Figures 2.6 and 2.15) is essential for the preliminary stages of water well siting. This enables the location, in combination with other parameters such as land-use, land-ownership and conditions of access, of favorable zones to carry out geophysical surveys for siting exploratory boreholes.

Geophysical methods, and particularly 2D-electric geophysical methods, generally cannot delineate the between different lithologies (Figure 2.16). Nevertheless, they do characterise the structure of the weathering profile (Figure 2.16). The typical response is, from top to bottom, (1) high resistivity (unsaturated saprolite, and or,

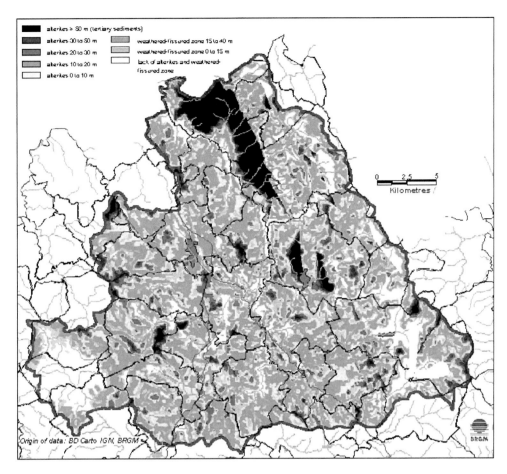

Figure 2.14a Examples of maps of the various layers constituting Hard Rock Aquifers. (a): geological map of the weathering cover on the Truyère, Lozère, France watershed: thickness of the saprolite (increasing thickness from yellow to red and black) and the fissured layer (increasing thickness from blue to green), white: weathering profile totally eroded (after Lachassagne *et al.*, 2001b). (*See colour plate section, Plate 6*).

where preserved, iron or bauxitic crust), (2) much lower resistivity, corresponding to the clayey saprolite, (3) regularly increasing resistivity within the sandy saprolite, and within the stratiform fissured layer, (4) the highest resistivity (up to more than 10.000 ohm.m) being reached within the unweathered and unfractured rock.

Such a high-resolution mapping (Figure 2.15) is also very suitable for the delineation of groundwater protection zones, but also for other applications such as siting a landfill (preferably on the upper part of the saprolite where it is clayey), locating a rock quarry (with a minimum cover cap, preferably on a place where the weathering profile is eroded, or at the boundary between two successive weathering profiles).

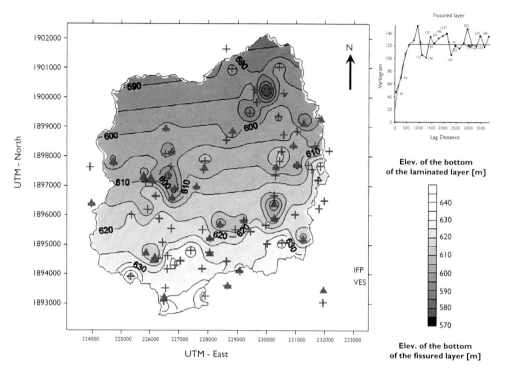

Figure 2.14b Isohypse map of the elevation (in m above sea level) of the base of the laminated layer Maheswaram, Andhra Pradesh, India (Dewandel *et al.*, 2006).

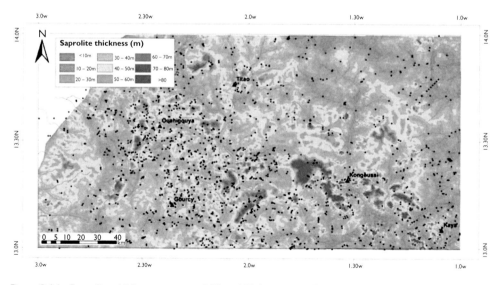

Figure 2.14c Saprolite-thickness over a ~250 × 100 km area in Burkina Faso (Courtois *et al.*, 2009). (*See colour plate section, Plate 7*).

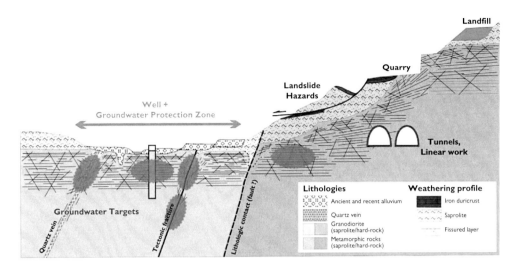

Figure 2.15 Various uses of a high resolution mapping of the weathering cover at the local scale (after Courtois et al., 2003). (See colour plate section, Plate 8).

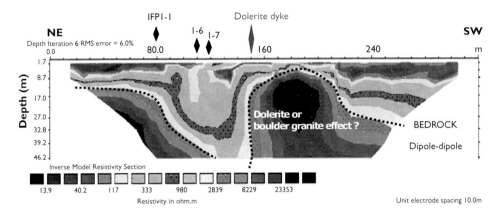

Figure 2.16 Examples of geophysical profiles in granitic weathered hard rock, Maheswaram, Andhra Pradesh, India.

2.3.3 Quantitative management and modelling of hard rock aquifer groundwater resource at the watershed scale

On the basis of this concept of stratiform layers and of the relative homogeneity of hard rock aquifers at a scale of few 100 meters (Dewandel et al., 2012), the use of water table fluctuation and groundwater budget methods at the watershed scale is appropriate (Maréchal et al., 2006). These techniques have already been extensively

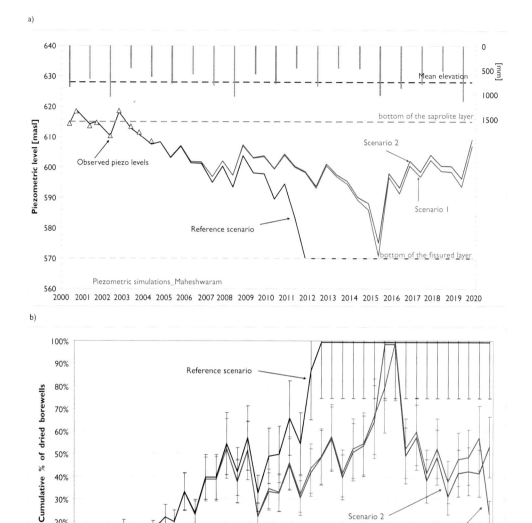

Figure 2.17 Example of long term resource management simulation at the watershed scale, Maheswaram, Andhra Pradesh, India. (a) annual rainfall (top right scale) and mean piezo-metric level (left scale). (b) number of dried wells.

used in the context of overexploited hard rock aquifers in India (Maréchal *et al.*, 2006; Dewandel *et al.*, 2007, 2010, Perrin *et al.*, 2011, 2012; Ferrant *et al.*, submit-ted). Decision support tools have been developed from these methodologies (Lachassagne *et al.*, 2001a; Dewandel *et al.*, 2007, 2011), and strategies built, some of them integrating climate change impacts (Ferrant *et al.*, submitted) (Figure 2.17).

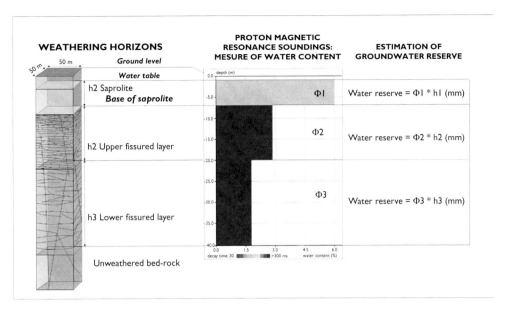

Figure 2.18 Principle of water reserve content mapping at watershed to regional scale.

2.3.4   Assessing hard rock aquifer groundwater reserves from the watershed to the regional scale

These new concepts allow assessment of hard rock aquifer groundwater reserves from the watershed to the regional scale on the basis of the combined use of the 3D geometry of the weathering profile, and the vertical distribution of the effective porosity determined, for instance, by Protonic Resonance Soundings (PRS) (Figure 2.18) (Wyns *et al.*, 2002, 2004; Baltassat *et al.*, 2005). PRS allows determining the mean effective porosity of each part of the aquifer, which is divided into three parts: saprolite, upper fissured layer, and lower fissured layer. For a given lithology, several PRS are used to determine the mean porosity for each layer, weighted by quality index of PRS. Then, the saturated thickness of each layer, multiplied by the porosity, gives the water content for each cell expressed in water height (Figure 2.19). The total water reserve is given by the addition of the water height of the three layers of the aquifer.

It also allows upscaling and regionalising hydrodynamic properties (Dewandel *et al.*, 2012). Taking into account the recharge of such aquifers, it is possible to evaluate the duration of non-point source pollutants such as diffuse nitrates.

2.3.5   Managing and protecting hard rock aquifer groundwater resources

The delineation of groundwater protection zones for water wells has to rely on a good knowledge of the geometry and hydrodynamic properties of hard rock aquifers. Methodologies for the optimisation of piezometric networks have also been developed (Zaidi *et al.*, 2007).

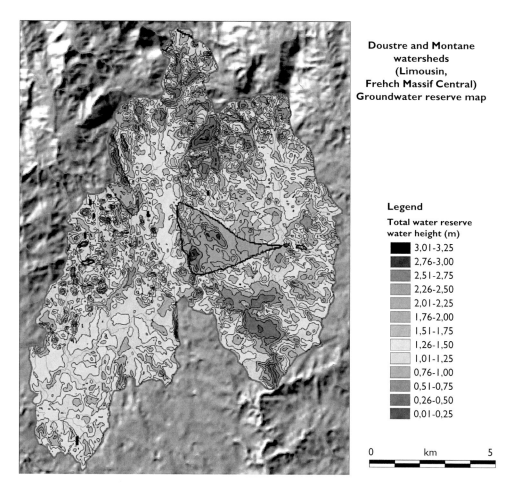

Figure 2.19 Example of map of water reserve on two watersheds totalizing 115 km² in mainly granitic context, Corrèze, French Massif Central. The triangular yellow, green to blue area in the middle of the figure corresponds to a biotite granite with higher groundwater reserves. The other lithological contours do not generate an enough groundwater reserve contrast to be easily identified on this map. (*See colour plate section, Plate 9*).

2.3.6   Computing the drainage discharge and the hydrogeological and hydrological impacts of tunnels drilled in hard rocks

The drainage discharge and the surface hydrogeological (piezometry in wells) and hydrological (discharge of streams and springs) impacts of shallow tunnels in hard rocks, with a depth below ground level between 0 and 300 m, have been forecasted on the basis of (i) the location of the tunnel within or below the various layers constituting the weathering profile, (ii) steady state groundwater discharge measured in existing tunnels, and (iii) analytical solutions for tunnel inflow (Lachassagne *et al.*, 2008). The actual discharge of the tunnels has later validated the methodology.

2.4   OTHER APPLICATIONS IN APPLIED GEOSCIENCES

Before being largely improved and developed for hydrogeological applications, the concepts about weathering profiles were first used as markers of geodynamic evolutions (Wyns, 1991) with geological mapping objectives (Wyns *et al.*, 1998) and radioactive waste storage safety analysis (Wyns 1991; 1997). Applications have also been developed in the field of soil and rock mechanics, notably for the forecast of the near subsoil mechanical properties for applications such as the burying of electrical and telephonic networks (Wyns *et al.*, 1999; 2005), and also for quarrying (Rocher *et al.*, 2003): evaluation of the thickness of the cover cap and determination of the optimal geo-mechanic quality of the materials. In the future more applications are expected to emerge such as preliminary surveys for highways and railways.

REFERENCES

Acworth R.I. (1987) The development of crystalline basement aquifers in a tropical environment. *Quaternary Journal of Engineering Geology* 20, 265–272.

Baltassat J.-M., Legtchenko A., Ambroise B., Mathieu F., Lachassagne P., Wyns R., Mercier J.L., Schott J.-J. (2005) Magnetic resonance sounding (MRS) and resistivity characterization of a mountain hard rock aquifer: the Ringelbach catchment, Vosges Massif, France. *Near Surface Geophysics* 3, 267–274.

Banfield J.F., Eggleton R.A. (1990) Analytical transmission electron microscope studies of plagioclase, muscovite, and K-feldspar weathering. *Clays Clay Minerals* 38(I), 71–89.

Bisdom E.B.A., Stoops G., Delvigne J., Curmi P., Altemuller H.-J. (1982) Micromorphology of weathering biotite and its secondary products. Contribution no.3 of the "Advisory Panel on Weathering Phenomena and Neoformations" of the "Sub-Commission on Soil Micromorphology of the I.S.S.S.". *Pedologie*, 32(2), 225–252.

Caine J.S., Evans J.P., Forster C.B. (1996). Fault zone architecture and permeability structure. *Geology* 24(11), 1025–1028.

Charlier J.B., Lachassagne P., Ladouche B. Cattan Ph., Moussa R., Voltz M. (2011). Structure and hydrogeological functioning of an insular tropical humid andesitic volcanic watershed: A multi-disciplinary experimental approach. *Journal of Hydrology* 398(2011), 155–170.

Chilton P.J., Foster S.S.D. (1995). "Hydrogeological Characterisation and Water-Supply Potential of Basement Aquifers In Tropical Africa." *Hydrogeology Journal* 3(1): 36.

Chilton P.J., Smith-Carington A.K. (1984) Characteristics of the weathered basement aquifer in Malawi in relation to rural water supplies. *Challenges in African Hydrology and Water Resources* (Proceedings of the Harare Symposium, July 1984). IAHS Publ. no. 144.

Cho M., Choi Y., Ha K., Kee W., Lachassagne P., Wyns R. (2003) Relationship between the permeability of hard rock aquifers and their weathering, from geological and hydrogeological observations in South Korea. *International Association of Hydrogeologists IAH Conference on "Groundwater in fractured rocks"*, Prague 15–19 September 2003, Prague.

Courtois N., Lachassagne P., Weng P., Theveniaut H., Wyns R., Joseph B., Laporte P. (2003) Détermination de secteurs potentiellement favorables pour la recherche d'eau souterraine à Cacao (Guyane) *Rapport BRGM/RM-52758-FR.*

Courtois N., Lachassagne P., Wyns R., Blanchin R., Bougaïre F.D., Some S., Tapsoba A. (2009) Large-Scale mapping of hard-rock aquifer properties applied to Burkina Faso. *Ground Water* 48(2), 269–283.

Dewandel B., Gandolfi J. M., Zaidi F.K., Ahmed S., Subrahmanyam K. (2007) A decision support tool with variable agro-climatic scenarios for sustainable groundwater management in semi-arid hard-rock areas. *Current Science* 92(8), 1093–1102.

Dewandel B., Lachassagne P., Boudier F., Al Hattali S., Ladouche B., Pinault J.L., Al-Suleimani Z. (2005) A conceptual model of the structure and functioning of the Oman ophiolite hard-rock aquifer through a pluridisciplinary and multiscale approach. *Hydrogeology Journal* 13, 708–726.

Dewandel B., Lachassagne P., Chandra S., Zaidi F.K. (2011) Conceptual Hydrodynamic model of a geological discontinuity in hard rock aquifers: example of quartz reef in granitic terrain in South India. *Journal of Hydrology* 405, 474–487.

Dewandel B., Lachassagne P., Wyns R., Maréchal J. C., Krishnamurthy N.S. (2006) A generalised 3-D geological and hydrogeological conceptual model of granite aquifers controlled by single or multiphase weathering. *Journal of Hydrology* 330(1–2), 260–284.

Dewandel B., Maréchal J-C., Bour O., Ladouche B., Ahmed S., Chandra S., Pauwels H. (2012) Upscaling and regionalizing hydraulic conductivity and efficient porosity at watershed scale in crystalline aquifers. *Journal of Hydrology* 416, 83–97.

Dewandel, B., Perrin, J., Ahmed, S., Aulong. S., Hrkal, Z., Lachassagne, P., Samad, M.S. Massuel (2010) Development of a Decision Support Tool for managing groundwater resources in semi-arid hard rock regions under variable water demand and climatic conditions. *Hydrological Processes* 24, 27884–2797.

Durand V., Deffontaines B., Léonardi V., Guérin R., Wyns R., Marsily de G., Bonjour J.-L. (2006) A multidisciplinary approach to determine the structural geometry of hard-rock aquifers. Application to the Plancoët migmatitic aquifer (NE Brittany, W France). *Bulletin de la Société Géologique de France* 177, 227–237.

Farmin R. (1937) Hypogene Exfoliation in Rock Masses. *The Journal of Geology* 45(6), 625–635.

Ferrant S., Caballero Y., Perrin J., Dewandel B., Aulong S., Ahmed S., Maréchal J.C. (Submitted). The likely impacts of climate change on farmer groundwater extraction from crystalline aquifer of south-India. Submitted to *Global Environmental Change*.

Folk R.L., Patton E.B. (1982) Buttressed expansion of granite and development of grus in central Texas. *Zeitschrift fuer Geomorphologie* 26(1), 17–32.

Foster S. (2012) Hard-rock aquifers in tropical regions: using science to inform development and management policy. *Hydrogeology Journal* 20, 659–672.

Genna A., Maurizot P., Lafoy Y., Augé T. (2005) Role of karst in the nickeliferous mineralisations of New Caledonia. *C. R. Geoscience* 337, 367–374.

Hill S.M. (1996) The differential weathering of granitic rocks in Victoria, Australia. *Journal of Australian Geology* 16, 271–276.

Hill S.M., Ollier C.D., Joyce E.B. (1995) Mesozoic deep weathering and erosion: an example from Wilson's Promontory, Australia. *Zeitschrift fuer Geomorphologie* 39, 331–339.

Holzhausen G.R. (1989) Origin of sheet structure, 1. Morphology and boundary conditions. *Engineering Geology* 27(1–4), 225–278.

Houston J.F.T., Lewis R.T. (1988) The Victoria Province drought relief project, II. Borehole yield relationships. *Ground Water* 26(4), 418–426.

Hsieh P.A., Shapiro A.M. (1996) Hydraulic characteristics of fractured bedrock underlying the FSE well field at the Mirror Lake Site, Grafton County, New Hampshire. Technical meeting, Colorado Springs, Colorado, *U.S. Geological Survey Water Resources Investigations Report* 94–4015.

Jahns R.H. (1943). Sheet structure in granites, its origin and use as a measure of glacial erosion in New England. *Journal of Geology* 51(2), 71–98.

Key R.M. (1992). An introduction to the crystalline basement of Africa. Hydrogeology of Crystalline Basement Aquifers in Africa. In Wright, E.P., Burgess W., *Geological Society Special Publication, London.* 66, 29–57.

Lachassagne P., Ahmed S., Golaz C., Maréchal J.-C., Thiery D., Touchard F., Wyns R. (2001a) A methodology for the mathematical modelling of hard-rock aquifers at catchment scale based on the geological structure and the hydrogeological functioning of the aquifer. *31st IAH Congress New approaches characterising groundwater flow. "Hard Rock Hydrogeology" session, Munich, Germany*, AIHS.

Lachassagne P., Lacquement F., Lamotte C. (2008) Assistance hydrogéologique dans le cadre des études préliminaires afférentes aux fonçages des tunnels de Violay, Bussière et Chalosset, sur le tracé de l'A89, section Balbigny-La Tour de Salvagny. Rapport de phase 3. *R. BRGM/RC-56085-FR*. Montpellier, BRGM.

Lachassagne P., Pinault J.-L. (2001) Radon-222 emanometry: a relevant methodology for water well siting in hard rock aquifer. *Water Resources Research* 37(12), 3131–3148.

Lachassagne P., Wyns R., Bérard P., Bruel T., Chéry L., Coutand T., Desprats J.-F., Le Strat P. (2001b) Exploitation of high-yield in hard-rock aquifers: Downscaling methodology combining GIS and multicriteria analysis to delineate field prospecting zones. *Ground Water* 39(4), 568–581.

Lachassagne P., Wyns R., Dewandel B. (2011) The fracture permeability of hard rock aquifers is due neither to tectonics, nor to unloading, but to weathering processes. *Terra Nova* 23, 145–161.

Le Borgne T., Bour O., de Dreuzy J.-R., Davy P., Touchard, F. (2004) Equivalent mean flow models for fractured aquifers: Insights from a pumping tests scaling interpretation. *Water Resources Research* 40, 12.

Le Borgne T., Bour O., Paillet F.L., Caudal J.-P. (2006) Assessment of preferential flow path connectivity and hydraulic properties at single-borehole and cross-borehole scale in a fractured aquifer. *Journal of Hydrology*, 328, 347–359.

Lelong F., Lemoine J. (1968) Les nappes phréatiques des arènes et des altérations argileuses; leur importance en zone intertropicale; les difficultés de leur exploitation. *Hydrogéologie Bulletin du Bureau de Recherches Géologiques et Minières. Deuxième série. Section III(2)*, 41–52.

Lenck P.-P. (1977) Données nouvelles sur l'hydrogéologie des régions à substratum métamorphique ou éruptif. Enseignements tirés de la réalisation de 900 forages en Côte d'Ivoire. *C.R. Acad. Sc. Paris* t. 285, 497–500.

Mabee S.B., Curry P.J., Hardcastle K.C. (2002) Correlation of Lineaments to Ground Water Inflows in a Bedrock Tunnel. *Ground Water* 40(1), 37–43.

MacFarlane M.J. (1992) Groundwater movement and water chemistry associated with weathering profiles of the African surface in Malawi. Hydrogeology of Crystalline Basement Aquifers in Africa. In Wright, E.P. and Burgess, W., *Geological Society Special Publication, London*. 66, 101–129.

Mandl G. (2005) *Rock joints. The Mechanical Genesis*. Berlin, Heidelberg, New York, Springer.

Maréchal J.-C., Ahmed S. (2003) Dark zones are Human-made. *Down to Earth* 12(4), 54.

Maréchal J.-C., Dewandel B., Subrahmanyam K. (2004) Use of hydraulic tests at different scales to characterize fracture network properties in the weathered-fractured layer of a hard rock aquifer. *Water Resources Research* 40(11), 17.

Maréchal J.C., Dewandel B., Ahmed S., Galeazzi L., Zaidi F.K. (2006) Combined estimation of specific yield and natural recharge in a semi-arid groundwater basin with irrigated agriculture. *Journal of Hydrology* 329(1–2), 281–293.

Michel J.P., Fairbridge R.W. (1992) *Dictionary of Earth Sciences/Dictionnaire des Sciences de la Terre. Anglais-Français/Français-Anglais*. Masson – Wiley, 304 p.

Neves M., Morales, N. (2007) Well productivity controlling factors in crystalline terrains of southeastern Brazil. *Hydrogeology Journal* 15(3), 471.

Oliva P., Viers J., Dupré B. (2003) Chemical weathering in granitic environments. *Chemical Geology* 202(3–4), 225–256.

Ollier C.D. (1988) The regolith in Australia. *Earth-Sciences Review* 25, 355–361.

Owen R., Maziti A., Dahlin T. (2007) The relationship between regional stress field, fracture orientation and depth of weathering and implications for groundwater prospecting in crystalline rocks. *Hydrogeology Journal* 15(7), 1231.

Perrin J., Ahmed S., Hunkeler D. (2011) The role of geological heterogeneities and piezometric fluctuations on groundwater flow and chemistry in hard-rock, southern India. *Hydrogeology Journal* 19, 6, 1189–1200.

Perrin J., Ferrant S., Massuel S., Dewandel B., Maréchal J.C., S. Aulong, Ahmed S. (2012) Assessing water availability in a semi-arid hard-rock regions watershed of southern India using a semi-distributed model. *Journal of Hydrology* 460–461(2012), 143–155.

Perrin J., Mascré C., Pauwels H., Ahmed S. (2011) Solute recycling: an emerging threat to groundwater quality in southern India. *Journal of Hydrology* 398(1–2), 144–154.

Pollard D.D., Aydin A. (1988) Progress in understanding jointing over the past one hundred years. *Geological Society of America Bulletin* 100, 1181–1204.

Razack M., Lasm T. (2006) Geostatistical estimation of the transmissivity in a highly fractured metamorphic and crystalline aquifer (Man-Danane Region, Western Ivory Coast). *Journal of Hydrology* 325(1–4), 164.

Rocher P., Mishellany A., Wyns R., Lacquement F., Gobron N., Greffier G., Germain Y. (2003) Identification et caractérisation des ressources en matériaux de substitution aux granulats alluvionnaires dans le département du Puy-de-Dôme: la zone des Combrailles. Report. *BRGM/RP52706-FR, BRGM:* 97.

Roets W., Xu Y., Rait L., El-Kahloun M., Meire P., Calitz F., Batelaan O., Anibas C., Paridaens K. (2008) Determining discharges from the Table Mountain Group (TMG) aquifer to wetlands in the Southern Cape, South Africa. *Hydrobiologia* 607, 175–186.

Roques C., Aquilina L., Bour O., Le Borgne T., Longuevergne L., Dauteuil O., Vergnaud V., Labasque Th., Lavenant N., Hochreutener R., Dewandel B., Mougin B., Schroetter J.M., Palvadeau E., Lucassou F., Jegou J.P. (2012) Hydrogeological and geochemical characterization of a deep hard-rock aquifer (Saint-Brice, French Britany). *Proceedings of the Int. Conf. on groundwater in Fractured Rocks, Prague, Czech Republic*, 21–24 May 2012, p. 39.

Sander P. (1997) Water-well siting in hard-rock areas: identifying promising targets using a probabilistic approach. *Hydrogeology Journal* 5(3), 32.

Shaw R. (1997) Variations in sub-tropical deep weathering profiles over the Kowloon Granite, Hong Kong. *Journal of the Geological Society of London* 154(6), 1077–1085.

Tardy Y., Roquin C. (1998) Dérive des continents, Paléoclimats et altérations tropicales. Orléans, *Editions BRGM*.

Taylor G., Eggleton R.A. (2001) *Regolith Geology and Geomorphology*. Chichester, John Wiley & Sons.

Taylor R.G., Howard K. (1999)Lithological evidence for the evolution of weathered mantles in Uganda by tectonically controled cycles of deep weathering and stripping. *Catena* 35(1), 65–94.

Taylor R.G., Howard K. (2000) A tectono-geomorphic model of the hydrogeology of deeply weathered crystalline rock: evidence from Uganda. *Hydrogeology Journal* 8(3), 279–294.

Theveniaut H., Freyssinet P. (1999) Paleomagnetism applied to lateritic profiles to assess saprolite and duricrust formation processes: the example of Mont Baduel profile (French Guiana). *Palaeogeography, Palaeoclimatology, Palaeoecology* 148(4), 209.

Theveniaut H., Freyssinet P. (2002) Timing of lateritization on the Guiana Shield: synthesis of paleomagnetic results from French Guiana and Suriname. *Palaeogeography, Palaeoclimatology, Palaeoecology* 178(1–2), 91.

Thiry M., Simon-Coinçon R., Quesnel F., Wyns R. (2005) Altération bauxitique associée aux argiles à chailles sur la bordure sud-est du bassin de Paris. *Bulletin de la Société Géologique de France* 176(2), 199–214.

Tieh T.T., Ledger E.B., Rowe M.W. (1980) Release of uranium from granitic rocks during in situ weathering and initial erosion (central Texas). *Chemical Geology* 29(1–4), 227.

Twidale C.R. (1982). *Granite landforms*. Amsterdam, Netherlands, Elsevier.

Wright E.P., Burgess W.G., Eds. (1992) *The Hydrogeology of Crystalline Basement Aquifers in Africa*. Geological Society Special Publication 66. London, The Geological Society.

Wyns R. (1991) Evolution tectonique du bâti armoricain oriental au Cénozoïque d'après l'analyse des paléosurfaces continentales et des déformations géologiques associées (in French: Structural evolution of the Armorican basement during the Cenozoic deduced from analysis of continental paleosurfaces and associated deposits). *Géologie de la France* 1991(3), 11–42.

Wyns R. (1997) Essai de quantification de la composante verticale de la déformation finie cénozoïque d'après l'analyse des paléosurfaces et des sédiments associés. *Proceedings des Journées Scientifiques CNRS – ANDRA. pp. 36–38. ANDRA, Poitiers*, pp. 36–38.

Wyns R. (2002) Climat, eustatisme, tectonique: quels contrôles pour l'altération continentale? Exemple des séquences d'altération cénozoïques en France. *Bull. Inf. Géol. Bass. Paris* 39(2), 5–16.

Wyns, R., BaltassatJ.M., Lachassagne P., Legtchenko A., Vairon J. (2004) Application of proton magnetic resonance soundings to groundwater reserves mapping in weathered basement rocks (Brittany, France). *Bulletin de la Société Géologique de France* 175(1), 21–34.

Wyns R., Clément J. P., Lardeux H., Gruet M., Moguedet G., Biagi R., Ballèvre M. (1998) Carte géologique de la France (1/50. 000), feuille Chemillé (483). Orléans.

Wyns, R., J.-C. Gourry, J.-M. Baltassat, Lebert F. (1999) Caractérisation multiparamètres des horizons de subsurface (0–100 m) en contexte de socle altéré. *PANGEA* 31/32: 51–54.

Wyns, R., Lacquement F., Corbier P., Vairon J. (2002) Cartographie de la réserve en eau souterraine du massif granitique de la Roche sur Yon. Orléans, *Rapport BRGM/RP-51633/FR*: 26.

Wyns R., Quesnel F., Lacquement F., Bourgine B., Mathieu F., Lebert F., Baltassat J.M., Bitri A., Mathon D. (2005) Cartographie quantitative des propriétés du sol et du sous-sol dans la région des Pays de la Loire. *R. BRGM/RP-53676-FR*, BRGM: 135.

Wyns R., Quesnel F., Simon-Coinçon R., Guillocheau F., Lacquement F. (2003) Major weathering in France related to lithospheric deformation. *Géologie de la France* 2003(1), 79–87.

Zaidi F., Ahmed S., Dewandel B., Maréchal J.C. (2007) Optimising a piezometric network in the estimation of the groundwater budget: a case study from a crystalline-rock watershed in southern India *Hydrogeology Journal* 15(6), 1131.

Chapter 3

Similarities in groundwater occurrence in weathered and fractured crystalline basement aquifers in the Channel Islands and in Zimbabwe

Jeffrey Davies[1], Nick Robins[1] & Colin Cheney[2]
[1]*British Geological Survey, Maclean Building, Wallingford, UK*
[2]*States of Jersey, Planning and Environment Department, Howard Davis Farm, La Route de la Trinité, Trinity, Jersey, Channel Islands*

ABSTRACT

The nature of weathered and fractured Precambrian-age crystalline basement aquifers was studied by workers in various parts of sub-Saharan Africa during the 1980s in response to the World Water Decade, and in the island of Jersey in the Channel Islands (Îles de la Manche) in the 1990s in response to increased water demand. Subsequently exploratory drilling in Jersey allows more detailed insight into the hydraulic properties of the weathered and fractured granite aquifer, sufficient to make comparison with development boreholes drilled into the Precambrian crystalline basement aquifer of sub-Saharan Africa, in this case using the Mutare area in eastern Zimbabwe as an example. Both cases offer similar groundwater occurrences in the upper weathered zone and lower fractured zones with regard to geology, geomorphology, water inflow zones, aquifer parameters, borehole yields, and groundwater quality.

3.1 INTRODUCTION

Groundwater in Precambrian basement rocks occurs in weathered, eroded and fractured zones (Singhal & Gupta, 1999). The degree of weathering and fracturing reflects the age and composition of the main lithologies, regional tectonism and volcanic intrusions, palaeoclimatology, palaeohydrology and land-surface erosion. The amount of groundwater available for supply is a reflection of the contemporary long-term average direct rainfall recharge.

The structured nature of the weathering profile was recognised by Ruxton & Berry (1957) while investigating deep weathering of granites in Hong Kong. The grade of weathering with depth was latterly formalised by Dearman *et al.* (1978) as outlined by Migon (2006). They recognised a gradual change in the degree of weathering with depth within the regolith from the soil layer or pedolith, through the saprolite zone of chemical weathering and into the basal saprock zone of fractured granite with numerous corestones. The weathering front at the base of the latter zone forms the interface between the solid rock and the overlying regolith. There may be an interlayering of weathered and fractured zones, the weathering relating to zones of water level oscillation reflecting water levels during specific palaeoclimatic events

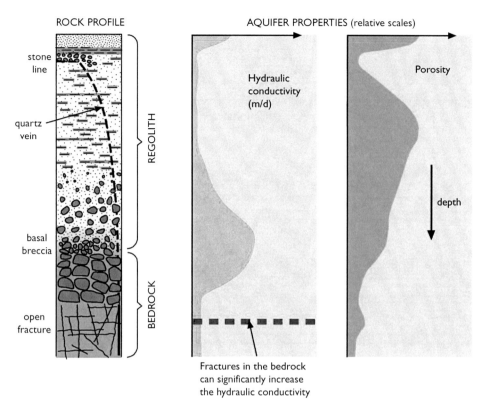

Figure 3.1 Graphic illustration of the weathered basement aquifer system (after Carl Bro, Cowiconsult & Kampsax-Kruger (1980); Jones (1985).

and palaeohydrology levels controlled by erosion (Ollier & Pain, 1996). Transmissivity and storativity decline as the degree of weathering diminishes with depth passing from granular regolith into fractured rock. This phenomenon and the classic graphic illustration (Figure 3.1) first described by Carl Bro, Cowiconsult & Kampsax-Kruger (1980) and Jones (1985) in south western Tanzania was latterly described by Davies (1984) in eastern Zimbabwe, Acworth (1987) in northern Nigeria, Grillot & Dussarrat (1992) in Madagascar, Dewandel *et al.* (2006) and Marechal *et al.* (2007) in India and latterly by Titus *et al.* (2009) in South Africa.

Hydrogeological properties in the shallow crystalline basement aquifers of sub-Saharan Africa have been described by Hydrotechnica (1985) and Interconsult (1985). These properties have been compared with similar aquifers in India (e.g. Reddy *et al.* 2009, Dewandel *et al.*, 2006; 2010) and in South America (e.g. Neves & Morales, 2006). The main difference between the African and South America savannah lands and India is the dominance and regularity of the monsoon rains over India which contrast with the drought cycles of southern Africa and the aridity within much of continental Brazil. Similar comparisons have been made with basement rocks in Europe (e.g. Banks *et al.*, 2006) and the commonality between Africa and the island of Jersey in the Channel Islands (Îsles de la Manche) where a shallow weathered granular regolith is preserved over Precambrian granite, described by Robins & Rose (2005). The main contrast is, again, rate of evapotranspiration and the availability of rainfall for aquifer recharge although the geological setting is strikingly similar.

This chapter focuses on the similarities in hydrogeological properties that occur within the shallow weathered and fractured granite aquifer on Jersey and Precambrian crystalline basement aquifer in sub-Saharan Africa, specifically eastern Zimbabwe. Although the two areas are vastly different in scale (Jersey has an area of just 117 sq km) and in modern and paleo-climates as well as tectonism the comparison tests the hypothesis that the shallow weathered basement aquifer is a recognisable global aquifer type. If the hypothesis is correct, the detailed and sophisticated exploratory work undertaken in Jersey can be used to inform understanding of the weathered crystalline basement aquifer elsewhere.

3.2   WEATHERED AND FRACTURED BASEMENT AQUIFER

In Jersey an early formal investigation of groundwater occurrence in the basement aquifer was undertaken by engineers belonging to the occupying German military forces during World War II (Robins & Rose, 2005). A second phase of investigation followed (T&C Hawksley, 1976) while a third modern phase of investigation supports the present understanding of the island's groundwater flow system (Robins & Smedley, 1994; 1998).

Studies of the distribution of crystalline basement aquifers in sub-Saharan Africa (Figure 3.2) and the occurrence of groundwater within the aquifers were consolidated in a thematic volume edited by Wright & Burgess (1992). Taylor and Howard (2000) further investigated the relationship of geomorphology and groundwater occurrence working in Uganda. Meanwhile in Francophone Africa, parallel studies of groundwater flow through weathered Precambrian Basement strata were being carried out, for example, the classic work reported by Dussarrat & Ralaimaro (1993) and detailed catchment investigations in Madagascar – Ambohitrakoho catchment, north of Antananarivo in central Madagascar by Grillot & Dussarrat (1992).

Precambrian crystalline basement granites, granitic gneisses and greenstones occur in eastern Zimababwe and in Jersey (Davies, 1984; Cheney *et al.*, 2006). Hydrogeological data were gathered in Zimbabwe and Jersey on project bases each carried out with different objectives and with access to different technological resources. Hydrogeological data were collected during a borehole drilling investigation in Jersey and during production drilling in Zimbabwe. At both locations the presence of fractures can be recognised from drill cuttings and their relationships to weathering and water occurrence derived from penetration logs and measured yield changes with depth. Coring was not undertaken.

The conceptual model of groundwater occurrence and transport in weathered and fractured Precambrian basement is well known (Figure 3.1) and essentially originates from the work of Jones and others in Tanzania in the late 1970s (Carl Bro, Cowiconsult & Kampsax-Kruger, 1980; Jones, 1985). The depth and degree of weathering

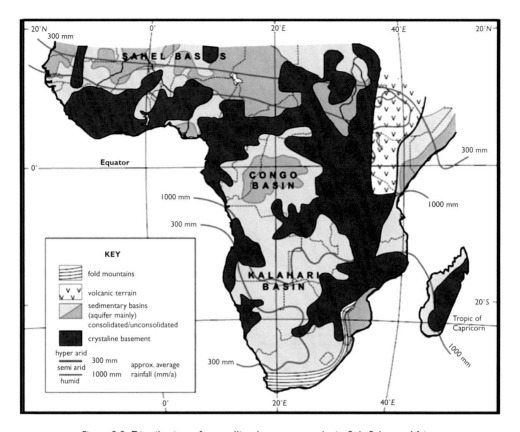

Figure 3.2 Distribution of crystalline basement rocks in Sub-Saharan Africa.

depend on a range of influences, for example, the basement in Africa is more weathered on the older African Erosion Surface than the Post-African Erosion Surface (Davies & Robins, 2007), with weathering commonly beyond a granular phase to an impermeable clay grade on the older surface. The weathering catalysts include:

- Lithology – occurrence of jointing within otherwise low permeability compact metamorphic and igneous rocks.
- Tectonism – occurrence of regional faulting and folding.
- Geomorphology – long term surface erosion and valley incision.
- Baseflow – groundwater flow to deepen the weathering front and lower the groundwater flow zone.
- Climate – effective rainfall and temperature.
- Soils – permeability, source of carbon and bacteria.
- Hydrology – drainage and run-off.

The work in Jersey was an investigation specifically aimed at the question of whether the water diviners better understood the groundwater system than did the hydrogeologists. This targeted project did, however, allow detailed borehole logging and investigation to be carried out and comparison of the shallow weathered zone to be made with the underlying fracture zone of weathering.

The project work in Zimbabwe was carried out during 1984 (Davies, 1984) as part of a refugee resettlement programme that coincided with a period of prolonged drought. Upper weathered zones were depleted due to declining water levels and only small volumes of water were obtained from lower fractured zones.

3.2.1   Jersey – La Rocque test site

Jersey is formed of ancient basement rocks including shales, granites, volcanic rocks and conglomerates, ranging in age from Precambrian to Cambro-Ordovician. Groundwater is abstracted for domestic, agricultural and industrial uses from a resource that is sustained by direct rainfall recharge over the island. The Jersey bedrock aquifer at the investigation site at La Rocque (49°10.089′ N: 2°1.818′ W) consists of weathered granite above hard fractured rocks, which possess minimal primary (intergranular) porosity or permeability. Most groundwater flow and storage occurs within the upper weathered fracture zone and borehole yields are dependent on the number, size and degree of lateral and vertical interconnection between the fractures. Under such conditions, failure to penetrate an adequate number of productive fractures results in a low yielding or dry borehole. In common with similar aquifers elsewhere, fractures are generally most common, larger and better interconnected at shallow depths, generally becoming fewer, less dilated and less well interconnected with increasing depth. Total borehole yields may increase cumulatively with depth, as a borehole penetrates an increasing number of productive fractures. Higher yields are normally obtained from the zone that occurs up to 25 m below the water table (Robins & Smedley, 1998).

The exploration borehole drilled at La Rocque penetrated superficial loessic silts and fluvial silty gravels above weathered granite. Between 10 and 43 m depth, weathered and fractured granitic rocks were penetrated. Fracture zones were identified by the presence of discoloured yellow or orange granite chippings. Rock chip samples of the superficial deposits and solid rocks penetrated were obtained at 0.5 m depth intervals during drilling and the borehole cleared of drill cuttings before the next 0.5 m section was drilled. This ensured the collection of a representative sample of cuttings for the 0.5 m section drilled. The locations of fracture zones were identified, using specific physical characteristics discussed below, and recorded. Drill penetration rates, related to specific hammer and bit assemblies, were measured at 0.5 m intervals. Drilling penetration rates varied in the granite reflecting hardness and weathering (Figure 3.3).

During drilling, air flush water discharge rates were measured at 3 m depth intervals within the upper section to be grouted and at 0.5 m intervals in the lower open hole section. Discharge rates were determined by capturing the air flush water from the borehole within a bund around the top of the borehole, while the water flowing out through an outlet pipe was timed into to a 14 l measuring tank. The cumulative air flush water yield was recorded to reach 7 l/s by 43 m depth (Figure 3.3). The borehole was pressure grouted to seal the borehole to surface. On drilling below 43 m to 55 m. harder granite was encountered, reflected by the slower drill penetration rate, with fewer water producing fractures being encountered. At 55 m depth the cumulative air discharge rate had reached about 1.5 l/s.

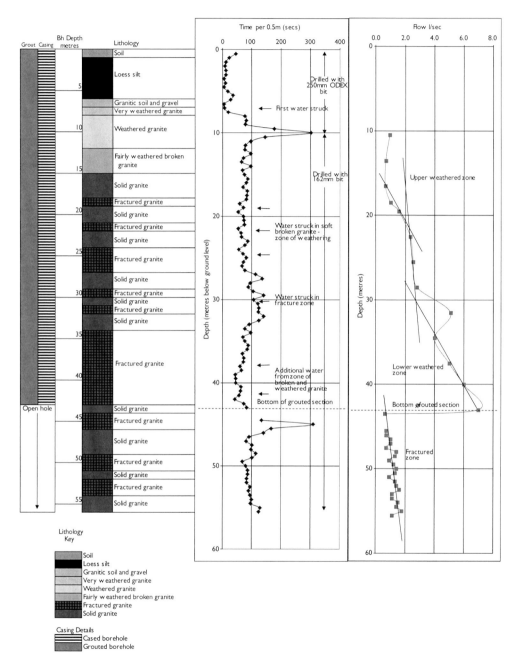

Figure 3.3 La Rocque borehole – construction, geological log, penetration rate and air flush yields (air flush yields are cumulative down the hole).

Test pumping of the lower open part of the borehole with yield step tests suggested a specific capacity of 15 m³/d/ m at yields up to 3.3 l/s. A 72 hour constant discharge test was carried out at a yield of 3.2 l/s and the data analysed to indicate a transmissivity of between 83 and 88 m²/d from an aquifer that reacted in a confined manner. An existing, but disused, borehole at the site served as a partially penetrating observation borehole.

The specific electrical conductivity of the water was brackish (1641 to 1666 µS/cm. Between 10 and 43 m depth, pH values lay in the range of 8.1 to 8.3, specific electrical conductivity varied between 1500 and 1620 µ/Scm and water temperature varied between 13.1 and 13.8°C. The rate of water discharge increased incrementally with depth from 0.6 l/s at 16.5 m bgl, to 2.5 l/s at 25.5 m depth, 4 l/s at 34.5 m and 7 l/s at 43 m depth. The increases in water inflow correlated with the occurrence of recognisable weathered and fracture zones. The rest water level in the 43 m deep borehole, before casing was installed, was 2.41 m. The borehole was cased and grouted to 43 m and all water inflow was successfully excluded.

Whilst drilling beyond the set grout, between 43.0 and 45.5 m depth, small quantities of water were encountered and the discharge rate increased gradually from 0.6 l/s at 43.5 m to 0.7 l/s at 45.5 m depth. Between 46.5 and 55 m, pH values lay in the range 8.3 to 8.7, conductivity varied between 1530 and 1670 µS/cm and water temperature varied between 12.9 and 13.5°C. The rate of water discharge increased incrementally with depth but at a much slower rate than in the upper section. The rate of discharge increased from 0.7 l/s at 45.5 m to 1.4 l/s at 50 m and to 1.75 l/s at 55 m, water inflow being from thin fracture zones in the granite (Figure 3).

As the shallow water bearing zone between ground level and 43 m was sealed off, the hydrogeological characteristics of the upper shallow and lower deep horizons could be defined and differentiated. The upper shallow horizon comprised two zones of softer weathered granite separated by a harder jointed granite. These zones correlate with increasing rates of water inflow and rates of drill penetration as determined during drilling. The lower deep horizon comprises harder jointed granite, with rates of groundwater inflow and drill penetration similar to those determined for the harder zone between the weathered zones above.

Analysis of step test results provides an aquifer loss coefficient (B) of 3.1×10^{-2} and a well loss coefficient (C) of 1.3×10^{-4}. Step tests suggest a sustainable specific capacity of around 15 m³/d/m at discharges not substantially greater than 3.25 l/s (280 m³/d). Analysis of constant rate tests indicate transmissivity values of 83 m²/d and 88 m²/d at the abstraction and observation boreholes respectively. Analysis of constant rate tests in the partially penetrating observation borehole suggest local storativity to be in the order of 1.3×10^{-5}. Analysis of recovery indicates transmissivity values of 84 m²/d and 86 m²/d at the abstraction and observation boreholes respectively. The close proximity of the transmissivity values derived from the two boreholes and different analytical methods suggests approximately 80 to 90 m²/d to be a reasonably accurate evaluation of local transmissivity. The closeness of rest water levels, recovery and derived transmissivity values in the two boreholes suggest both the deep abstraction and shallow /observation boreholes are intercepting the same aquifer and that the deep and shallow fracture systems are closely linked.

The storativity value is indicative of confined conditions; the first water strike was at a depth of 10 m and the water rose to 2.4 m. The confining layer is likely to comprise the silty and clayey loess, present to a depth of 5.5 m and underlying heavily weathered granites, in which any fractures are likely be filled with relatively impermeable material.

In summary, the cement grout sealed the shallow main water bearing horizons, the borehole remained dry after the grout had been removed from within this upper cased section of the borehole. The total discharge of 7 l/s measured in the shallow upper zone was successfully closed out and only 1.5 l/s was obtained from the lower deep zone. Cheney *et al.* (2006) concluded that 'A fractured granite aquifer exists in the La Rocque area, which appears capable of providing significant borehole yields. Yields are better at shallow depth than in the deeper granite and are cumulative with increasing depth, being related to the number and productivity of the fractures penetrated, rather than the yield obtained from any one single large fracture…'

3.2.2 Eastern Zimbabwe

Rural communities in Eastern Zimbabwe are almost entirely dependent on groundwater as surface supplies are ephemeral and many are unsafe. A village water point typically comprises a hand pump in a borehole or well with a sustainable yield of barely 0.3 l/s. As water levels decline in dryer years water points may fail because of mechanical problems caused by increased usage and stress or, as has been shown to be the case in parts of neighbouring Malawi (Davies *et al.*, 2011), because the groundwater resource is incapable of matching demand. A project within a drought relief drilling programme illustrates how the shallow weathered basement aquifer performs: the Nyamazura Resettlement Scheme involving 400 families in 12 villages (Davies, 1984).

A set of boreholes were drilled and test-pumped within Precambrian granites and granitic gneisses with associated dolerites at Nyamazura (53°53.11' S: 32°24.74' E) in Mutare District in Manicaland, eastern Zimbabwe. These boreholes were installed to supply groundwater to a number of villages demarcated a rural refugee resettlement scheme. Between July and September 1984 thirteen boreholes were drilled and three cleaned out. The drainage pattern is rectilinear and follows structural lineations trending north west to south east, north east to south west and north-south. Annual precipitation varies from 370 and 1300 mm (long term average 700 mm). However, between1982 and 1984 the area experienced drought with <550 mm/a. Actual evaporation exceeds most rainfall day amounts and recharge only occurs during scarce prolonged intensive periods of rain. The natural vegetation of bush scrub has been replaced over much of the area by open grassland. Sludge chipping samples were collected during drilling (Figure 3.4, Table 3.1) and each borehole was test pumped and water samples collected for chemical analysis.

Precambrian Basement Strata include granitic gneisses of the Older series and the granodiorite adamalite series intruded by dolerite dykes and sills of various ages. The coarse-grained granites and gneisses are composed of white quartz and feldspar with mafic minerals. These rocks often contain pink feldspathic veins. Schistic strata form hills in the north east of the area. The schistose bands of dark green micaeous amphibolitic schists are interbedded with white to grey coarse-grained gneiss. Precambrian

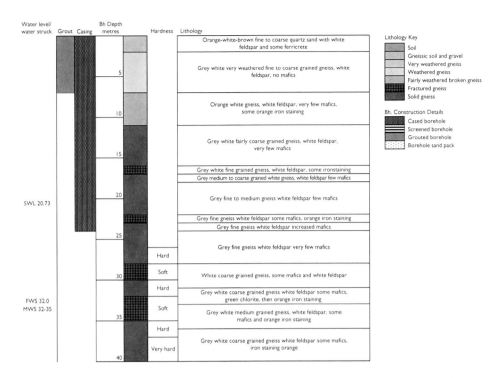

Figure 3.4 Details of a typical borehole at Nyamazura (SWL – depth to static water level, FWS – depth of first water strike, MWS – depth of main water strike).

Table 3.1 Borehole details Nyamazura Resettlement Scheme.

Bh No.	Elevation	Bh Depth	Bedrock	Soil	Regolith	SWRL	First WS	Main WS	Yield	Transmissivity	Sp Cap
	m	m	type	m	m	m	m	m	l/sec	m²/day	m³/day/m
NZ1	1022	42	gneiss	2	35	20.73	32	32			
NZ2	1018	30	granite	4	27	10.15	14	14			
NZ3	1062	40	gneiss	6	38	24.5	26	26			
NZ4A	1075	13	granite		13	dry					
NZ4B	1080	22	gneiss	3	11	5.7	10.5	10.5	0.83	25.7	4
NZ5	1100	25	gneiss	4	25	3.6	4	11	0.70	2.2	
NZ6A	1068	26	dolerite	1	18	14.1	18	18	0		
NZ6B	1055	35	gneiss	3	24	2.7	8	11	0.62	5.9	
NZ6Bbail	1055	35	gneiss	3	24	2.7	12	12	0.24	1.3	9
NZ7A	1040	44	dolerite	5	36	27.7	34	34	0.37	0.6	
NZ7Bbail	1045	32	gneiss	10	29	24.78	28	29	0.60	10.3	33.9
NZ8	1060	54	gneiss			12.2			0.67	8.7	14.1
NZ9	1045	32	gneiss	10	22	19.4	21	21	0.68	24.3	24.4
NZ10	1035	26	gneiss			6.51			0.59	20.2	19.7
NZ11A	1077	56	gneiss			4.17			0.83	5.7	
NZ11Bbail	1075	20	dolerite	9	12	6.1	11	11	0.07	0.02	6.3
NZ12A	1057	52	gneiss	5	25	14.5	26	26	0.52		
NZ12Bbail	1067	23	gneiss	8	22	6.53	12	12	0.03	5.7	20.3

Sp Cap – specific capacity, WS depth of water strike.

age dolerites are intruded into the granites and gneisses within two east-west trending zones of the central and southern parts of the area. They are characterised by rounded dark green dolerite blocks and a bright brick red soil.

Twelve boreholes were test pumped. At NZ4B (Table 3.1), a 6-hour constant yield/drawdown test was undertaken followed by a 1 hour recovery test. A standard constant yield/draw down test of 3 hours with a one hour recovery test was carried out at NZ5, NZ8, NZ9, NZ10 and NZ6B. Reduced time period tests were conducted at NZ7 A, NZ11 A and NZ12 A as yields were poor in these boreholes. A short bail test, bailing the borehole using cable tool percussion equipment for a period of 30 minutes, recording the number of bails (volume of water abstracted), the maximum drawdown level and recovery over a 60 minute period was carried out at NZ6B, NZ11B and NZ12B. Only the recovery data from the bail tests were used for analysis. The constant yield/drawdown data were analysed using the Jacob Straight Line Approximation Method and the recovery data from these and bail tests analysed using the Theis Recovery Method.

The transmissivity values obtained from the bail tests were usually low, there being insufficient 'late' data available to allow adequate analysis of the results (Table 3.1). The dolerite sills and dykes have low transmissivities with values <0.1 m^2/d. In the weathered basement complex aquifers transmissivity was lowest in watershed areas, e.g.at NZ5 where transmissivity is <10 m^2/d increasing down the hydraulic gradient to >20 m^2/d in the valleys.

Groundwater recharge off dolerite hills creates pockets of weakly mineralised groundwater with specific electrical conductance increasing to 400 µs/cm towards the valleys. The groundwater is alkaline with pH up to 7.5 increasing in parallel with alkalinity down the hydraulic gradient.

The weathered zone is thickest beneath interfluves (up to 30 m) where the zone of water table oscillation is greatest. The thinnest zones of weathering are developed adjacent to the main rivers where the zone of water table oscillation is least. In the area between the ridges and the rivers the zone of weathering is commonly 20 m thick. The weathered Precambrian Basement aquifer has a transmissivity of between 10 and 20 m^2/d. The depth of weathering is enhanced along fault and prominent joint lineation zones.

Lateral groundwater flow through the Basement Complex aquifer is controlled by topography, structural lineations and impermeable dolerite dykes. Groundwater throughflow in the north west of the area is of the order of maximum 0.30 Mm^3/a falling to 0.16 Mm^3/a in drought periods. Within the central part of the area groundwater through flow is estimated to be maximum 0.60 Mm^3/a with a minimum flow of 0.30 Mm^3/a.

3.3   DISCUSSION

The findings from the projects in Zimbabwe and Jersey are summarised in Table 3.2. The aquifers comprise:

- Low permeability weathered and fractured granite aquifers with limited storage capacity.

Table 3.2 Comparison of summary findings from Zimbabwe and Jersey with those of studies in India and Brazil.

	Zimbabwe	Jersey	India	Brazil
Climate	Semi-arid to subtropical – long droughts	Temperate, occasional short dry periods	Semi-arid, monsoonal rains June to October	Sub-tropical to temperate occasional droughts
Lithology	Weathered and fissured granite/gneiss	Weathered and fractured granite	Weathered and fractured granite	Weathered and fractured crystalline basement
Tectonism	Some north-south faults	Some ESE – WNW faults	Some faulting	NW-SE and E-W faulting
Geomorphology	Post-African surface eroded	Eroded incised plateau sloping gently to south	Eroded incised plateau sloping gently to north	Eroded block faulted piedmont
Aquifer	Confined to semi-confined	Semi-confined to confined	Unconfined to semi-confined	Semi-confined to confined
Depth of weathering	11 to 38 m	43 m	40 m	20–60 m
Depth to water table	3 to 28 m	2 m	12–28 m depressed by over-pumping	>5 m
Borehole yield	0.03 to 0.8 l s^{-1}	0.1 to 4 l s^{-1} with depth	1–10 l/s	0.1–1.7 l/s
Specific capacity	4 to 34 m^3 d^{-1} m^{-1}	0.8 to 15 m^3 d^{-1} m^{-1} with depth	Not specified	Mean 1.4 m^3 d^{-1} m^{-1} fractured Mean 6 m^3 d^{-1} m^{-1} weathered
Transmissivity	1 to 25 m^2 d^{-1}	0.3 to 90 m^2 d^{-1} with depth	0.1 to 155 m^2 d^{-1}, mean 30 m^2 d^{-1}	Not specified
Data source	Davies (1984)	Cheney et al. (2006)	Marechal et al. (2007), Dewandel et al. (2006) & Dewandel et al. (2010)	Neves & Morales (2006) Meju et al. (2001)

- Pumping from deeper fractured granite causes water levels to fall in shallow weathered/fractured granite, i.e. the two are connected.
- They have reduced permeability with depth towards the weathering front.
- Rate of baseflow diminishes with depth (Figure 3.5).

These findings are comparable with those obtained from studies of similar weathered and fractured low permeability Basement aquifers in India (Marechal *et al.*, 2007; Dewandel *et al.*, 2006; Dewandel *et al.*, 2010) and Brazil (Meju *et al.*, 2001; Neves & Morales, 2010) (Table 3.2). The similarities in groundwater occurrence in these four areas suggest that the weathered basement aquifer is indeed a global aquifer type

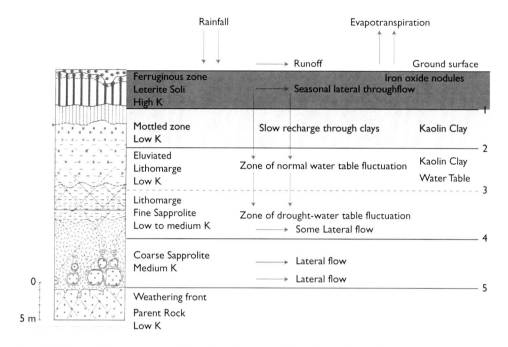

Figure 3.5 Deeper 'fracture' zones that reflect lowering of baseflow and weathering front levels. Note vertical percolation of ingressing rainwater to the water table minus evapotranspiration and any runoff and interflow laterally according to the topographical gradient. In the saturated zone of the aquifer groundwater may flow laterally according to the prevailing hydraulic gradient.

with reasonably uniform properties. Work reported from elsewhere in Africa, India and South America also support this contention (Hydrotechnica, 1985; Interconsult, 1985; Reddy *et al.*, 2009). Figure 5, therefore, becomes a universal representation of the weathered crystalline type aquifer. In addition, the detailed understanding obtained at the Jersey test site can sensibly be applied to the aquifer on a global basis, despite the different geotectonic, climatological and geomorphological histories that have been applied to the aquifer worldwide.

3.4   CONCLUSIONS

Investigation in Jersey was carried out to a high technical standard while in Zimbabwe it was primarily undertaken to find water supplies for rural communities. Nevertheless the two studies offer little differences in the nature of the weathered basement aquifer; one study carried out in a small island setting, the other in a continental setting, but the observed weathering products are similar in both. Looking further afield the same pattern of weathering has been reported from India and Brazil.

The Jersey study is a unique study in weathered basement rocks that provides detailed data that quantify the distribution of the depth to yield relationship of groundwater occurrence within the upper weathered aquifer sequence and the underlying fractured aquifer. The Zimbabwe case study characterises the hydraulic features of the weathered and fractured Precambrian crystalline basement aquifer showing the relationship between borehole depth with yield and permeability. The distribution of groundwater occurrence within the Jersey aquifer sequence is comparable with that found in Zimbabwe while these same characteristics are also found in India and Brazil. This supports the hypothesis that the weathered crystalline basement aquifer is a universal aquifer type which possesses common hydraulic properties. Therefore, the detailed knowledge gained at Jersey can be used to inform understanding of basement rock aquifers elsewhere. In addition the Jersey experience explains why yields decline in prolonged dry periods as, for example, was the case during the drought in Zimbabwe, i.e. water tables had already declined into the lower permeability fractured zone during the drought relief phase project work.

REFERENCES

Acworth R.I. (1987) The development of crystalline basement aquifers in a tropical environment. *Quarterly Journal of Engineering Geology* 20, 265–272.

Banks D., Gundersen P., Gustafson G., Maleka J. Morland G. (2006) Regional Similarities in the distribution of well yield from crystalline rocks in Fennoscandia. *Norges geologiske undersøkelse Bulletin* 450, 33–47.

Carl Bro, Cowiconsult & Kampsax-Kruger (1980) Water master plans for Iringa, Ravuma and Mbeya regions, the geomorphological approach to the hydrogeology of the basement complex. Report Carl Bro, Cowiconsult & Kampsax-Kruger, report for Danish International Development Agency (www.sadcgwarchive.net last accessed September 2012).

Cheney C.S., Davies J., Darling W.G., Rukin N. Moon B. (2006) Jersey Deep Groundwater Investigation. British Geological Survey BGS Commissioned Report, CR/06/221C. 151pp.

Davies J. (1984). Short Term Hydrogeological Assignment to Zimbabwe, 3rd May–15th September 1984. Volume 2 – The Hydrogeology of the Nyamazura Area, Zimbabwe. BGS Technical Report WR/OS/84/16 (www.sadcgwarchive.net, last accessed September 2012)

Davies J. Robins N.S. (2007) Groundwater occurrence north of the Limpopo: are erosion surfaces the key? *Geological Society of South Africa Groundwater Division Groundwater Conference*, Bloemfontein (CD).

Davies J., Farr, J.L. Robins N.S. (2011) Is the Southern African Archean Basement Aquifer up to the job? *Geological Society of South Africa/IAH Conference on Groundwater, Our Source of Security in an Uncertain Future*, Pretoria (CD).

Dearman W.R., Baynes F.J. Irfan T.Y. (1978). Engineering grading of weathered granite. *Engineering Geology* 12, 345–374.

Dewandel B., Lachassagne P., Wyns R., Maréchal J.C, Krishnamurthy N.S. (2006) A generalized 3-D geological and hydrogeological conceptual model of granite aquifers controlled by single or multiphase weathering. *Journal of Hydrology* 330, 260–284.

Dewandel D., Perrin J., Ahmed S., Aulong S., Hrkal Z., Lachassagne P., Samad M. Massuel S. (2010) Development of a tool for managing groundwater resources in semi-arid hard rock regions: application to a rural watershed in South India. *Hydrological Processes* 24, 2784–2797.

Dussarrat B., Ralaimaro J. (1993) Caracterisation hydrogeologique de bassins versants emboites sur socle altere en zone tropicale d'altitude: exemple des Hautes Terres de Madagascar. *Hydrogeologie* 1, 53–64.

Grillot J.C., Dussarrat B. (1992). Hydraulique des unites d'interfluves et de bas-fond tourbeux: un example en zone de socle altere (Madagascar). *Journal of Hydrology* 135, 321–340.

Hydrotechnica (1985) Accelerated Drought Relief Programme, Victoria Province, Zimbabwe, Final Report, Volume 2 Scientific Text, Chapter 9 Hydrogeology.

Interconsult (1985) National Master Plan for Rural Water Supply and Sanitation, Republic of Zimbabwe, Volume 2, part 2. (www.sadcgwarchive.net, last accessed September 2012)

Jones M.J. (1985) The weathered zone aquifers of the basement complex areas of Africa. *Quarterly Journal of Engineering Geology* 18, 35–46.

Marechal J.-C., Dewandel B., Ahmed S. & Lachassagne P. (2007) Hard rock aquifers characterization prior to modelling at catchment scale: an application to India. In: Krásný J., Sharp J.M. (editors) *Groundwater in fractured rocks*. IAH Selected papers on Hydrogeology 9, 227–242.

Meju M.A., Fontes S.L., Ulugerger E.U., La Terra E.F., Germano C.R., Carvalho R.M. (2001) A Joint TEM-HLEM Geophysical Approach to Borehole Siting in Deeply Weathered Granitic Terrains. *Ground Water* 39, 554–567.

Migon P. (2006) *Granite Landscapes of the World*. Oxford University Press, 384 pages.

Neves M.A., Morales N. (2006) Well productivity controlling factors in crystalline terrains of southeastern Brazil. *Hydrogeology Journal* 15, 471–482.

Ollier C., Pain C. (1996) *Regolith, Soils and Landforms*. John Wiley, Chichester, UK, 316 pages.

Reddy D.V., Nagabhushanam P., Sukhija1 B.S., Reddy A.G.S. (2009) Understanding hydrogeological processes in a highly stressed granitic aquifer in southern India. *Hydrological Processes* 23, 1282–1294.

Robins N.S., Smedley P.L. (1994) Hydrogeology and hydrogeochemistry of a small, hard rock island – the heavily stressed aquifer of Jersey. *Journal of Hydrology* 163, 249–269.

Robins N.S., Smedley P.L. (1998) The Jersey groundwater study. British Geological Survey Research Report RR/98/5. 48pp.

Robins N.S., Rose E.P.F. (2005) Hydrogeological investigation in the Channel Islands: the important role of German military geologists during World War II. *Quarterly Journal of Engineering Geology and Hydrogeology* 38, 4, 351–362.

Ruxton B.P., Berry L. (1957) Weathering of granite and associated erosion features in Hong Kong. *Geological Society of America* 68, 1263–1282.

Singhal B.B.S., Gupta R.P. (1999) *Applied Hydrogeology of Fractured Rocks*. Kluwer Academic Publishers.

Taylor R.G., Howard K.W.F. (2000) A tectono-geomorphic model of the hydrogeology of deeply weathered crystalline rock: Evidence from Uganda. *Hydrogeology Journal* 8, 279–294.

T&C Hawksley (1976) Report on water resources. Unpublished technical report T&C Hawksley, Aldershot, UK.

Titus R., Friese A., Adams S. (2009) A tectonic and geomorphic framework for the development of basement aquifers in Namaqualand – a review. In: Titus R, Beekman H, Adams,S & Strachan L (editors) *The Basement Aquifers of Southern Africa*. Water Research Commission, Pretoria, Report No. TT 428–09, pp. 5–18.

Wright E.P., Burgess W.G. (1992) *The hydrogeology of crystalline basement aquifers in Africa*. Geological Society, London, Special Publications 66, 264 pp.

Chapter 4

Outcrop groundwater prospecting, drilling and well construction in hard rocks in semi-arid regions

António Chambel
Geophysics Centre of Évora, University of Évora, Department of Geosciences, Portugal

ABSTRACT

This chapter presents some recommendations for prospecting, drilling and well construction in hard rocks in semi-arid regions. Considering that these conditions are present in many countries where technology is not always available, the chapter concentrates on the most basic and simple methods to plan where best to drill and maximize success through the direct observation of rock types, weathering and fracturing. The advantage for the geologist and hydrogeologist in an arid or semi-arid environment is that vegetation is normally scarce and the weathering layer thin, allowing a direct view of the rock in circumstances impossible in other climate regions of the world. The close observation of the weathering material, and especially of the fracture network, mainly the fracture density, dip, extension and interconnection, can provide important information for a field hydrogeologist who can then plan the best place for drilling. The most appropriate drilling technique, if available in the area, is rotary percussion, also designated as down-the-hole drilling, with drilling rates that can achieve 100 m per day in normal circumstances. This allows a well to be constructed in about two days, essential in the case of disaster relief. Finally, some information is given about well construction, careful planning of the work, protection to preserve the water quality, avoiding problems of partial or total collapse of the hole during construction or of the well after completion, and how to avoid direct contact between the surface or sub-surface waters with the aquifer along the walls of the well to protect the well and the aquifer against contamination.

4.1 INTRODUCTION

Groundwater prospecting is a basic task wherever there is a need to improve drilling results for water supply purposes. But in many countries of the developing World prospecting is not a priority or even possible due to the lack of professional hydrogeologists and the lack of financial resources (sometimes donors send money just for drilling with no payment for data gathering and study).

This chapter is targeted at those hydrogeologists and geologists who have less experience with hydrogeological field studies, in order to provide them with a guide to quickly identify features that can help them to make decisions. The semi-arid zones are present in many developing countries and the techniques can be applied in these conditions. Geophysics is of great importance in many places. For geophysical guidance see general books, some more theoretical, some focused more on hydrogeology, such as, for example, Telford *et al.* (1990); Parasnis (1996); Roy (2001); Lowrie (2007); Kirsch (2009); Milsom & Eriksen (2011); Chandra (2014) or practical examples presented in many other publications totally or partially dedicated to geophysics (e.g. Kellet & Bauman, 2004; Mohamed *et al.*, 2012; Kouadio *et al.*, 2013; Venkateswaran & Jayapal, 2013).

The quality of groundwater prospecting has to do with knowledge. It depends on the available information: geological or hydrogeological maps and information, geological, geophysical and hydrogeological studies, drilling reports, and other data. But reliable information is scarce in many countries or in specific parts of some countries. Groundwater is present in most environments on the Globe. It normally has good quality, it is reasonably well protected from contamination and is often present in places where no permanent surface water is available (e.g. deserts and other semi-arid areas).

In many of the less developed areas of the World, technology may be lacking, financial resources may be inadequate and there are difficulties even with simple chemical water analyses. Sanitary conditions are an overwhelming problem causing groundwater contamination. With few resources, simple prospecting processes are essential, in order to get the maximum consistency between investment and output. Many of the low income countries include extensive hard rock areas in semi-arid climate zones, while others are situated in tropical or mountainous areas. This chapter helps to address the issue of outcrop prospecting, using knowledge, common sense, and some technology when available. The main focus is on practical field observations and less expensive methods that will help hydrogeologists make the best decisions for these geological and hydrogeological settings. Drilling and technical construction of wells and groundwater protection are also presented.

4.2   OUTCROP GROUNDWATER PROSPECTING IN SEMI-ARID HARD ROCK ENVIRONMENTS

Krásný (1996) and Rubbert *et al.* (2006) demonstrate that hard rocks have three to four layers of interest for hydrogeology. Krásný (1996) describes three layers that represent the geological environment of a hard rock aquifer (Figure 4.1):

- Upper weathered bedrock
- Middle fissured bedrock
- Unweathered bedrock

Thus the productivity of a well is dependent on:

- The groundwater level
- The thickness of the two first layers
- The resulting weathering characteristics of the upper two layers: the presence of clay materials can make the layers impermeable and unproductive, whereas a granular composition can be beneficial both for infiltration and storage
- Number, length, dip, fracture intersection, dilation, and composition and filling of the fractures in the second and third layers.

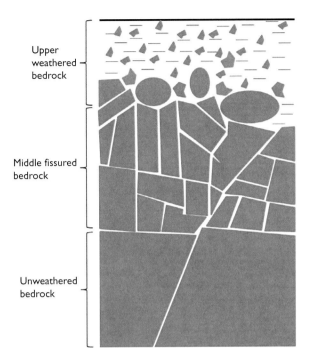

Upper
weathered
bedrock

Middle fissured
bedrock

Unweathered
bedrock

Figure 4.1 Schematic profile of a hard rock environment, with the three characteristic vertical zones (after Krásný, 1996).

If the first layer has a low clay content and a high porosity and the second layer is represented by the same kind of weathering materials as the matrix between boulders or rock fragments, a 10 m thick saturated zone is of interest for prospecting, for water supply to small rural communities. If the thickness exceeds 20 m, it has the potential to be used for irrigation. One characteristic of this kind of aquifer is the potential to overexploit in dry periods, as this can induce a higher rate of infiltration, when much of the infiltration capacity is controlled by a piezometric level near the soil surface (Figure 4.2). This rapid replenishment of the aquifer is much more effective if the weathered layer is porous, and less effective when fine-grained particles are present or the fractured layer is close to the surface.

The kind of rocks that can best represent these circumstances are varied, but the most acid rocks (granites and similar rocks) are those whose weathering layer can better mimic a granular porous media in the upper layer. The weathering of basic rocks, like gabbros or diorites, are more likely to result in a fine-grained clayey weathered layer, which retains the infiltration water for a longer time and reduces the recharge potential and the storage capacity. The hardest of the metamorphic rocks offer the best potential for recharge (i.e., quartzites or greywackes), and the clay rocks (e.g. shales) produce clay-rich weathered layers. The depth of the weathered layer in basic rocks is normally deeper than in acid rocks, and the deeper weathered layer can form excellent aquifers in semi-arid areas, beneath superficial clayey layers, with water accumulating in the fissured bedrock. This is the situation, for example, in the Gabbros of Beja, in Portugal (Duque, 2005).

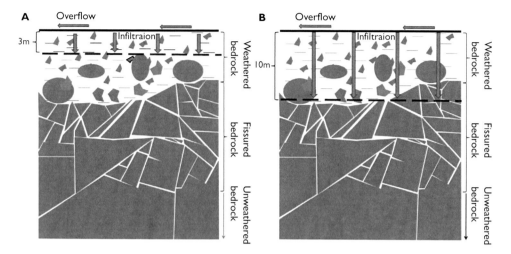

Figure 4.2 Explanation of how groundwater levels can affect the recharge capacity of a shallow aquifer in hard rocks. In case A, in natural conditions, the overflow can occur after the replenishment of the first 3 m of the unsaturated zone with water; in case B, the overexploitation permits a much higher recharge, by filling the 10 m length space between the artificial water level and the surface, before overflow happens.

Direct observation of the fractured zone can be problematic, as it is normally covered by the weathered zone. In many of the semi-arid regions, however, the weathered layer is thin or non-existent, and the fractured zone is at least partially exposed. The best way to get more information on the fracture zone is through the application of geophysics. When direct observation is possible, attention must be paid to:

- Fracture grading (highly fractured to non-fractured)
- Fracture dip
- Fracture intersection (interconnection)
- Grade of fracture opening (dilation)
- Filling of fractures (clay, quartz, calcite, etc.) and the internal fracturing of the filling materials
- Depth of fracturing (reduced dilation with depth).

If the fractures are interconnected, an assembly of inclined and horizontal fractures (Figure 4.3A and 4.3B), or vertical and horizontal (Figure 4.3C and 4.3D), then the prospects for drilling vertical boreholes are good although the conditions in A and B are the most favorable.

Where the fractures are not all interconnected (Figure 4.4, showing 3 different clusters in each case), the chance of striking water is much higher when inclined fractures are present (Figure 4.4A and 4.4B), compared with the clusters of vertical and horizontal fractures (Figure 4.4C and 4.4D). Two of the boreholes fail to hit the target in this last case. The analysis of the fracture network will help decide where to drill. In addition the actual relief will assist in locating the optimum drilling site as intense fracturing may coincide with lower-lying ground.

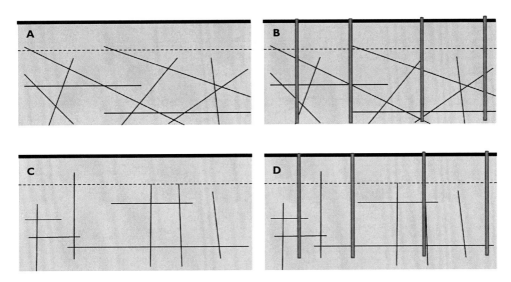

Figure 4.3 Example of two possible fracture patterns where fractures are all interconnected (on a tilted fracture network, down a vertical and horizontal fracture pattern). The wells will always be successful in both patterns.

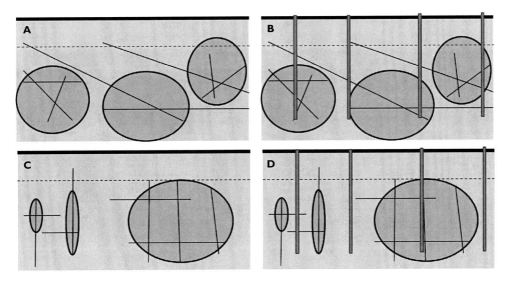

Figure 4.4 Example of another two possible fracture patterns where fractures are disconnected in some places (there are three clusters of fractures each one marked by a darker circle). The set of wells will always be successful in the top pattern (A, B), but not in the lower one (C, D).

The facture dilation and the material filing the fractures (if it exists and what kind of geological materials are present) also need to be considered. Depending on the water pressure, open fractures with 1 mm dilation can sustain wells producing 1 l/s, and are important for small supplies and use of hand pumps. The best condition is where the fractures are open and clean (with no infill material). Clay will close a fracture and make it impermeable. Calcite, quartz and oxides can also close a fracture, but, if the infill material is also fractured, it can still be productive.

Figure 4.5 to Figure 4.8 show examples of some features that can be seen in the field and what kind of interpretation can be drown from each situation. The views in Figure 4.5 are taken from natural outcrops or in previous cuts for engineering works (road construction for example).

The quartzite's of Figure 4.5A and 4.5B (Castelo de Vide, Portugal) form good aquifers in hard rocks in Portugal. Figure 4.5B shows the high level of fracturing and its opening, which permits abstraction, and raid recharge during rainy events. Figure 4.5C and 4.5D show two examples of migmatites (Helsinki, Finland) where the quartz veins are sin-metamorphic (4.5C) or fill totally the fractures (4.5D), closing all the fractures to the circulation of water.

Figure 4.5E and 4.5F show a granite area in Portugal (Portalegre region) and metamorphic rocks cut by volcanic dykes in Jordan (Aqaba region). Both have a fracture network where large blocks are separated by fractures of limited extension. It is possible to verify that an open fracture closes in just 2 or 3 m depth, and the links between the open fractures are rare. In this case, the expectation of getting water from these rocks is low. In the Portuguese case, these rocks correspond to the less productive sector of all the igneous and metamorphic Paleozoic rocks of South Portugal (ERHSA, 2001; Chambel et al., 2007).

Figure 4.5G and 4.5H show shale formations in the Mértola region of Portugal. In Figure 4.5H an intercalation of a quartzite fractured layer is identified. Stratification is clearly visible in Figure 4.5G and the shales are cut by subvertical fractures, but these fractures are neither continuous nor permeable, sometimes closed, sometimes with fracture space filled with clay gouge materials. The close observation of the behaviour of shales around the fractured quartzite in Figure 4.5H show that the fractures crossing the quartzite layer are closed by material resulting from crushing of clay minerals from the shale layers surrounding the quartzite. Wells are only productive in this area when more recent fractures are filled with quartz veins. In the conditions of both Figure 4.5G and Figure 4.5H, the yields are normally low (less than 1 l/s, with more than 50% of the wells dry).

A vertical exposure of at least 5 m height is ideal to try to preview the fracture network with depth. Road and railway cuts are normally excellent areas of observation. Cuttings can answer the questions:

- Is the fracture pattern the same up to down the exposed wall and with good features for water flow (fracture opening, interconnections, favorable dip of the fractures)?
- Is the fracture pattern changing slowly with depth (less intense fracturing, less connection) although still worthy of drilling?
- Is the fracture pattern changing rapidly with depth? How does it change? Is the fracture network disappearing with depth?

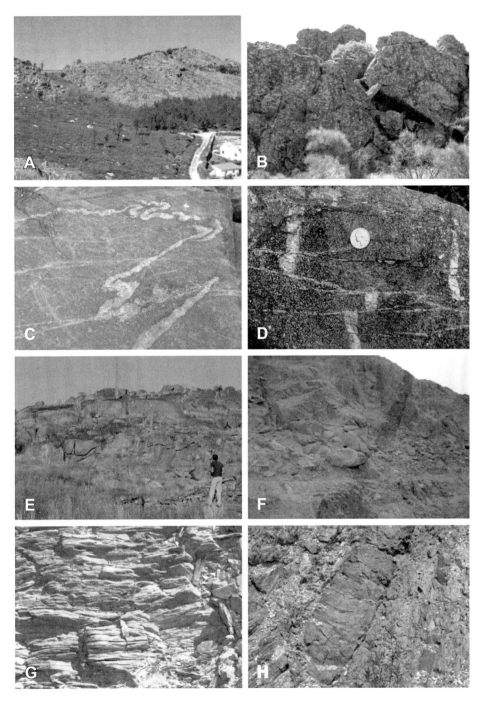

Figure 4.5 A–B: Quartzites, Castelo de Vide, Portugal; C–D: Migmatites, Helsinki, Finland; E: Granites, Portalegre, Portugal; F: Metamorphic rocks cut by a volcanic dyke, Aqaba, Jordan; G–H: Shales, Mértola, Portugal, with the intercalation of a quartzite layer in case H. (*See colour plate section, Plate 10*).

Figure 4.6 Excavation for geotechnical purposes, important for the selecting drilling sites (the objective was to create an industrial area in this area and both studies were occurring at the same time). Figures B, C and D show more favourable features for water than Figure A (fractures filled with highly fractured quartz veins). (*See colour plate section, Plate 11*).

In the first case it is likely that the pattern will persist for 20–30 m and is favorable for drilling. In the second case, the situation is not so favorable, although inclined fractures are more favorable than vertical ones. In the third case, the groundwater potential is negligible.

It is sometimes possible to observe the rock from existing works in the area. In Figure 4.6, the objective was to plan an industrial area. During the planning, some excavations were made for geotechnical purposes and it was possible to observe in situ the kind, structure and fracturing of the rock. This assisted in planning the drilling positions for wells to supply the site. The excavations were only 1 to 1.5 m deep, in an area where the weathered layer was just a few centimeters thick. The rocks are metamorphic shales, schists and greywackes. The photos of Figure 4.6B, 4.6C and 4.6D show fractures with better potential than Figure 4.6A, due to the filling of fractures with highly fractured veins of quartz and the presence of greywackes intercalated with shales and schists. In the case of Figure 4.6A, the more thin sediments dominate (shales and schists) and the fractures are thinner, without quartz inside the fractures.

Figure 4.7 shows a shale formation crossed by a fractures filled with quartz veins, some highly fractured. The position of the fractures (highly tilted, with many of them

Figure 4.7 Area in metamorphic rocks (shales, schists, greywackes) in Granja, Mourão (Portugal). With such quantity, position and fracturing of quartz veins, it would be a good place for drilling. In all the other visible cuts in this village the quartz veins are absent. (*See colour plate section, Plate 12*).

horizontal), and the links between them, make this an excellent position to get water. This place was selected after field work around a village of 900 inhabitants (Granja, Mourão, Portugal), and, in the same geologic formations around the village there were no other areas with similar features. Nevertheless, in this village, a group of productive wells on the opposite side of the village is supplying water to the population. The water is abstracted from hard rocks overlaid by a sedimentary formation. In this geologic environment, direct observation of the aquifer is not possible and the only way to get information on the underlying rocks is to use geophysical methods. The wells, some of them more than 30 years old, were probably drilled without any technical support. As the area proved to be productive in the past, all the subsequent wells were drilled in the same area.

Figure 4.8 shows a well that was also drilled many years ago, into shales and schists. In a cut on a nearby dirt road a 100 m long 5 m high vertical rock wall is visible. The uppermost weathered layer is thin, and there are just a few centimeters of soil (which is characteristic of other semi-arid areas). The metamorphic rocks are cut by fractures, some of them filled with quartz veins, mainly the sub-horizontal fractures, which are otherwise ideal for yielding water. This well has a yield of 8 l/s, obtained by a flow test of 8 h, and it supplies a village with about 70 inhabitants. A new well was drilled during a campaign in 2004–2005, in a similar geologic setting 80 m away from this one, and has a 20 l/s yield.

Figure 4.8 Example of a well in metamorphic rocks (shales, schists, greywackes) in central Portugal (area of Nisa). Quartz veins and fractures cut the rock in this specific position, where there is a productive well (yield of 8 l/s) which supplied a population of 70 inhabitants (the well is no more in use for water supply). (*See colour plate section, Plate 13*).

4.3   DRILLING METHODS

See Driscoll (1986), Repsold (1989) or Misstear *et al.* (2006) for a review of drilling methods. In hard rock environments, the advantages of using mechanical drilling rather than digging are described by Larsson (1987):

- Drilled wells need less space than excavated ones
- It is easier to protect a drilled well against pollution than an excavated one
- A well of 40–60 m depth can be drilled in one to two days, while an excavated 15–20 m deep well dug by hand may take six months
- The drilled wells are more reliable due to their greater penetration of the aquifer.

But there are disadvantages:

- The drilled wells need pumps, which are more expensive than equipment that can be used in hand dug wells
- For irrigation or bigger supply systems for domestic purposes, submersible pumps are needed and these require electricity, which is not always available
- The costs of water pumping is higher in drilled wells than in the excavated ones, especially when the water levels are low

Figure 4.9 Drilling rig for rotary percussion. The compressed air enters the drill tubes (in the image near the top of the tower), and drives the percussion of the down-hole-hammer. The rotation of the hammer is given by the engine on the top of the drilling pipes.

- Usually, and due to the yields obtained in drilled wells in hard rocks, the water cannot be used for agriculture without a storage reservoir
- The drilling rigs are not available in some less developed regions or it may be too expensive to use them.

The most suitable mechanical drilling method in hard rocks is rotary percussion, also called down-the-hole hammer (Figure 4.9). From all the techniques, rotary percussion is the most efficient, quick and easy to use method, and applies rotation of the hammer and percussion at the bottom of the well. The transmission of power is hydraulic (rotation) and compressed air (percussion). The cooling fluid is the air and it can be, in some circumstances, combined with water, foam or other liquids when the well is nearly dry to avoid dust and cuttings from the drilling, forming a mud that sticks to the hammer, slowing or even stopping the drill. Rotary methods can also be used, but are only at an advantage in special circumstances when a thick weathered or a sedimentary layer is present on top of the aquifer.

Using compressed air as the drilling fluid allows the water strikes to be detected while drilling, as the cuttings and the water rise to the surface in just a few seconds. This does not happen in the rotary system, as the use of water as the circulation fluid masks the water strikes.

The decisions that need to be made are:

- Drilling technique
- Well depth (according to the field observation and the expected thickness of the weathered and fractured zone)
- Drilling inclination (normally vertical, but can be angled depending on the inclination of fractures)
- Drilling diameter (depending on the expected yield and calculated abstration rate).

A strong and close control of the drilling is advisable and must include:

- Analysis of the cuttings (rock type, weathering conditions, fracturing)
- Water strike levels and volumes (measured using simple methods, such as measuring the time to fill a container with known volume)
- Rate of drilling
- The noise of the drilling equipment (a change of the kind of noise indicates a change in the drilling conditions, which can represent a defect in the drilling equipment or a change in lithology or in the fractured rock network).

4.4   WELL CONSTRUCTION

The construction of a well in a hard rock environment involves risks, including the siting of the well. During drilling, some other issues must be carefully evaluated (see for example Huffman & Miner, 1996; Tomlinson, 2008; IGI, 2013). In reality, the first few metres are crucial to the success, protection and durability of a well. The initial drilling is normally easy, as the hammer is cutting soft (weathered) rock material in the unsaturated zone. Only in a few cases, when groundwater level is near the surface, will this zone be hard to drill with the rotary percussion method, and the bit can be clogged with clay material. Adding water to the basic drill fluid (air is the normal) can help at this stage.

4.4.1   The need for conductor casing during drilling

It is essential to case off the upper weathered zone before drilling on. The problem otherwise will occur when a water strike in the fractures begins to cause erosion in the upper part of the well. The compressed air used as circulation fluid rises at high speed with water and debris from the cuttings and begins to erode the upper

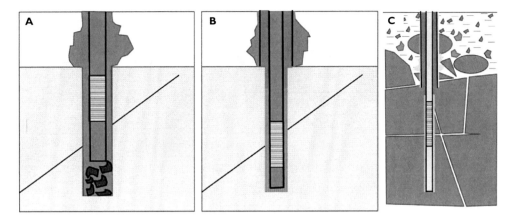

Figure 4.10 Schematic representation of a well in an area with a weathered layer and one productive fracture (A and B). In case A the position of the screen is incorrect, due to the presence of debris in the bottom of the well. In case B the screen is located in the right position, due to the protection given by the casing from the weathered layer. Case C shows a good positioning of the well screen in a more realistic case, with well casing in the upper part of the well, preventing debris entering the well.

layer in the unsaturated zone. Figure 4.10A shows an example of erosion, in this case with just one productive fracture, and the debris from the upper weathered zone is falling down the well. The cavity may develop such that the rig is in danger of collapse. The final position of the screen may be misplaced, as can be seen in Figure 4.10A in this case caused by the fall of debris between removal of the drill pipes and the insertion of the casing. The right calculation of the position of the screen pipe (in front of the productive fracture) is camouflaged by the debris falling from the weathered layer, leading to the final position of the screen away from the fracture.

It is important to spend some time thinking how to protect the weathered layer before drilling the fractured hard rock mass. The easiest way is to drill the first few metres with a larger diameter hammer (for example 254 mm) until the firm hard rock is detected, and to stop drilling and insert an plain, protective conductor pipe with the same dimension (normally made of steel and cemented into the rock) and then to drill inside this pipe with a hammer of 216 or 178 mm (see example in Figure 4.10B). The upper layer is now protected and erosion will not occur during drilling. If some weathered rock falls during the drilling (as it is the case in Figure 4.10B), the drilling space will be protected by the conductor pipe.

In more than 90% of hard rock boreholes, collapse is related to the upper weathered layer and not with the fractured zone. If, however, the fracture zone is causing problems, it may be necessary, before inserting the completion casing, to clean out the hole with the drill string and bit.

After the completion casing is inserted, the conductor casing may either be left in place or removed. Sometimes, when the condutor casing is more than 6 m long, it can be difficult to remove it, as the rig may have insufficient power to pull it.

If the water use is for domestic supply, and if the protection tube is made of iron, it must always be removed in order to avoid contamination by iron oxides; if a long conductor pipe is needed it should be made of steel or PVC which can be left in place. In most cases, when the well is complete the conductor pipe is not needed anymore provided the sanitary seal between the completion casing and the well walls is properly installed, using clay or preferably cement grouting.

4.4.2   Completion casing and well screens

The casing and well screen may be made of PVC, and must support at least 10 kg/cm^2 of pressure in order to avoid damage from overpressure along the well walls. Steel casing may be preferred in a highly fractured borehole. The screen sections must be inserted in the correct position (see Figures 4.9B and 4.9C). The opening of the screens must be about 1.5 mm. The casing diameter reflects the production yield and pump bowl diameter, the drilled well diameter and the fracture density and fill material. There must be an annular space between the pump and the inner part of the casing sufficient to permit water transport to the pump intake.

The enlargement of wells due to high pumping yields is rare in hard rocks. The annular difference between drilling and casing diameters must be bigger if the fracture network is extensive and the fracture infill includes clay. If there are only a few clean dilated fractures, an annular space between a drilling diameter of 216 mm and a 140 mm completion casing is sufficient to avoid problems.

4.4.3   Annular filter and grouting

A gravel pack must be inserted in the space between the casing and the well walls, from the bottom up to the weathered zone (as can be seen in Figure 4.10B). From the top of the gravel to the surface, the space must be filled with impermeable material (clay or cement grout), in order to seal the upper part of the well from surface and near surface contaminants.

The diameter of the gravel must be compatible with the screen slot size. For a screen slot of 1.5 mm, the graded gravel pack must be between 3 and 7 mm in diameter in order to prevent it passing through the slot.

4.4.4   Examples how to drill and construct a well in different geological settings

Figure 4.10C shows a well-designed borehole, with protection in the upper part, where the drilling penetrated a weathered zone with boulders. A protecting steel pipe is set into the firm hard rock, and drilling from that point on is done inside the pipe. After completion, the casing is inserted, with the screen in front of the productive fractures, and the well is completed with gravel pack between the casing and the well walls and sealed to surface.

If the hard rock is under a layer of loose sedimentary material (see Figure 4.11), drilling with the rotary percussion can result in an enlargement of the originally planned diameter (Figure 4.11A), especially if water is contained in the sediments. If dry, this problem is less likely to occur.

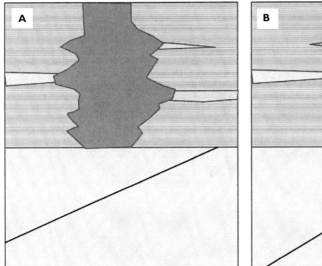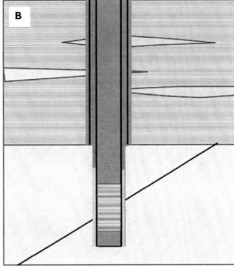

Figure 4.11 Example of a well in hard rocks under a loose sedimentary formation (clay, sand). The drilling was by rotary percussion in the sedimentary layer and this can cause an uncontrolled enlargement of the well (case A). In this geologic setting, the drilling on the sedimentary layer can be done using the rotary process, the well must then be isolated by a conductor pipe (case B) and the drilling on the hard rocks behind can be done using the rotary percussion method. Finally, the casing with the screen is inserted, the gravel filter pack installed and, on top of it, the seal of impermeable material is emplaced (case B).

The rotary method is better suited to drilling the sediments (direct circulation if the sediments are fine-grained, reverse circulation if the sediments are coarser, especially if coarse gravel or cobbles are present). After drilling the sediments, the walls must be protected by conductor casing, and drilling of the hard rock below may be continued with the rotary percussion method to a diameter that permits the hammer to pass inside the conductor pipe. This arrangement will allow the quickest time to drill the borehole. When drilling is finished, the casing with the screen must be inserted in the right position (with the screen in front of the productive fractures) and the well finished with the gravel pack and sanitary seal (Figure 4.11B).

4.4.5 Data collection during drilling

During drilling (Figure 4.12A), valuable information can be obtained from the cuttings (Figure 4.12B to 4.12E). The cuttings must be collected with a bucket (at least every 3 m drilled depth), and kept in a row later to be described and logged (Figure 4.12B). The final drilling report must contain this information, which can be done in the field or in a laboratory. The samples can be stored in plastic bags identified with the borehole name and number, location and depth interval. Figure 4.12C shows a possible fracture in the central zone with coarser grains indicating the likelihood of a fracture (Figure 4.12D and 4.11E). Confirmation comes from flat surfaces in the drill cuttings particularly

Figure 4.12 Drilling in hard rocks (A). Steps during the construction of the well: drill cuttings taken every 3 m (B), a section showing material which indicates a probable fracture (C), close analysis of the cuttings directly picked from the well during drilling (D), or after washing the cuttings with water to clean away dust (E), and observing the flow coming from the well during drilling, using compressed air directed to the bottom of the well (F). (*See colour plate section, Plate 14*).

those that have quartz, calcite, iron or manganese oxides on the surface. Non-fractured homogeneous rocks are represented by fine-grained cuttings. Figure 4.12F shows the flow coming from drilling with compressed air. With a bucket of known capacity it is possible to measure the time to fill it and calculate the discharge to give an idea of the likely final abstraction yield. After completion, the yield must be confirmed by test pumping. In Figure 4.12F the yield corresponds to approximately 1 l/s.

4.5   CONCLUDING STATEMENT

In semi-arid and arid areas rock weathering is normally shallower than in humid areas. Dryer environments generate less vegetation, which helps inspection of rock outcrop, fracturing and weathering. Aquifers in hard rocks require careful prospecting before drilling. Hard rocks have three distinct layers: the upper shallow weathered bedrock, the middle fissured bedrock, and the fresh bedrock below. Water is stored in both fractures and the rock matrix, but is transported mainly through the fractures. The capacity to transmit water is related to the interconnection between the fractures rather than the overall storage capacity of the fracture system.

Vertical fractures are more difficult to intersect with a vertical well than an inclined well, and horizontal fractures will be intersected by all wells, except the horizontal ones. Best conditions occur where vertical, horizontal and inclined fractures are present, they interconnect and are present over a thick zone, their distribution is regular and the space inside the fractures is not filled with clay or other impermeable minerals. An indication of the thickness of the two upper zones is given by geophysics, either electrical resistivity or electromagnetic methods.

Rotary percussion is the preferred drilling method. It is essential to record the lithologies, the fracture levels and the water strikes. Observation of the penetration rate can help identify fractures.

Good planning, the right construction skills and a sustainable exploration plan are essential to create a durable sustainable working well. Isolation, protection, casing, gravel packs, sanitary seals and borehole caps are the essential components of a successful well. Finally, if the abstraction rate is high a water balance calculation is needed to ensure sustainability of supply.

ACKNOWLEDGMENT

The author acknowledges the funding provided by the Évora Geophysics Centre, Portugal, under the contract with FCT (the Portuguese Science and Technology Foundation), PEst-OE/CTE/UI0078/2014.

REFERENCES

Chambel A., Duque J., Nascimento J. (2007) Regional study of hard rock aquifers in Alentejo, South Portugal: methodology and results. In: Krásný J. & Sharp J. (editors), *Groundwater in Fractured Rocks* IAH Selected Papers on Hydrogeology 9, 73–93.

Chandra P.C. (2014) *Groundwater geophysics in hard rock*. Taylor & Francis, London, 350 pp.

Driscoll F.G. (1986) *Groundwater and wells*. Johnson Division, USA, 1089 pp.

Duque J. (2005) Hydrogeology of the aquifer system of the Gabros of Beja [in Portuguese]. PhD. Thesis, University of Lisbon, Lisbon, 419 p.

ERHSA (2001) Study of groundwater resources of Alentejo region [in Portuguese]. CCR-Alentejo: Coordination Commission of Alentejo Region, Évora, Portugal.

Huffman R.L., Miner D.L. (1996) *Your water supply well construction and protection*. North Carolina Cooperative Extension Service, Publication Number: AG 469. http://www.bae.ncsu.edu/programs/extension/publicat/wqwm/ag469.html

IGI (2013) *Water well construction*. The Institute of Geologists of Ireland. http://www.igi.ie/assets/files/Water%20Well%20Guidelines/Guidelines.pdf, consulted September 2013.

Kirsch R. (2009) Groundwater geophysics: A toll for hydrogeology. Springer, 548 p.

Krásny J. (1996) Hydrogeological environment in hard rocks: An attempt at its schematizing and terminological considerations. In: *Proceedings of the First Workshop on Hardrock Hydrogeology of the Bohemian Massif, Acta Universitatis Carolinae, Geologica*, 40, 2, 115–122.

Kellet R., Bauman P. (2004) Mapping groundwater in regolith and fractured bedrock using ground geophysics: a case study from Malawi, SE Africa. Canadian Society of Exploration Geophysicists (CSEG) Recorder Focus Article, pp. 24–33.

Kouadio K.E., Konan-Waidhet A.B., Koffi K., Lasm T., Savane I., Nagnin S. (2013) Interpretive approach to hydrogeological and geophysical prospection data for the choice of the best boreholes sites in area of fractured rocks in Ivory Coast. *International Journal of Scientific & Engineering Research* 4(7), 1963–1970.

Larsson I. (1987) Les eaux souterraines des roches dures du Socle. Project 8.6 du Programme Hydrologique International, UNESCO, 282 p.

Lowrie W. (2007) *Fundamentals of geophysics*. Cambridge University Press, Cambridge, 390 p.

Milsom J., Eriksen A. (2011) *Field geophysics. Series geological field guide* 28, Wiley, 304 p.

Misstear B., Banks D., Clark L. (2006) *Water wells*. John Wiley & Sons, UK, 498 p.

Mohamed N.E., Brasse H., Abdelgalil M.Y., Kheiralla K.M. (2012) Geoelectric and VLF electromagnetic survey on complex aquifer structures, Central Sudan. *Comunicações Geológicas* 99(2), 95–100.

Parasnis D.S. (1996) *Principles of applied geophysics*. Springer, 456 p.

Repsold, H. (1989) *Well logging in groundwater development*. Verlag Heinz Heise, 136 p.

Roy K.K. (2001) Potential theory in applied geophysics. Springer, 651 p.

Rubbert T., Miesler T., Bender S. (2006) Hydrogeological Modeling in the Combined Porous-fractured Aquifer System of the Bavarian Forest. In Chambel A. (editor) *Proceedings of the 2nd Workshop of the IAH Iberian Working Group on Hard Rock Hydrogeology*, AIH-GP, Évora, Portugal, 31–39.

Telford W.M., Geldart L.P., Sheriff R.E. (1990) *Applied geophysics*. Cambridge University Press, Cambridge, 770 p.

Tomlinson C.A. (2008) Private Well Guidelines. Drinking Water Program, Bureau of resource Protection, Department of Environmental Protection, Commonweal of Massachusetts, 96 p. http://www.mass.gov/dep/water/laws/prwellgd.pdf

Venkateswaran S., Jayapal P. (2013) Geoelectrical Schlumberger investigation for characterising the hydrogeological conditions using GIS in Kadavanar Sub-basin, Cauvery River, Tamil Nadu, India. *International Journal of Innovative Technology and Exploring Engineering* 3(2), 196–202.

Chapter 5

Sustainable yield of fractured rock aquifers: The case of crystalline rocks of Serre Massif (Calabria, Southern Italy)

Antonella Baiocchi[1], Walter Dragoni[2],
Francesca Lotti[1] & Vincenzo Piscopo[1]
[1]*Dipartimento di Scienze Ecologiche e Biologiche,*
Università degli Studi della Tuscia, Viterbo, Italy
[2]*Dipartimento di Scienze della Terra, Università degli Studi di Perugia,*
Perugia, Italy

ABSTRACT

Potential water supply aquifers in many countries are composed of crystalline rocks. In Calabria, southern Italy, crystalline aquifers are tapped by low-yield wells. However, the long-term effects are not well known. The objective was to determine whether the groundwater development of the Serre Massif is 'sustainable' or not. Results are similar to those obtained for crystalline rocks in the Mediterranean region elsewhere. Climate, granitoid covers and topography influence the amount of aquifer recharge. Hydrostratigraphy reveals a shallow weathered layer above a fractured layer that could be treated as a porous-equivalent. A groundwater model simulates the effects of pumping under different scenarios. Sustainable management could be achieved in the following ways: an intermittent and/or alternated use of more wells; a constant, continuous pumping rate assuming prescribed limits of drawdown; and the use of horizontal boreholes or radial collector wells. This last solution appeared more efficient for long time frames.

5.1 INTRODUCTION

Crystalline rocks are found in water supply aquifers in many regions of Africa, South America and Asia. These aquifers are generally characterised by low yield. Several studies have shown that crystalline aquifers are comprised of weathered mantle and fractured bedrock where groundwater flow takes place. The hydraulic conductivity and storage of the weathered mantle and underlying fractured bedrock are derived from geodynamic and geomorphic processes (e.g., Taylor & Howard, 2000; Dewandel *et al.*, 2006; Foster, 2012). From top to bottom, the hydrogeological profile of the crystalline rocks typically includes a saprolite layer from the in situ decomposition of the bedrock (a few tens of meters thick), and a fissured layer that is characterised by dense fracturing in the first few tens of meters and by decreasing fracture density with depth. These fractures have been explained by several processes, such as lithostatic decompression, tectonic activity, cooling stress and weathering processes.

These two layers make up a composite aquifer below which is the fresh basement which is only locally permeable where tectonic fractures are present. Saprolite typically constitutes the capacitive function of the composite aquifer but the fractured layer also has a transmissive function (e.g., Chilton & Smith-Carington, 1984; Acworth, 1987; Rushton & Weller, 1985; Barker *et al.*, 1992; Wright, 1992; Banks & Banks, 1993; Stober & Bucher, 2000; Taylor & Howard, 2000; Cho *et al.*, 2003; Dewandel *et al.*, 2006; Krásný & Sharp, 2007). In addition, these studies have shown that these aquifers (including the saprolite and fractured layer) generally have low transmissivities of less than $6 \times 10^{-4} \, m^2/s$ (e.g., Uhl & Sharma, 1978; Chilton & Smith-Carington, 1984; Houston & Lewis, 1988; Howard *et al.*, 1992; Chilton & Foster, 1995; Taylor & Howard, 2000; Maréchal *et al.*, 2004; Foster, 2012). Therefore, a high incidence of well failure occurs and well yields are generally low. Thus, defining the sustainable pumping rate is more difficult in hard rock terrain than in other cases. In addition, determining which areas are suitable for groundwater abstraction is difficult. It is common practice to maintain a discharge rate that will not allow the water level in the well to drop below a prescribed limit. This prescribed limit is identified from the nature and thickness of the aquifer and the depth of the well (Van Tonder *et al.*, 2001).

In Italy, the hydrogeological properties of crystalline rocks are not well known. These aquifers are characterised by lower groundwater yields than the carbonate and alluvial aquifers that are widely used for water supply (Civita, 2005; 2008). Thus, the limited available knowledge regarding the hydrogeology of these rocks results from the data obtained during digging underground galleries (e.g., Maréchal, 1999) or from investigations conducted for the construction of dams (e.g., Calcaterra *et al.*, 1993; Celico *et al.*, 1993). Although the interest in crystalline rock hydrogeology is scarce relative to the interest in other higher yielding aquifers, the large extent of these rocks and the scarcity of water resources in Sardinia and Calabria provide reasons for thorough analysis of these rocks (Barrocu, 2007). The objective of the chapter is to examine the hydrogeological characteristics of a representative area of crystalline rocks in Calabria and to determine efficient and sustainable groundwater development methods.

5.2 STUDY AREA

The area under examination lies 15 km south of Catanzaro (Figure 5.1) and belongs to the north-eastern portion of the Serre Massif (an extended relief of the Calabrian Arc in southern Calabria).

In this portion of the Serre Massif, a Paleozoic basement outcrop exists that is composed of rocks with variable composition between tonalite and granodiorite. A transgressive Late Miocene conglomerate-calcarenite-clay-evaporite succession and an Early Pliocene conglomerate-sand-clay succession unconformably overlay the crystalline basement. Middle Pliocene to Middle Pleistocene deposits, including a conglomerate-sand-sandstone-clay marine succession, fill the main tectonic depressions. Late Pleistocene fluvial terraced deposits, marine terraces and a Late Pleistocene-Holocene alluvial fan also outcrop in the coastal plain and locally on higher relief (Ogniben, 1973; Di Nocera *et al.*, 1974; Amodio Morelli *et al.*, 1976; Romeo & Tortorici, 1980; Bonardi *et al.*, 1984; Schenk, 1984).

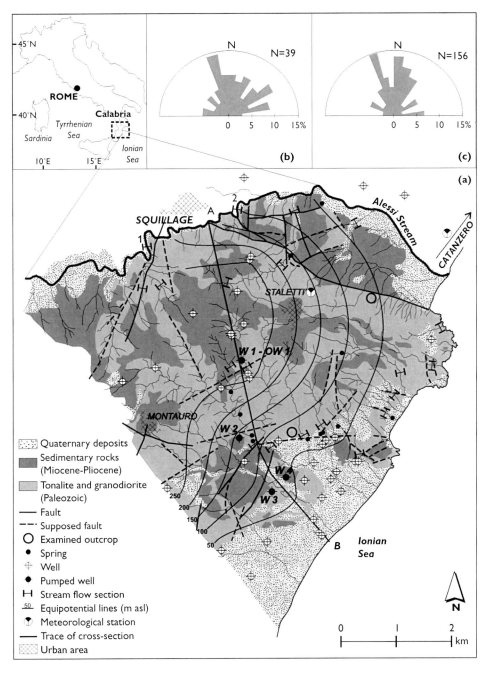

Figure 5.1 Surface geology and hydrogeology of Staletti Hill (a), rose diagrams of the faults (b), rose diagram of the crystalline basement discontinuities (c).

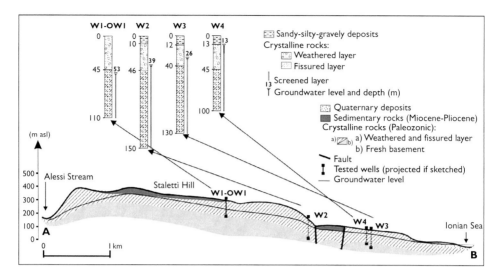

Figure 5.2 Hydrogeological cross section through Stalettì Hill and the main characteristics of the tested wells.

The present tectonic setting of the Calabrian Arc is a result of pre-Hercynian to Cenozoic events, which favoured the exhumation of thick Hercynian crustal sections (Schenk, 1990; Thomson, 1994; Caggianelli *et al.*, 2000; Festa *et al.*, 2003). Tectonic processes detected within the area are generally recent and linked to an extensional stage, such as the normal NW-SE – and NE-SW-oriented faults (Ghisetti, 1979; Van Dijk *et al.*, 2000; Tansi *et al.*, 2007). These faults divide the crystalline basement and its sedimentary cover into several blocks that form the coastal plain (Figure 5.2).

The area is characterised by a typical Mediterranean climate with maximum rainfall in the autumn and winter and minimum rainfall during the summer. The mean annual precipitation and air temperature are 932 mm and 16.2°C, respectively.

The crystalline rocks of the Calabrian Arc form a low-permeability aquifer (average yield of 4–7 l/s per km^2) where groundwater circulation occurs in the saprolite and fractured zones close to the surface (on average the first few tens of meters). In the most heavily fractured zones (i.e., near the faults), groundwater flow is more active and can occur at a greater depth. Groundwater flows into the valleys and feeds the streams, numerous low-discharge springs (generally less than 0.01 m^3/s) and alluvial coastal aquifers (Celico *et al.*, 2000; Allocca *et al.*, 2006).

In Calabria, the groundwater is widely used for drinking and irrigation. For example, approximately 2000 sources (springs and wells) are used for drinking water (REGIONE CALABRIA, 2009). The alluvial and carbonate aquifer yield the most water for drinking. However, many scattered wells are present in the crystalline rocks that discharge less than 3 l/s. Despite their low productivity, the wells in crystalline rocks are important local sources of drinking water, especially in the little towns and villages located in the hilly and mountain areas of the region (REGIONE CALAB-RIA, 2009). The hydrogeological behaviour of the crystalline rocks is not thoroughly understood.

5.3   INVESTIGATIONS AND DATA COLLECTION

Field investigations were conducted in an area of approximately 30 km^2 (Figure 5.1). These investigations included a geological survey, stratigraphy, a preliminary rock mass fracturing analysis, stream and spring discharge measurements during low-flow conditions, water level measurements and pumping tests. Other data were acquired from the literature (ISPRA, 2012).

The surface geology of the area was based on a geological map at a scale of 1:50,000 (ISPRA, 2010) and on detailed geological survey interpretations. The lithological logs of 19 wells were obtained from the records of private companies and from direct observation during the cutting of drilled wells. The information gathered included stratigraphy, aquifer formations and water level.

The faults and fractures reported in the geological maps were verified in the field, and then, their azimuth direction frequencies were determined. Only faults and fractures with lengths greater than 250 m were considered in the statistical analysis. In addition, two outcrops of the crystalline basement were examined to determine the orientation of fractures in each outcrop.

Flow measurements were conducted in the dry season. Measurements were taken using a tank or current meter for the rate of discharge depending on the type of emergence point. Data were collected at 11 springs and 15 stream sections. The accuracy of the flow measurements ranges from 5 to 10%. In the same year water level measurements were conducted in 13 wells with depths of between 80 and 150 m. The water level of the other 18 wells that are located in or near the investigation area were acquired from an existing database.

Pumping tests at four wells that penetrate the crystalline aquifer were carried out at a constant rate to measure the drawdown of the tested well (W1, W2, W3 and W4 in Figure 5.1). For one pumping test, the drawdown was measured in the tested well (W1) and also in an observation well (OW1). The pumping data were interpreted with analytical techniques (Kruseman & de Ridder, 1994) and Aquifer Test 4.1 software.

Rainfall and air temperature data for the area were obtained from the Staletti and Catanzaro meteorological stations, respectively, for the period between 1940 and 2001 (ARPACAL, 2009). The Staletti Meteorological Station and Catanzaro Meteorological Station are located within the investigation area at 330 and 343 m asl, respectively. The Catanzaro Meteorological Station is the nearest station with available temperature records (approximately 15 km) (Figure 5.1). Data were processed to statistically analyse their homogeneity, the cumulative residuals method was applied, and data gaps were identified to complete the dataset (Haan, 1977). Missing data were reconstructed on a monthly basis. The mean monthly data were used to calculate the potential and actual evapotranspiration with the Thornthwaite-Mather method (1955). It was not possible to use more modern or reliable methods because other data, such as wind speed and radiation, were unavailable.

5.4 HYDROGEOLOGICAL INVESTIGATION RESULTS

The area includes Stalettì Hill and its surroundings. The surface geology of the area is shown in Figure 5.1. The geological cross-section (Figure 5.2) shows that the crystalline basement is divided in the coastal plain through NNE-SSW – and ENE-WSW-oriented faults. The lithostratigraphy within the boreholes highlights the presence of sandy-silty-gravely deposits in the first few tens of meters, which overlay the crystalline basement. First, the Paleozoic rocks are heavily weathered to form a soil with a high percentage of silt. Next, these rocks are weathered to form a sandy soil. Finally, the Paleozoic rocks are weathered to create fractures (Figure 5.2). The discontinuity directions of the fissured layers are generally consistent with those of the faults (as shown in the rose diagrams of Figure 5.1). The rock outcrop surveys highlight a variable frequency of persistent discontinuity (varying from 0.1 to 2 discontinuities per meter) and opening widths of less than 1 mm (Figure 5.3). Through the drilled boreholes, the maximum depth of the heavily fractured granitoid bedrock was verified to be 100 m.

A potentiometric map of the crystalline aquifer was drawn by interpolating the groundwater levels and Alessi Stream head (Figure 5.1). The hydraulic gradient varies from 0.04 to 0.15, and the steepest gradients are located in the steepest hill slopes. These high gradients are rare but not unique. Similar gradients have been reported in crystalline formations by several authors (e.g., Bauer *et al.*, 1999; Gustafsson & Morosini, 2002; Maréchal *et al.*, 2006). The crystalline aquifer (sometimes unconfined and sometimes confined by weathered mantle or sedimentary cover) discharges towards the Alessi Stream in the northern slope and feeds the coastal aquifer in the southern slope. The evidence for this hydrogeological scheme are as follows: increasing flow rate along the Alessi Stream during the dry season (approximately 0.06 m^3/s between stream sections 1 and 2, shown in Figure 5.1), the absence of basal flow in any other stream and the reduced discharge of springs at different heights on the hillslope (overall, approximately 0.005 m^3/s).

A detailed quantification of the crystalline aquifer resources is not possible because continuous measurements of streamflow are unavailable and little is known regarding the flow towards the coastal plain aquifers. A first evaluation of the aquifer recharge was conducted based on climatic data and soil characteristics.

The mean annual precipitation and temperature are 932 mm and 16.2°C, respectively. The actual evapotranspiration was estimated by the Thornthwaite-Mather method by considering a total available water-holding capacity of 150 mm based on the soil texture, field capacity, permanent wilting point and land cover of the area. The mean effective infiltration was estimated from the hydrologic analyses that were carried out by other authors in neighbouring morphologically and geologically comparable basins. These studies showed that approximately 40% of the water surplus (i.e., effective rainfall or the difference between precipitation and actual evapotranspiration) constitutes the groundwater recharge (REGIONE CALABRIA, 2009). Table 5.1 contains the water budget of the area. This budget shows that the recharge from precipitation is approximately 166 mm/yr. Figure 5.4 shows the monthly distribution of the precipitation, actual evapotranspiration and water surplus.

Figure 5.3 A granitoid outcrop showing the discontinuity network of the fissured layer. *(See colour plate section, Plate 15).*

Table 5.1 Water budget according to the Thornthwaite-Mather method.

Term	Value
Mean annual precipitation	+ 932 mm/yr
Mean annual actual evapotranspiration	−516 mm/yr
Water surplus: *Infiltration (−166 mm/yr)*	−416 mm/yr
Surface runoff (−250 mm/yr)	

The wells used for the pumping tests are between 100 and 150 m deep (W1, W2, W3 and W4 in Figures 5.1 and 5.2), have a diameter of 210 mm and are screened in the fractured bedrock. In addition, W3 and W4 drain the overlying weathered mantle (Figure 5.2). W1 well is located in the hilltop, W2 at the foot of the hill near some faults, and W3 and W4 in the lowered and faulted crystalline block (Figure 5.1).

The pumping tests were carried out at a constant rate (from 3.4 to 5.8 l/s, Table 5.2) for approximately 34 hours. Drawdown and recovery data were measured in the pumped wells. For W1, the pumping test included the measurement of draw-

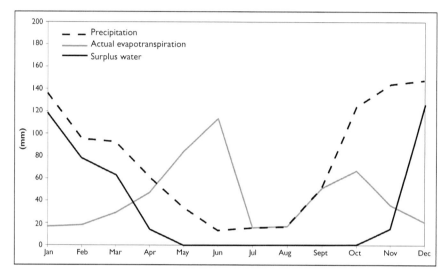

Figure 5.4 Plot of the monthly soil water budget.

Table 5.2 Results of the pumping tests: H$_{sat}$ saturated thickness of the aquifer intercepted by well (m); Q constant rate of pumping (l/s); T$_d$ transmissivity determined by the drawdown data (m²/s); T$_r$ transmissivity determined by the recovery data (m²/s); S storativity (ad.); BB boundary barrier; DP double porosity; WE well effect; LA leaky aquifer.

Well	H$_{sat}$	Q	Drawdown model	T$_d$	Recovery model	T$_r$	S
W1	57	4.2	Theis	4.80×10^{-4}	Theis	2.50×10^{-4}	
			Theis BB	4.80×10^{-4}	DP	2.82×10^{-4}	
OW1	57	4.2	Theis	3.61×10^{-4}	Theis	3.90×10^{-4}	2.23×10^{-3}
			Theis BB	3.61×10^{-4}	DP	3.50×10^{-4}	2.23×10^{-3}
W2	104	4.8	DP	1.00×10^{-4}	Theis WE	1.83×10^{-4}	
			DP BB	8.97×10^{-5}	DP	1.98×10^{-4}	
W3	90	3.4	LA	3.07×10^{-5}	DP	3.80×10^{-5}	
			DP	2.56×10^{-5}			
W4	76	5.8	DP	9.96×10^{-5}	Theis WE	6.35×10^{-5}	
			Theis	9.96×10^{-5}	DP	5.36×10^{-5}	

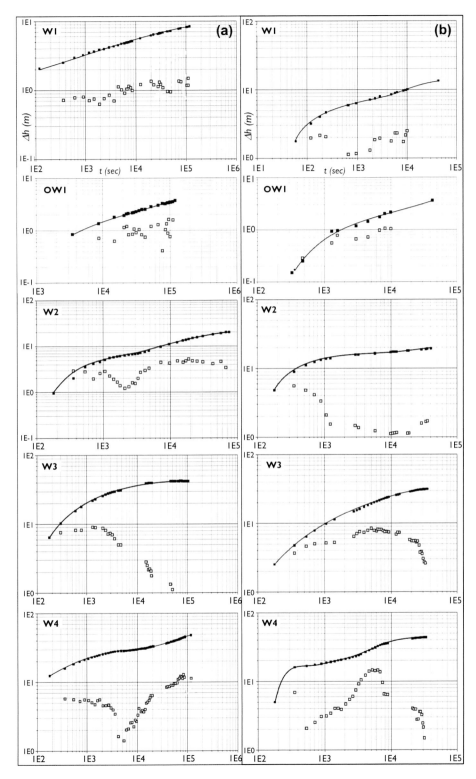

Figure 5.5 Drawdown-time (Δh-t) (solid line) and first derivative of drawdown (blank squares) plots resulting from the pumping (a) and recovery tests (b).

down and recovery both in the pumped well (W1 in Table 5.2) and in an observation well (OW1 in Table 5.2) located 30 m away.

The pumping test data diagnosis included the following methods: comparison of drawdown and recovery plots (semi-log and log-log plots) with theoretical models, analysis of the first derivative and determination of the aquifer parameters by considering the best-fit theoretical model. The preliminary recovery data were transformed by using the Agarwal (1980) method. The data were also smoothed to estimate the general slope of the derivative, and the signal was re-sampled at regularly spaced time intervals (using a logarithmic scale) (Renard *et al.*, 2009).

The log-log plots of the tests are shown in Figure 5.5. The comparison of drawdown-time curves and first derivative curves identified the type of aquifer model and verified the well-bore storage effect and the presence of boundaries. This analysis shows that different theoretical models can explain the measured data, including Theis, double porosity and leaky aquifer models, as summarised in Table 5.2. Using the drawdown and recovery data, similar transmissivity values result from the two best-fit models for the same test. Transmissivity values determined in W1 are comparable with those determined in the observation well (OW1). The overall transmissivity varies from 2.6×10^{-5} to 4.8×10^{-4} m²/s. In addition, the storativity value, determined from data measured in the observation well (OW1) during the W1 test, is 2.2×10^{-3} (Table 5.2).

5.5 GROUNDWATER FLOW MODEL

A simplified numerical model was constructed to analyse the long-term response of the aquifer to pumping and to obtain suggestions for potential aquifer development methods.

5.5.1 Conceptual model

Hydrostratigraphy shows a shallow porous layer, including the weathered hard rock or its sedimentary cover (up to 40 m), followed by a fissured layer that is up to 100 m deep. The main aquifer is the fissured rock that has a saturated thickness of tens of meters. This aquifer appears confined, leaky or unconfined depending on the thickness and grain size of the material that cover it. The average amount of aquifer recharge was estimated to be 170 mm/yr.

The potentiometric map of the fissured aquifer (Figure 5.1) highlights its continuity at the study area scale (approximately 20 km²). The simultaneous response to pumping W1 and observing OW1 agree with the drawdown-time plots, which confirms the continuity of the fissured aquifer at this scale (Figure 5.5). The transmissivity values are comparable for the well and observation well in the W1-OW1 pair (Table 5.2). Therefore, the transmissivity values from the other pumped wells are significant for the aquifer characterisation.

Generally speaking, all pumping test results show a low aquifer transmissivity and comply with the drawdown-time and first derivative curves of different theoretical models. The aquifer response can result from the heterogeneity of the fissured aquifer. Although this aquifer is fissured and not Darcian (in a strict sense), the aquifer can be treated as a porous equivalent medium at the scale of the pumping test because measured data fit the Theis and leaky aquifer theoretical models. This interpretation appears to be consistent with the continuity of the fissured layer and the density and orientation of the discontinuities observed in the outcrops. In addition, the transmissivity values are not dependent on well depth. This could result from the heterogeneous permeability of the fractured layer, which is higher in the first tens of meters than in the deeper wells that have lower transmissivity (Table 5.2). Thus, the degree of fissuring in the crystalline rocks decreases with depth.

Groundwater in the investigated area mainly discharges into the Alessi Stream at the northern boundary and into the coastal aquifer at the southern sector. The lowest crystalline aquifer hydraulic heads were found in the southern sector. Inflow from the western sector and outflow towards the eastern sector of the Serre Massif occur in a single portion of the crystalline aquifers (Figures 5.1 and 5.2). Secondary discharge towards springs with low flow occurs but is insignificant in the numerical model.

5.5.2   Model set up

A finite difference transient model was constructed. The MODFLOW2000 (McDonald & Harbaugh, 1988) code was selected, and the Groundwater Vistas user interface was used. Based on the conceptual hydrogeological model, a three-dimensional reconstruction of the aquifer was built. Because of the dense degree of fracturing in the investigated layer and the scale of the model, the model likened the aquifer to an equivalent porous medium. This approach is often used when velocity and transport phenomena are not considered (i.e., when only the flow is considered) (e.g., Angelini & Dragoni, 1997; Tiedeman et al., 1997; Shapiro, 2002; Scanlon et al., 2003).

A grid consisting of one layer, 159 rows and 100 columns of variable size covered the 21.35 km^2 area from the Alessi Stream to the coast was used. The lateral extension was defined on the available data and on the local boundary conditions (Figure 5.6). A telescopic grid refinement was used around the wells that were tested (from 100×100 m to 10×10 m). The grid was oriented with the main directions of the flow, faults and discontinuities. The thickness of the layer was determined by the topography and by the thickness of the fractured layer of the crystalline rocks. This thickness was slightly adjusted during the calibration process.

The northern boundary was simulated by the MODFLOW drain package by assigning different elevations to the Alessi Stream valley bottom. The southern boundary corresponds to the coast, which was simulated by a constant head boundary. The eastern and western limits are represented by general head boundaries, which were based on groundwater levels in the wells and the discharge areas located outside of the model domain (Figure 5.1).

The initial recharge rates were assigned to agree with the water budget and were varied slightly within the margins of uncertainty during the calibration process. Initial heads were obtained from the equipotential map that was reconstructed for the crystalline aquifer (Figure 5.1). Initial hydraulic conductivity values and their distributions were deduced from the transmissivity values that resulted from the pumping tests and the thickness of the weathered and fissured granitoids. The aquifer was considered horizontally isotropic and vertically anisotropic (ratio of 10). The spatial distribution

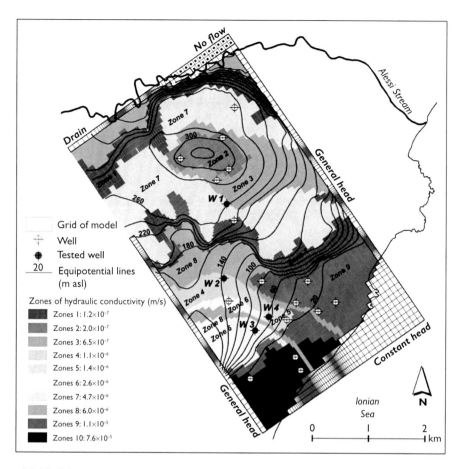

Figure 5.6 Model area, boundary conditions and results of the steady-state calibrated model showing the hydraulic conductivity zones and potentiometric contour map. (*See colour plate section, Plate 16*).

of the hydraulic conductivity was modified during the calibration process. This value was adjusted within physically reasonable upper and lower bounds.

5.5.3   Calibration and results

The first stress period of the transient model was run until equilibrium was reached. Hydraulic heads, discharge rates of the systems and topography constraints were controlled during the calibration process. The horizontal and vertical hydraulic conductivities, drain conductance and recharge were varied during this process. A sensitivity analysis was performed to determine the importance of the different parameters.

The transient model was calibrated by using the drawdown and discharge data that were measured during the pumping tests. Time discretisation considered 9 stress periods. Recharge was only applied in the first steady-state period, which

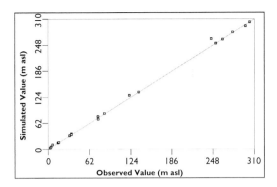

Figure 5.7 Comparison of the observed and simulated heads.

Table 5.3 Model groundwater budget.

Term	Value (m³/s)	Value (mm/yr)	Percentage of total inflow or outflow
Recharge	+0.114	+168	42
Inflow from western boundary	+0.155	+228	58
Outflow to sea	−0.127	−188	47
Outflow to Alessi Stream	−0.027	−40	10
Outflow to eastern boundary	−0.115	−170	43
Budget discrepancy = −0.0036%			

represents the six month period of water surplus that results from the soil water budget (Figure 5.4).

New sensitivity analyses, trial-and-error and a moderate use of automatic calibration reduced the residuals between the observed and calculated values. The results of the calibrated steady-state model are shown in Figure 5.6 as hydraulic conductivity zones and as a potentiometric contour map and in Figure 5.7 as a scatter plot of the hydraulic heads. Additionally, the mean error, mean absolute error and root mean square error between the computed and measured heads were 0.96 m, 2.06 m and 3.85 m, respectively. The water budget of the steady-state period is reported in Table 5.3. These results are consistent with field measurements and/or data elaboration. The transient stress period results (2–9) are shown in Figure 5.8 with reference to the measured versus computed heads of the 4 tested wells.

The final sensitivity analysis in the transient state (Figure 5.9) shows that the model is most sensitive to horizontal hydraulic conductivity and to recharge. Only a small influence was derived from the storage coefficient. The most sensitive hydraulic conductivity zones are those where pumping wells are located, beneath the alluvial plain and the steep slopes around the crystalline massif.

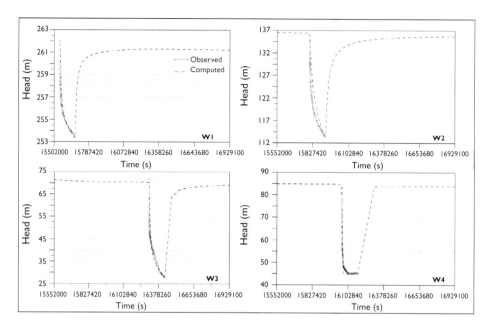

Figure 5.8 Comparison of the observed and simulated heads for the 4 tested wells.

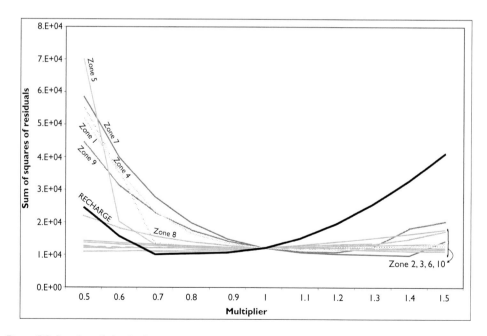

Figure 5.9 Results of the final parameters sensitivity analysis, in the transient state. Curves of the different zones refer to the hydraulic conductivity zones in Fig. 5.5.6.

Table 5.4 Results of the different simulated pumping scenarios: H_{sat} saturated thickness of the aquifer intercepted by well (m); Q constant rate of pumping (l/s); Δh drawdown (m); E percentage of emptying, i.e. $\Delta h/H_{sat}$ (%); *d* dried up; *(126)* emptying time (days); Q_{tot} total discharge at the end of 6 months of simultaneous pumping (l/s).

Case	W1 (H_{sat} = 57)			W2 (H_{sat} = 68)			W3 (H_{sat} = 97)			W4 (H_{sat} = 92)			Q_{tot}
	Q	Δh	E	Q	Δh	E	Q	Δh	E	Q	Δh	E	
A	4.2	27.4	48	4.8	*d*	*(11)*	3.4	16.4	17	5.8	*d*	*(22)*	7.6
	3.0	20	35	3.0	34	50	3.4	16.4	17	3.0	39	42	12.4
	5.4	40	70	3.2	38	56	8.0	43	44	4.0	56	61	20.6
	6.0	*d*	*(126)*	3.5	*d*	*(126)*	10.0	52	54	4.5	50	54	14.5
B	2.2	15	26	1.7	16	24	5.0	25	26	1.7	25	27	10.6
C	2.2	12	21	1.7	11	16	3.4	12	12	1.7	18	20	9.0
	4.0	15	26	3.0	17	25	5.0	16	16	2.4	25	27	14.4
	4.2	19	33	4.8	27	40	8.0	25	26	5.8	39	42	22.8
	8.0	32	56	6.0	35	51	10.0	34	35	7.0	51	55	31.0

5.5.4   Scenarios

The calibrated model was used to simulate different pumping scenarios from the tested wells.

The first simulation (case A) evaluated the drawdown following 6 months of simultaneous and continuous abstraction from the four tested wells at different flow rates. Table 5.4 presents the results and the percentage of the saturated thickness that was emptied in each well (i.e., the ratio between drawdown and the initial saturated thickness of the aquifer intercepted by the well). For some wells, a considerable drawdown occurred or the well dried up at a discharge rate that was less than the maximum simulated discharge rate. The maximum total discharge rate from simultaneous and continuous pumping was approximately 21 l/s, with significant drawdown (percentage of emptying of the well varied between 44 and 70%). This total discharge is less than the outflow of the aquifer towards the sea (127 l/s in Table 5.3).

A second simulation (case B) included the evaluation of discharge after 6 months of simultaneous abstraction from the wells at the end of the wet season. A prescribed limit for drawdown was maintained in the wells. The percentage of the saturated thickness that was emptied in the wells was between 24 and 27%. In this case, a lower total discharge results than in case A (Table 5.4).

The last simulation (case C) included the evaluation of drawdown after 6 months of simultaneous abstraction at the end of the wet season and at different discharge rates after replacing the vertical wells with horizontal 130 m long wells. There is a lower drawdown during simultaneous abstraction, a higher total discharge (up to 31 l/s) and a moderate percentage of emptying of the wells (from 35 to 56%) (Table 5.4).

5.6   CONCLUDING REMARKS

Investigation of the crystalline rocks in part of Calabria highlights the typical hydrogeological features that occur also in other parts of the world in similar terrain. These aquifers consist of an integrated system of weathered and fractured layers that overlay fresh unweathered crystalline rock. The aquifer thickness varies between 50 and 100 m. A dense network of fracturing was caused by specific and intense tectonic events that characterise the Calabrian Arc. This feature and the significant precipitation rate permit considerable aquifer recharge (approximately 170 mm/yr). The continuity of the aquifer was determined from a potentiometric map of measured groundwater levels and the response of the wells to pumping. The aquifer discharges to just one stream. However, the high hydraulic gradients and low transmissivity values produce only low well yields.

The derived hydrogeological parameters should be viewed as a preliminary estimate. Nevertheless, the results help to describe the complex pumping response of the crystalline aquifer. All drawdown-time plots show that the aquifer is extremely heterogeneous. The heterogeneity of the aquifer relates to the degree of fracturing, the hydrogeological effect of the faulted zones and the influence of the weathered layer on the fractured zone. However, analysis of the time-drawdown and first derivative of drawdown curves can be carried out by using the porous equivalent medium model because the volume of the pumped aquifer is more than the representative elementary volume (according to the orientation and spacing of the discontinuities).

The simplified groundwater model that was developed is a valuable tool for verifying the hydrogeological behaviour of the crystalline rock aquifer. The calibrated model was also useful in transient conditions for simulating different pumping style scenarios. Together, the different simulation and pumping test results suggested a way forward for sustainable groundwater development in the aquifer.

Groundwater flow in the crystalline rock aquifer is towards alluvial plain, the sea and the stream. General criteria to assess a sustainable yield should be used to tap groundwater flow that discharges into the sea. These criteria should be applied to assure strict control of sea water intrusion and to maintain a stream base flow for ecosystem needs. This solution is consistent with the sparse distribution of the built-up areas located on the crystalline massifs in Calabria Region.

Second, a suitable method for tapping groundwater that reflects the low transmissivity of the aquifer and the likelihood of the well drying up is needed. The following three possibilities result from the simulations and pumping tests: i) assign the maximum tested pumping rate that considers an intermittent and alternated use of more wells to prevent them from drying up; ii) define a constant and continuous pumping rate that assumes drawdown as a percentage of well depth; and iii) use horizontal boreholes or radial collector wells. This last solution appears to be more suitable because the increased drainage surface limits drawdown for the same discharge. Thus, one single horizontal or collector well can replace many vertical wells. However, other factors, economic and social, should also be considered.

ACKNOWLEDGEMENTS

This study was supported by the "Hydrogeology of Fractured and/or Karst Systems: theoretical research and application to groundwater management under current climate change" research that was sponsored by the Italian Ministry of Education, Universities and Research (grant PRIN 2008YYZKEE). The authors appreciate the review and suggestions made by A.L. Mayo and J.M. Sharp and another anonymous reviewer.

REFERENCES

Acworth R.I. (1987) The development of crystalline basement aquifers in a tropical environment. *Quarterly Journal of Engineering Geology* 20, 265–272.

Agarwal R.G. (1980) A new method to account for producing time effects when drawdown type curves are used to analyze pressure buildup and other test data. *55th SPE Annual Technical Conference and Exhibition*, SPE Paper 9289. Sept. 21–24, Dallas, TX.

Allocca V., Celico F., Celico P., De Vita P., Fabbrocino S., Mattia C., Monacelli G., Musilli I., Piscopo V., Scalise A.R., Summa G., Tranfaglia G. (2006) *Carta Idrogeologica dell'Italia Meridionale*. Istituto Poligrafico e Zecca dello Stato, Roma.

Amodio Morelli L., Bonardi G., Colonna V., Dietrich D., Giunta G., Ippolito F., Liguori V., Lorenzoni S., Paglionico A., Perrone V., Piccarreta G., Russo M., Scandone P., Zanettin Lorenzoni E., Zuppetta A. (1976) L'arco Calabro-Peloritano nell'orogene appenninico Maghrebide (The Calabrian-Peloritan Arc in the Apennine-Maghrebide orogen). *Memorie della Società Geologica Italiana* 17, 1–60.

Angelini P., Dragoni W. (1997) The problem of modeling limestone springs: the case of Bagnara (North Apennines, Italy). *Ground Water* 35(4), 612–618.

ARPACAL Centro Funzionale Multirischi (2009) Banca dati meteorologici, URL: http://cfd.calabria.it, last visited 03/04/2012.

Banks, S.B., Banks, D. (eds.) (1993). *Hydrogeology of Hard Rocks Vols. I & II*. Proc. 24th Congress of International Assoc. Hydrogeologists, June/July 1993, Ås, Norway, 1206 pp.

Barker R.D., White C.C., Houston J.F.T. (1992) Borehole siting in an African accelerated drought relief project. In: Wright E.P., Burgess W.G. (editors) *Hydrogeology of crystalline basement aquifers in Africa*. Geological Society, London Special Publication 66, 183–201.

Barrocu G. (2007) Hydrogeology of granite rocks in Sardinia. In: Krásný J, Sharp JM (editors) *Groundwater in fractured rocks*. IAH Selected Papers on Hydrogeology 9, 33–44.

Bauer P.W., Peggy S.J., Kelson K.I. (1999) *Geology and hydrogeology of the Southern Taos Valley, Taos County, New Mexico*. New Mexico Bureau of Mines and Mineral Resources, New Mexico Tech Socorro, New Mexico.

Bonardi G., Messina A., Perrone V., Russo S., Zuppetta A. (1984) L'unità di Stilo nel settore meridionale dell'Arco Calabro Peloritano. *Bollettino della Società Geologica Italiana* 103, 279–309.

Caggianelli A., Prosser G., Del Moro A. (2000) Cooling and exhumation history of deep-seated and shallow level, late Hercynian granitoids from Calabria. *Geological Journal* 35, 33–42.

Calcaterra D., Ietto A., Dattola L. (1993) Aspetti geomeccanici e idrogeologici di ammassi granitoidi (Serre calabresi). *Bollettino della Società Geologica Italiana* 112, 395–422.

Celico P., Piscopo V., Berretta G. (1993) Influenza di un invaso sulla circolazione idrica sotterranea in rocce scistose. *Geologia Applicata e Idrogeologia* 28, 253–261.

Celico F., Celico P., De Vita P., Piscopo V. (2000) Groundwater flow and protection in the Southern Apennines (Italy). *Hydrogéologie* 4, 39–47.

Chilton P.J., Foster S.S.D. (1995) Hydrogeological characteristics and water–supply potential of basement aquifers in tropical Africa. *Hydrogeology Journal* 3, 3–49.

Chilton P.J., Smith-Carington A.K. (1984) Characteristics of weathered basement aquifer in Malawi in relation to rural water-supplies. IAHS Publication 144, 57–72.

Cho M., Ha K-M, Choi Y-S, Kee W-S, Lachassagne P., Wyns R. (2003) Relationship between the permeability of hard-rock aquifers and their weathered cover based on geological and hydrogeological observation in South Korea. In: *Poceedings IAH Conference on 'Groundwater in fractured rocks'*, 15–19 September, Prague.

Civita M. (2005) Idrogeologia applicata ed ambientale. CEA, Milano.

Civita M. (2008) L'assetto idrogeologico del territorio italiano: risorse e problematiche. *Quaderni della Società Geologica Italiana* (3).

Dewandel B., Lachassagne P., Wyns R., Maréchal J.C., Krishnamurthy N.S. (2006) A generalized 3-D geological and hydrogeological conceptual model of granite aquifers controller by single or multiphase weathering. *Journal of Hydrology* 320, 260–284.

Di Nocera S., Ortolani F., Russo M., Torre M. (1974) Successioni sedimentarie messiniane e limite Miocene-Pliocene nella Calabria settentrionale (Messinian sedimentary successions and Miocene-Pliocene boundary in northern Calabria). *Bollettino della Società Geologica Italiana* 93, 575–607.

Festa V., Di Battista P., Caggianelli A., Liotta D. (2003) Exhumation and tilting of the late-Hercynin continental crust in the Serre Massif (Southern Calabria, Italy). *Bollettino della Società Geologica Italiana* Special Volume 2, 79–88.

Foster S. (2012) Hard-rock aquifers in tropical regions: using science to inform development and management policy. *Hydrogeology Journal* 20, 659–672.

Ghisetti F. (1979) Evoluzione neotettonica dei principali sistemi di faglie della Calabria Centrale. *Bollettino della Società Geologica Italiana* 98, 387–430.

Gustafsson E., Morosini M. (2002) In-situ groundwater flow measurements as tool for hardrock site characterisation within the SKB programme. *NGU Bulletin* 439, 33–44.

Haan C.T. (1977) *Statistical Methods in Hydrology*. The Iowa State University Press, Ames.

Houston J.F.T., Lewis R.T. (1988) The Victoria Province drought relief project, II. Borehole yield relationships. *Ground Water* 26(4), 418–426.

Howard K.W.K., Hughes M., Charlesworth D.L., Ngobi G. (1992) Hydrogeologic evaluation of fracture permeability in crystalline basement aquifers of Uganda. *Applied Hydrogeology* 1, 55–65.

ISPRA (2012) URL http://sgi2.isprambiente.it/indagini/; last visited 03/04/2012

ISPRA (2010) URL http://www.isprambiente.it/Media/carg/580_SOVERATO/Foglio.html; last visited 10/05/2012.

Kalf R.P., Woolley D.R. (2005) Applicability and methodology of determining sustainable yield in groundwater systems. *Hydrogeology Journal* 13, 295–312.

Krásný J., Sharp J.M. (editors) (2007) *Groundwater in fractured rocks*. Taylor & Francis, London.

Kruseman G.P., de Ridder N.A. (1994) *Analysis and Evaluation of Pumping Test Data*. ILRI, Wageningen, The Netherlands.

Maréchal J.C. (1999) Observation of Alpine crystalline massifs from underground galleries. 1. Hydraulic conductivity at the massif scale. *Hydrogéologie* 1, 21–32.

Maréchal J.C., Ahmed S., Engerrand L., Galeazzi L., Touchard F. (2006) Threatened groundwater resources in rural India: an example of monitoring. *Asian Journal of Water, Environment and Pollution* 3(2), 15–21.

Maréchal J.C., Dewandel B., Subrahmanyam K. (2004) Use of hydraulic tests at different scales to characterise fracture network properties in the weathered-fractured layer of a hard rock aquifer. *Water Resources Research* 40, W11508, 1–17.

McDonald M.G., Harbaugh A.W. (1988) *A modular three-dimensional finite-difference ground-water flow model*. US Geological Survey Open-File Report, Washington.

Ogniben L. (1973) Schema geologico della Calabria in base ai dati odierni. *Geologica Romana* 12, 243–585.

Romeo M., Tortorici L. (1980) Stratigrafia dei depositi miocenici della Catena Costiera calabra meridionale e della media valle del F. Crati (Calabria) (Stratigraphy of Miocene deposits of southern Calabrian Coastal Chain and of middle R. Crati valley (Calabria)). *Bollettino della Società Geologica Italiana* 99, 303–318.

REGIONE CALABRIA (2009) URL http://www.regione.calabria.it/ambiente Piano di Tutela delle Acque della Regione Calabria. Dipartimento di Politiche dell'Ambiente, Regione Calabria, Catanzaro; last visited 03/04/2012.

Renard P., Glenz D., Mejias M. (2009) Understanding diagnostic plots for well-test interpretation. *Hydrogeology Journal* 17, 589–600.

Rushton K.R., Weller J. (1985) Response to pumping of a weathered-fractured granite aquifer. *Journal of Hydrology* 80, 299–309.

Schenk V. (1984) Petrology of felsic granulite, metapelites, metabasics, ultramafics, and meta-carbonates from Southern Calabria (Italy): Prograde metamorphism, uplift and cooling of a former lower crust. *Journal of Petrology* 25, 255–298.

Schenk V. (1990) The exposed crustal cross section of southern Calabria, Italy: structure and evolution of a segment of Hercynian crust. In: Fountain D.M., Salisbury M.H. (editors) *Exposed Cross-Sections of the Continental Crust*, Kluwer, Dordrecht, The Netherlands.

Scanlon B.R., Mace R.E., Barrett M.E., Smith B. (2003) Can we simulate regional groundwater flow in a karst system using equivalent porous media models? Case study, Barton Springs Edwards aquifer, USA. *Journal of Hydrology* 276, 137–158.

Shapiro A.M. (2002) Characterizing fractured rock for water supply: From well filed to watershed. In: *Fractured-Rock Aquifers*, Proceeding National Ground Water Association, Denver, Colorado, 6–9.

Stober I., Bucher K. (editors) (2003) *Hydrogeology of crystalline rocks*. Kluwer Academic Publishers, Dordrecht, The Netherlands.

Tansi C., Muto F., Critelli S., Iovine G. (2007) Neogene-Quaternary strike-slip tectonics in the central Calabrian Arc (southern Italy). *Journal of Geodynamics* 43, 319–414.

Taylor R., Howard K. (2000) A tectono-geomorphic model of the hydrogeology of deeply weathered crystalline rock: Evidence from Uganda. *Hydrogeology Journal* 8, 279–294.

Thomson S.N. (1994) Fission track analysis of the crystalline basement rocks of the Calabrian Arc, southern Italy: evidence of Oligo-Miocene late-orogenic extension and erosion. *Tectonophysics* 238, 331–352.

Thornthwaite C.W., Mather J.R. (1955) The water balance. *Publication in Climatology*, 8(1), Drexel Institute of Technology, Laboratory of Climatology, Centerton, New Jersey.

Tiedeman C.R., Goode D.J., Hseih P.A. (1997) Numerical simulation of ground water flow through glacial deposits and crystalline bedrock in the Mirror Lake Area, Grafton County, New Hampshire. *US Geological Survey Professional Paper* 1572 pp.

Uhl V.W., Sharma G.K. (1978) Results of pumping tests in crystalline-rock aquifers. *Ground Water* 16(3), 192–203.

Van Dijk J.P., Bello M., Brancaleoni G.P., Cantarella G., Costa V., Frixa A., Golfetto F., Merlini S., Riva M., Torricelli S., Toscano C., Zerilli A. (2000) A regional structural model for the northern sector of the Calabrian Arc (southern Italy). *Tectonophysics* 324, 267–320.

Wright E.P. (1992) The hydrogeology of crystalline basement aquifers in Africa. In: Wright EP, Burgess WG (editors), *Hydrogeology of crystalline basement aquifers in Africa*, Geological Society, London Special Publication 66, 1–127.

Van Tonder G.J., Botha J.F., Chiang W.H., Kunstmann H., Xu Y. (2001) Estimation of the sustainable yields of boreholes in fractured rock formation. *Journal of Hydrology* 241, 70–90.

Chapter 6

From geological complexity to hydrogeological understanding using an integrated 3D conceptual modelling approach – insights from the Cotswolds, UK

S.H. Bricker, A.J.M. Barron, A.G. Hughes,
C. Jackson & D. Peach
British Geological Survey, Kingsley Dunham Centre, Nottingham, UK

ABSTRACT

Adequate hydrogeological conceptualisation of structurally complex fractured aquifers requires the support of detailed geological mapping and three dimensional understanding. With a geological framework in place uncertainties in hydrological understanding and irregularities in hydraulic observations may be rationalised. Using the Cotswold of southern England, which are underlain by the ooidal limestone-dominated Middle Jurassic Inferior Oolite and Great Oolite groups, 3D modelling software GSI3D and Geographical Information Systems (GIS) have been used to integrate observed hydraulic behaviours with the 3D geological framework. In this way a conceptual model is developed to assist simulation of groundwater flow and the predicted response of groundwater levels and river flows to climatic extremes. The structural and lithological complexity of the bedrock results in sub-catchments which exhibit individual hydraulic responses and a hydrogeological setting dominated by shallow rapid fracture pathways and copious spring discharge.

6.1 INTRODUCTION

Geological maps have always been an expression of the geologist's 3D understanding of the structure and composition of the earth. In recent years 3D geological 'framework' models describing the subsurface have begun to supersede the traditional 2D geological map. Initially these tended to be of shallow superficial deposits but are increasingly of complex faulted bedrock systems (Aldiss *et al.*, 2012). The capability for 3D geological characterisation is being driven by multi-faceted applications e.g. for engineering (Merritt *et al.*, 2007), land-use planning (Campbell *et al.*, 2010) and hydrogeology (Robins *et al.*, 2005, Royse *et al.*, 2009) such that 3D models provide the foundation for many integrated modelling approaches. Within the discipline of hydrogeology, 3D geological models have been applied to both hydrogeological and groundwater modelling e.g. for permeability mapping, as a framework for risk screening tools (Marchant *et al.*, 2011), to delineate aquifer volumes and boundaries within groundwater flow models and to visualise model outputs (Wang *et al.*, 2010). A detailed 3D geological model may also be applied where geology is structurally complex. The model supports hydrogeological conceptualisation where uncertainties in hydrological understanding and irregularities in hydraulic observations may be rationalised (Royse *et al.*, 2010).

The Jurassic sequence of the Cotswolds, in which lie the headwaters of the River Thames, is complex both in terms of lithology, where units are thin and laterally variable, and structure. The strata are highly fractured, disrupted by steeply dipping normal faults and cambering and dissected by deep river valleys. By resolving the geological configuration in 3D it is easier to visualise and understand the hydrogeology and subsurface flow processes.

The limestones of the Middle Jurassic within the Cotswolds form a principal aquifer. Licensed groundwater abstraction nears 50 Ml/d much of which is used for public water supply serving 250 000 people. Non-consumptive abstractions for fish farming and mineral workings are also common within the region. Base flow derived from the Jurassic sequence of the Cotswolds is reputed to have sustained river flows further downstream within the Thames basin during recent droughts (e.g. 2004–2006) (pers. comm. Jones, 2011). Rivers draining the Cotswolds have high base flow indices (BFI) all year round; the BFI for the River Churn at Cirencester varies from 0.71–0.89, while the BFI for the River Coln at its downstream gauging station varies from 0.89–0.98 seasonally (CENTRE FOR ECOLOGY AND HYDROLOGY, 2012). Base flow within the River Thames downstream of the Cotswolds remains above 50 Ml/d during a typical summer recession and exceeds 25 Ml/d during some of the more extreme drought periods (period of record 1992–2008) (CENTRE FOR ECOLOGY AND HYDROLOGY, 2012). However, the sustainable management of water resources in the Cotswold area is challenging particularly under low flow conditions where groundwater resources are already fully allocated (ENVIRONMENT AGENCY, 2007). Public water supply licences have been subject to review, both within the Cotswolds and further downstream in the Thames basin, in an effort to reduce the impacts of groundwater abstraction on river flows. Meanwhile climate change predictions (UK Climate Impacts Programme) suggest reductions in summer river flows typically between 20–50% by 2050 within the Thames Basin (ENVIRONMENT AGENCY, 2012).

This chapter describes the integration of hydrogeological data within a geological framework model as a means to further our understanding of the catchment hydrogeology and controls on groundwater flow. The mid-Jurassic series of the Cotswolds is used as a case study. Using the 3D modelling software and Geographical Information Systems (GIS) the influence of lithology and structure on groundwater flow and discharge processes is examined along with the importance of perched water tables and shallow flow paths for river base flow contribution. The relationship between the Middle Jurassic Great and Inferior Oolite limestone successions controls the potential for hydraulic interconnectivity between the Great Oolite and Inferior Oolite aquifers. The refined conceptual model provides a basis for onward numerical simulation of groundwater flow to assess the response of hydrological system to climate change and extreme events.

6.2   3D GEOLOGICAL MODELLING

A 3D platform for geological mapping using the Geological Surveying and Investigation in three dimensions tool (GSI3D, ©Insight GmbH) has been available to geologists since the late 1990s. Its successful uptake since then is largely due to the intuitive

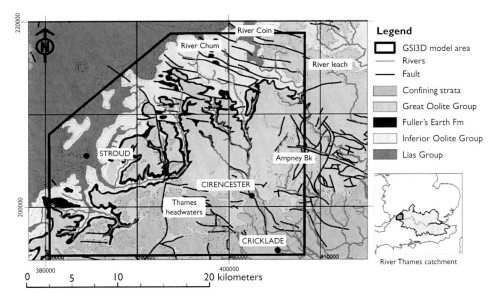

Figure 6.1 Location of the study area showing extent of the Cotswolds 3D geological model and 250k mapped geology with location of faults and drainage network. The position of the River Thames catchment (grey) and study area (green) is indicated on the smaller location map. Contains Ordnance Survey data © Crown Copyright. (*See colour plate section, Plate 17*).

geologically-based methodology that allows the incorporation of a geologist's expert knowledge (Kessler *et al.*, 2009). This approach is similar to 3D geology methodologies adopted elsewhere (e.g. Kaufmann & Martin, 2008). Such 3D models are able to encapsulate the implicit and tacit knowledge from the mapping geologist and convey the degree of uncertainty in the depth profile which is otherwise absent in the interpretation of 2D maps and cross-sections by the non-expert viewer (Howard *et al.*, 2009). The end product from such models may, therefore, be viewed as a geological interpretation as opposed to a modelled output. This more qualitative knowledge-driven approach has advantages over interpolation-based software packages, where geological data are unevenly distributed and the complexity of the geological structure requires significant extrapolation (Kessler *et al.*, 2009).

The confidence in GSI3D and other similar modelling approaches is manifest in the wide range of applications and in the variety of end-users which includes geological surveys, regulators, the water industry and local government (Kessler *et al.*, 2009). For the Cotswolds the environmental regulator provided the original impetus for the development of a 3D geological framework model to assess groundwater catchment divides for better groundwater resource management (Maurice *et al.*, 2008). Recognising the complexity of the geological setting, and the implications for successful resource management the original 3D framework model was refined to provide the basis for an enhanced hydrogeological conceptual model.

The 3D geological model developed in GSI3D covers some 600 km² of the Cotswolds between Cricklade in the south-east and Gloucester in the north-west (Figure 6.1).

While the model does not provide complete coverage of the Cotswolds the lithological sequence and geological features present within the study area are representative of the wider Cotswolds area. Inferences made with respect to the integration of the 3D geology with hydrogeological characteristics are, therefore, more widely applicable. The 3D geology was modelled by the regional geologist using 139 borehole records within 42 cross-sections with a combined length of nearly 600 km in conjunction with digital surface bedrock maps at the 1:50 000 scale. 35 bedrock units from the Whitby Mudstone Formation at its base through to the Oxford Clay Formation are incorporated into the 3D model with strata mapped at the formation or member scale where appropriate. The model also includes approximately 80 faults for which the geological displacements have been resolved.

The model provides 3D surfaces for the base of each of the geological units, synthetic cross-sections and information about fault displacements and river bed geology. It is the integration of this geological information with hydrogeological datasets, both within the GSI3D model and within inter-operable software packages such as GIS that considerably advances the hydrogeological conceptualisation.

6.3   GEOLOGY

The bedrock geological strata of the Cotswolds are of Jurassic age and comprise an alternating sequence of limestones and mudstones which were laid down in a shallow marine environment (Sumbler *et al.*, 2000). The geological sequence of the study area is summarised in Table 6.1 and mapped in Figure 6.1. The Early Jurassic age was marked by a period of global sea-level rise during which the predominant deposition was of thick marine mudstones of the Lias Group. Following global sea-level recession, most of the Mid Jurassic is characterised by relatively modest changes in sea level (both eustatic and related to regional uplift or subsidence) and moderate subsidence of the shelf areas (Barron *et al.*, 2012). Consequently, many of the formations are difficult to map at regional scales being relatively thin, and laterally variable and impersistent especially where facies belts have migrated laterally through time (Barron *et al.*, 2012). This was when the ooidal limestones of the Inferior and Great Oolite Group aquifers were laid down. These two aquifer units are separated by the Fuller's Earth Formation, a predominantly land-derived silicate mud deposit which marks a brief period of subsidence (Barron *et al.*, 2012). Sea-level rise and coastal retreat during the late stages of the Jurassic age led to the widespread deposition of silicate mudstones of the Kellaways and Oxford Clay formations which serve to confine the Great Oolite aquifer.

The study area comprises rolling upland sloping gently to the south-east towards the upper reaches of the River Thames and dissected by incised river valleys. Regionally the bedrock dips about 1° to the south-south-east, although locally dips may reach 5° or more mainly as a result of faulting (Sumbler *et al.*, 2000). The alternating sequence of permeable limestones and low permeability mudstones gives rise to significant spring discharge and has also led to widespread cambering along escarpments and hill slopes in which (largely under periglacial conditions) the weaker mudstone units have deformed and squeezed out causing disruption in the overlying more competent limestones (Sumbler *et al.*, 2000). The fractured limestone aquifers of the mid-Jurassic strata are of most hydrogeological interest.

Table 6.1 Geological sequence present within the study area along with the aquifer classification as defined by the environmental regulator; adapted from Neumann *et al.* (2003).

Age	Lithostratigraphy	Rock Type	Aquifer classification
Quaternary	Alluvium	Variable limestone gravel, loam and clay	Secondary
	River terrace deposits	Mainly limestone gravel	Secondary A
Middle to Upper Jurassic	Oxford Clay Formation	Grey mudstone	Unproductive
	Kellaways Formation	Grey mudstone overlain by fine-grained sand	Unproductive
Middle Jurassic	*Great Oolite Group:* Cornbrash Formation	Rubbly, shell-detrital limestone	Secondary A
	Forest Marble Formation	Mudstone with beds of shell-detrital, ooidal limestone	Principal/ Secondary A
	White Limestone Formation	Limestone with minor mudstone beds	Principal
	Hampen Formation	Sandy and ooidal limestone with mudstone beds	Principal
	Taynton Limestone Formation	Shell-detrital ooidal limestone	Principal
	Fuller's Earth Formation	Grey mudstone with limestone beds	Unproductive
	Inferior Oolite Group: Salperton Limestone Formation	Shelly, ooidal limestone	Principal
	Aston Limestone Formation	Shelly, sandy limestone	Principal
	Birdlip Limestone Formation	Ooidal limestone	Principal
Lower Jurassic	*Lias Group (part):* Bridport Sand Formation	Sandy mudstone and fine-grained sandstone	Principal
	Whitby Mudstone Formation	Mudstone with limestone beds at base	Unproductive
Principal	Geological units with high intergranular and/or fracture permeability.		
Secondary A	Permeable layers capable of supporting water supplies at a local rather than strategic scale.		
Unproductive	Rock layers or superficial deposits with low permeability that have negligible significance for water supply or river base flow.		

The Middle Jurassic succession comprises the Great Oolite and Inferior Oolite groups. For the purpose of this, and previous studies (e.g. Maurice *et al.*, 2008; Royse *et al.*, 2010) the lower aquifer comprises the Inferior Oolite Group, composed entirely of limestone, and the fine-grained sandstone beds of the underlying Bridport Sand Formation (uppermost Lias Group). The upper aquifer is comprised of a succession of limestone strata from the upper part of the Fuller's Earth Formation, through to the limestone beds at the base of the Forest Marble Formation (all part of the Great Oolite Group; Table 6.1; Figure 6.1). Colloquially, these are referred to as the Great Oolite and Inferior Oolite aquifers, which range in thickness from 32 to 55 m and 30 to 80 m respectively. The mudstone beds of the intervening Fuller's Earth Formation are between 20 and 50 m thick and the Whitby Mudstone Formation underlying the Bridport Sand comprises 30 to 95 m of low permeability mudstone. The upper part of the Forest Marble is dominated by silicate mudstone beds, and although these include lenses of limestone, and are capped by the thin limestone succession of the Cornbrash Formation, they form an effective confining layer above the Great Oolite aquifer, together (in the extreme south-east) with the thick mudstone-dominated succession of the Kellaways and Oxford Clay formations, overlying the Cornbrash.

The limestone units which dominate these aquifers range in lithology from cross-bedded coarse-grained shelly and ooidal to fine-grained and sandy types with subordinate and laterally impersistent mudstone layers (Barron et al., 2012). In terms of their gross lithologies the Great Oolite and Inferior Oolite aquifers are similar although the Great Oolite contains a significant proportion of interbedded mudstone. The aquifers units are highly dissected by fracturing which occurs both along and across bedding planes and by faulting where displacements of over 50 m cause the juxtaposition of different geological units (Maurice et al., 2008). The fracture control on surface water drainage is also evident with river networks often aligned with major lineaments.

6.4   HYDROGEOLOGY OF THE MID-JURASSIC GREAT OOLITE AND INFERIOR OOLITE AQUIFERS

The limestone-dominated Great Oolite and Inferior Oolite aquifer units are of principal importance for groundwater resources. Several investigations into the hydrogeology of these aquifers were completed in the 1980s and 1990s, driven largely by groundwater resource exploitation and accompanying legislation. The investigations invariably focussed on a specific local issue or aquifer unit e.g. the Great Oolite (Rushton et al., 1992) and only rarely (Rushton et al., 1992) appear in the peer-reviewed literature. A notable exception is a review of the hydrogeochemistry of the Jurassic limestones provided by Morgan-Jones & Eggboro (1981). Findings from these investigations are presented in Maurice et al. (2008) while the hydrogeology is summarised in Allen et al. (1997) and Neumann et al. (2003).

The limestones of the Great Oolite and Inferior Oolite Groups are generally well-cemented with low inter-granular permeability and low storage potential (Morgan-Jones & Eggboro, 1981). Aquifer productivity derives from secondary deformation with rapid groundwater flow occurring along fracture pathways. Karstic flow, where fractures are enhanced by dissolution, is also suggested (Allen et al., 1997). The combination of low storage and high transmissivity means that both aquifers are 'flashy' with large seasonal variations in groundwater levels and a rapid response to rainfall (Rushton et al., 1992). Shallow pathways which provide rapid groundwater flow to rivers prior to the recovery of aquifer storage are evident as well as deeper pathways for groundwater which recharge the confined aquifer over longer timescales (Rushton et al., 1992). While separated from each other by the mudstones of the Fuller's Earth Formation, there are places where the Great Oolite and Inferior Oolite are in hydraulic continuity as a result of faults which connect the aquifers (Maurice et al., 2008). The impact of faulting and fracturing on groundwater systems also extends to cross-catchment groundwater flow (Maurice et al., 2008) and migrating and intermittent river reaches where rivers recharge the aquifer (Rushton et al., 1992) as well as limiting the extent of drawdown due to abstraction (Morgan-Jones & Eggboro, 1981). Where information about the geological structure and faulting in the third dimension is often limited, particularly for confined or concealed aquifers, 3D geological models may be used to interrogate the detailed geology and explore the relative importance of structure versus lithology on groundwater flow.

6.4.1   Integrating hydrogeological data with geology models

The integration of hydrogeological data with the geological interpretation has always been essential for the development of conceptual and numerical groundwater models. However the development of a 3D geological framework model brings advantages by capturing more explicitly the geologist's structural interpretations such that the geology is not oversimplified in subsequent process modelling. Some novel approaches to integrate spatial geology and hydrogeology datasets in a 3D model for both visualisation and scientific understanding have been explored within this project and are summarised in Table 6.2.

Table 6.2 Description of the hydrogeological datasets integrated with the geological framework model.

Hydrogeological data	Source	Format	Purpose
Permeability classification	BGS permeability mapping	Colour ornamentation within 3D model	Distinguishes more permeable formations and the mechanisms of groundwater flow.
Springs	BGS springs dataset	Location and elevation (X,Y,Z data) illustrated within 3D model and GIS.	Provides an indication of perched groundwater, conductive faults, spring lines and spring typology.
Observation borehole (OBHs) time series data	Borehole completion taken from BGS borehole records. Groundwater time series from environmental regulator	Long-term time series data for multiple OBHs. Coded into 3D model as X,Y, Z data visible in synthetic cross-sections.	Illustration of observed groundwater data and representative seasonal variations to assess against e.g. geology, structural features, cross-aquifer flow, groundwater contours.
Groundwater contour surfaces	Environmental Agency (England and Wales)	Digitised raster surfaces for display in 3D model and GIS.	Indicates groundwater flow directions, zones of discharge, groundwater-surface water interaction, cross-aquifer groundwater flux potential, spatial extent of aquifer confinement.
River flow data River-bed geology	Environmental Agency (England and Wales) BGS geological maps	Long-term time series flow data from gauging stations, spot-flow gauging and stream-head migration data. Not directly integrated in 3D model or GIS. Riverbed elevations and geology interrogated within GIS.	Identify and examine intermittent river sections and zones gaining and losing river flows. Explore relationship between groundwater levels, river flows and geological setting.

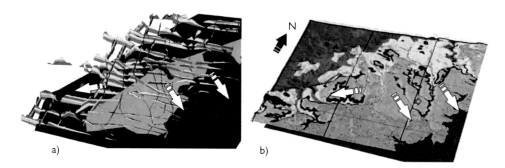

Figure 6.2 Outputs from the GSI3D model showing a) the geological cross-sections with maximum groundwater level surfaces for the Inferior Oolite aquifer (dark blue) and the Great Oolite aquifer (light blue), and; b) the mapped geology at surface (1:250 000 scale) draped over the digital terrain model along with the groundwater level surface to highlight where artesian groundwater levels exist in the Inferior Oolite aquifer under maximum groundwater level conditions (dark blue). White arrows highlight the regional groundwater flow direction; note the catchment divide between the River Thames to the east and the River Severn to the west. Contains Ordnance Survey data © Crown Copyright. (*See colour plate section, Plate 18*).

6.4.2   Interaction between the Great Oolite and Inferior Oolite aquifer

The Great Oolite and Inferior Oolite aquifers are hydraulically independent aquifer units with different water table elevations despite evidence of localised connectivity e.g. via faults (Maurice *et al.*, 2008). To gain an initial overview of the aquifer systems, groundwater level contour surfaces for maximum and minimum groundwater level conditions, for each of the aquifers were imported into the geological model and viewed in 3D (Figure 6.2). Under maximum groundwater level conditions in the west of the model area groundwater levels in the Inferior Oolite are below those of the Great Oolite i.e. potential downward flow of groundwater between the aquifers while to the east the groundwater levels in the Inferior Oolite aquifer are above those of the Great Oolite hence a potential upward flow of groundwater. By introducing a digital terrain model (DTM) as a capping surface to the geology model the interaction of the Great Oolite and Inferior Oolite water tables with surface water systems is apparent and artesian groundwater levels can be seen (Figure 6.2). Under minimum groundwater level conditions, the Inferior Oolite aquifer appears to contribute only to the River Coln, a tributary of the River Thames (Figure 6.1). Under high groundwater level conditions, there again appears to be greater contribution from the Inferior Oolite aquifer within the upper reaches of the River Coln than within the River Churn, another tributary of the River Thames (Figure 6.1). This may suggest that groundwater contributions to the headwaters of the River Churn are from perched springs following shorter pathways elevated above the main aquifer as a result of well-developed fracture zones. The more pronounced accretion in river flows down the valley of the River Coln, compared with the River Churn, where initial spring flow contribution in the headwaters appears to be lost further downstream, might support this hypothesis (Parades, 2012). The interaction of the Great Oolite aquifer with surface water streams

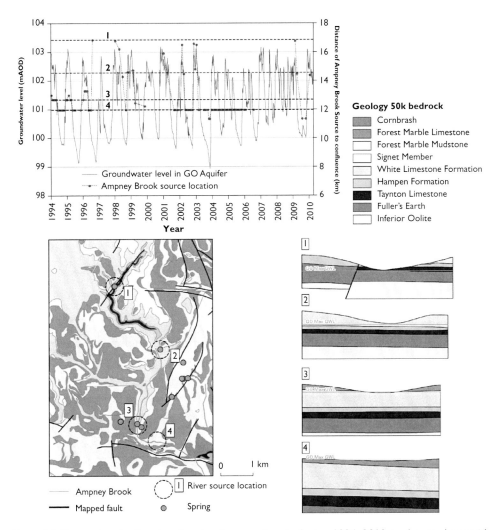

Figure 6.3 Stream head migration for the Ampney Brook during 1994–2010 is shown along with groundwater level variations within a Great Oolite observation well. Four locations to which the Ampney Brook migrates to are identified. The surface and sub-surface geology at each of these locations is shown. Groundwater level and river source location data © Environment Agency copyright and/or database rights 2012. All rights reserved. (*See colour plate section, Plate 19*).

appears to be more restricted under maximum groundwater level conditions. However, there is an apparent contribution from the Great Oolite aquifer at the head of the River Thames where several springs emanate and within the lower sections of the Ampney Brook (Figures 6.1 and 6.3). There may also be some contribution from the Great Oolite aquifer at its boundary with the Inferior Oolite aquifer within in the valley of the River Churn where the channel is more heavily incised than its neighbouring tributaries.

The presence and effect of faulting on the aquifer units is not always obvious, particular where they are confined. As a result groundwater contour maps constructed within a 2D mapping environment, albeit with hydrogeological expertise, are unlikely to encapsulate satisfactorily the 3D geological setting, the role of faults in compartmentalisation of the aquifer and the potential cross-connection of aquifers. Inspection of the fault sections within the GSI3D model suggests that while throws may on occasion be large (over 50 m) there are actually very few instances within the area where the Inferior Oolite or Great Oolite are juxtaposed as a result of the displacement. This implies that groundwater flow between the two aquifer systems is a function of conductivity along the fault zone. Viewing existing groundwater contour maps in combination with groundwater levels observed in boreholes within a 3D geological model resolves groundwater level contours around the 3D geology and fault networks and for example, faults are categorised into those which are conductive and those which act as barriers.

6.4.3 Structural control on hydrogeology

Morgan-Jones & Eggboro (1981) suggest that groundwater flow is less controlled by the stratigraphy and more by the structure yet spring lines occur at the junction of specific stratigraphical units such as the Fuller's Earth Formation or are associated with marl bands within units (Allen *et al.*, 1997). To assess the influence of structure versus lithology further, spring locations were correlated with a) mapped geological units using the digital geological map at the 1:50 000 scale (DiGMapGB50), and b) mapped faults all within a GIS format. Correlating springs with stratigraphical units yields predictable results which are in keeping with previous investigations (Morgan-Jones & Eggboro, 1981). For example numerous springs are observed at the base of Great Oolite aquifer at its junction with the Fuller's Earth Formation and at the base of the Inferior Oolite aquifer at its junction with the underlying Lias mudstones. Spring discharge is also coincident with the Hampen Formation, an interbedded limestone and marl unit of the Great Oolite and the Harford Member, a sand and clay unit within the Birdlip Limestone Formation of the Inferior Oolite (Figure 6.3).

To the north of the area, coincident with the outcrop of the Inferior Oolite aquifer, is a strong association between springs and mapped faults (Figure 6.4). These springs are not rising along faults under artesian pressure, they are perched and discharge at an elevation above the main water table within the Inferior Oolite. Enhanced deformation and higher fracture density within the limestone units around the fault zone offers a plausible explanation for the discharge of groundwater in the shallow zone emanating at the perched springs. There are faults located on the outcrop of the Great Oolite aquifer, to the east of the area, within the lower Coln valley (Figure 6.1) and adjacent River Leach catchment which don't have any springs associated with them despite expected artesian groundwater conditions in the Inferior Oolite aquifer. Either the throws on the faults in this area are insufficient to provide a pathway for artesian Inferior Oolite groundwater to reach surface or that any upward flow of groundwater from the Inferior Oolite aquifer contributes to the overlying Great Oolite aquifer which itself is not artesian and is expected to have available storage during low flow

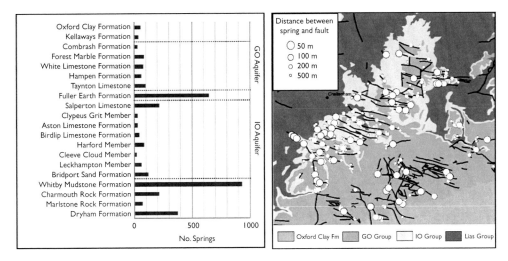

Figure 6.4 The number of springs associated with each of the geological unit mapped at surface (1:50 000), and; the extent to which springs are associated with mapped fault. (*See colour plate section, Plate 20*).

periods. The intermittent nature of flows in the River Leach along this faulted section offers further evidence that the Great Oolite aquifer in this area becomes depleted during summer periods where river flows are lost to Great Oolite aquifer storage.

The influence of structure on river flows is further illustrated by examining stream-head migration of the rivers along their intermittent sections. Using the Ampney Brook (Figure 6.1) as an example (Figure 6.3) we see that there are four principal locations to which the Ampney Brook rises to, all of which are coincident with geological features or the presence of springs; the first lies on a mapped fault; the second occurs at the boundary of the Hampen Formation with the overlying White Limestone Formation; the third occurs at the boundary of the Great Oolite limestones with the overlying confining layer of the Forest Marble mudstones; the fourth occurs at the confluence of a dry valley with the main channel of the Ampney Brook. An association between stream-head migration and groundwater levels in the Great Oolite aquifer is also evident, while the additional influence of groundwater abstraction from the Great Oolite aquifer has also observed (James, 2011).

Further information about the geological controls on the hydrological observations may be elucidated by viewing the river flow accretion profiles in combination with the river bed geology extracted from the GIS and geological profiles across river sections derived from GSI3D. For example, river flow accretion profiles along the River Churn exhibit a reduction in flow over both the Inferior Oolite and the Great Oolite limestones with the river recharging the underlying aquifer (Figure 6.5). Meanwhile there are significant increases in river flows over the mudstone units of the Lias Group, Fuller's Earth Formation and the upper Jurassic clays, reflecting the re-emergence of groundwater at the boundary of these less permeable units with the more permeable limestones at well-developed spring lines. The significance of the

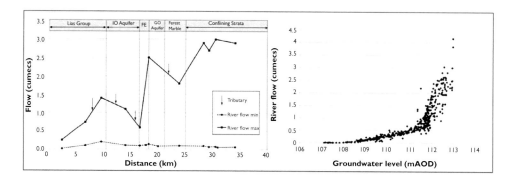

Figure 6.5 Flow accretion profile for the River Churn showing the influence of bedrock geology on gaining and losing river sections and; the relationship between groundwater levels in the Great Oolite at observation well SP00/142 and flows in the River Churn at the same location. River flows measured manually at spot locations and reported to an accuracy of one litre per second. Groundwater level and river flow data © Environment Agency copyright and/or database rights 2012. All rights reserved.

activation of these spring lines and groundwater discharge points as aquifer storage is replenished is evident when correlating river flows in the Churn at Cirencester (Figure 6.1) against groundwater levels within the Great Oolite aquifer at a nearby observation well (SP00/142). A groundwater level threshold is observed, controlled by the discharge elevation of the spring line, above which increases in groundwater level are small compared to increases in river flow (Figure 6.5).

6.4.4   Conceptual understanding using spring type

Spring discharge, whether stratigraphically or structurally controlled, exerts a strong influence on the hydrological observations within the Cotswolds and characterisation of the spring discharge mechanisms may be used to define broad hydrogeological domains (Figure 6.6). For example, in areas where perched groundwater is present and shallow and rapid pathways for groundwater discharge are expected: conversely areas can be defined where groundwater discharge, via springs, occurs at the intersection of the main aquifer water table. A simple approach is adopted whereby the ground elevation of the spring is compared with groundwater level elevations in both the Great Oolite and the Inferior Oolite aquifers. Where the spring elevation lies well above the groundwater level elevation we consider the spring to be perched. Where there is good agreement between the spring elevation and the groundwater level in one of the aquifers there is likely to be potential groundwater discharge from that aquifer unit. Where there is good agreement between the spring elevation and the groundwater level in both of the aquifers groundwater discharge from either or both aquifer units is plausible and the two aquifers may be in hydraulic continuity at that point. The rules used to define the hydrogeological domains are described in Table 6.3.

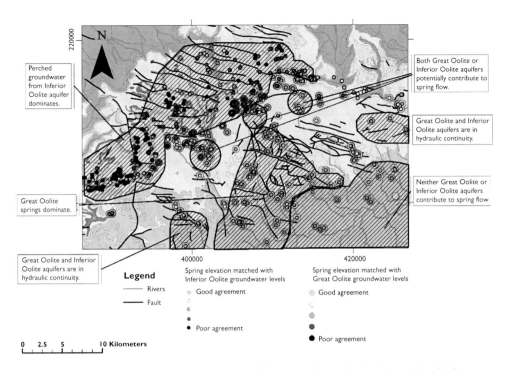

Figure 6.6 Hydrogeological domains for the Cotswold aquifers derived using the relationship between spring elevation and groundwater levels within the underlying Great Oolite and Inferior Oolite aquifer as an indicator of groundwater discharge processes and aquifer inter-connectivity. Contains Ordnance Survey data © Crown Copyright. (*See colour plate section, Plate 21*).

Table 6.3 Hydrogeological significance of elevation agreement between spring elevations and groundwater level elevations.

Elevation agreement	Hydrogeological significance
The spring elevation shows no correlation with groundwater levels in either the Great Oolite or Inferior Oolite aquifer.	The spring is discharging groundwater from a different aquifer unit OR the spring is perched above the main water table.
The spring elevation is correlated with the groundwater level in one of the aquifers.	The Great Oolite and Inferior Oolite aquifers are hydraulically independent at this location with different groundwater levels AND there are perched springs associated with upper aquifer OR artesian springs associated with the lower aquifer.
The spring elevation correlates with groundwater levels in both the Great Oolite and Inferior Oolite aquifer.	The Great Oolite and Inferior Oolite are in hydraulic continuity at this location OR the similarity between the Inferior Oolite and Great Oolite groundwater level is coincidental.

The hydrogeological domains may be verified using more localised understanding derived from the 3D geological model and groundwater observations or from anecdotal information, for example throws on the faults may be examined in the 3D geological model to confirm or exclude the potential cross-aquifer connection.

6.5   DISCUSSION AND CONCLUSIONS

The accuracy of a 3D geological framework model may be constrained by both the particular approach and the quality of data available so care must be taken not to over-interpret the results. For example, much of our understanding of an aquifer system derives from groundwater contour maps; while compiled using high quality groundwater level data and expert judgement they were delineated using a 2D approach and interrogation of these contours in 3D may show them to be deficient in some areas. Equally important, the 3D geology is an interpretation and is limited by the availability, spatial density and quality of borehole logs and the accuracy and scale of geological mapping available. Thus the integration of geological and hydrogeological information should be viewed as an iterative process where new observations may inform understanding developed in the other.

The 3D geological framework is built by extrapolation and interpolation from and between point information to produce an upscaled more generalised understanding. The same is true of developing a groundwater or hydrogeological conceptual model in three dimensions. Point observations are taken and interpolated spatially to reveal spatial trends. When these approaches are integrated both the hydrogeological model and the 3D geological framework are subjected to a new set of constraints that must be resolved. Local hydrogeological detail can be examined against the background of the geological setting to reveal errors or hopefully agreement in the datasets and conceptual understanding. The success of this process, demonstrated here in the Cotswolds lies in the enhanced ability to resolve many of the detailed hydrological observations with what can be elucidated from the 3D geology. This results in the refinement of a broader conceptual understanding. In this case, for example, such that hydrogeological domains may be delineated over a wider area. This ability to use detailed local information and data to refine conceptual understanding at the greater scale, i.e. from local to subcatchment to catchment and even regional level, introduces a level of refinement and new science hitherto not available.

In the Cotswolds this approach has allowed a multi-scale approach by using a combination of 3D geological framework modelling, geospatial hydrogeological data and information manipulation in this integrated fashion. At the local scale the model was used to understand specific hydraulic observations such as spring emergence at fault zones but also more broadly to assess catchment groundwater flow and hydrogeological domains. There are significant geological controls on the hydrogeological characteristics, exerted by both the stratigraphy and by the structure. The limestone aquifer units being highly deformed by intense fracturing and faulting, provide shallow pathways in the unsaturated zone for rapid groundwater discharge to springs elevated above the main zone of aquifer storage. The highly fractured unsaturated zone around incised river valleys is equally important for directed aquifer recharge and river losses. While fracturing provides the dominant control on aquifer recharge and groundwater conveyance, groundwater discharge is often controlled by the presence of mudstones in the stratigraphical sequence along which prominent spring lines have developed. Faulting also appears to provide a localised mechanism for groundwater discharge and the exchange of groundwater between the two aquifers.

While many of the observations about the Great Oolite and Inferior Oolite aquifers are in agreement with previous investigations, the 3D process demonstrates more clearly the relationship between the 3D geology and how that gives rise to the hydraulic responses observed and the hydrogeological conceptual model has been considerably improved on a local catchment scale.

ACKNOWLEDGEMENTS

The authors thank NERC for funding this work within the Changing Water Cycle – Hydrological Extremes and Feedbacks (HydEF) project and members of the HydEF project team for their advice. We also thank the reviewers for their constructive comments, Holger Kessler and Ian Cooke for their support with the 3D geological modelling and Henry Holbrook for the illustrations. The contributions of visiting MSc students Katy James, Cristian Parades and Fiona Marks are also gratefully acknowledged. This paper is published with the permission of the Executive Director of the British Geological Survey (Natural Environment Research Council).

REFERENCES

Aldiss D.T., Black M.G., Entwisle D.C., Page D.P., Terrington R.L. (2012). Benefits of a 3D geological model for major tunneling works: an example from Farringdon, east-central London, UK. *Quarterly Journal of Engineering Geology and Hydrogeology* 45(4), 405–414.

Allen D.J., Brewerton L.J., Coleby L.M., Gibbs B.R., Lewis M.A., MacDonald A.M., Wagstaff S.J., Williams A.T., Barker J.A., Bird M.J., Bloomfield J.P., Cheney C.S., Talbot J.C., Robinson V.K. (1997). The physical properties of major aquifers in England and Wales. British Geological Survey, Hydrogeology Series, Technical Report WD/97/34.

Barron A.J.M., Lott G.K., Riding J.B. (2012) Stratigraphical framework for the Middle Jurassic strata of Great Britain and the adjoining continental shelf. British Geological Survey report RR/11/006.

Campbell S.D., Merritt J.E., O Dochartaigh B. E., Mansour M., Hughes A.G., Fordyce F.M., Entwisle D.C., Monaghan A.A., Loughlin S.C. (2010) 3D geological models and their hydrogeological applications: supporting urban development: a case study in Glasgow-Clyde, UK. *Zeitschrift der Deutschen Gesellschaft fur Geowissenschaften* 161(2): 251–262.

CENTRE FOR ECOLOGY AND HYDROLOGY (2012). National River Flow Archive, Centre for Ecology and Hydrology (CEH). Online data access Sept 2012. http://www.ceh.ac.uk/data/nrfa/data/time_series.html?39129

ENVIRONMENT AGENCY (2007). The Cotswolds Catchment Abstraction management Strategy Document, Environment Agency (EA), 2007. Online publication GETH1007BNME-E-E.

ENVIRONMENT AGENCY (2012). Climate change and river flows in the 2050s, Environment Agency (EA) Science Summary SC070079/SS1. Online publication SCHO1008BOSS-E-P, Sept 2012.

Howard A. S., Hatton W., Reitsma F., Lawrie K.I.G. (2009). Developing a geoscience knowledge framework for a national geological survey organisation. *Computers and Geosciences* 35: 820–835

James K., (2011). An overview of the Upper Thames Catchment Hydrogeology, and the Development of Neural Network Modelling for Baseflow Simulation. MSc Thesis. Cardiff University, UK.

Kaufmann O., Martin T. (2008). 3D geological modelling from boreholes, cross-sections and geological maps, application over former natural gas storages in coal mines. *Computers and Geosciences* 34(3): 278–290.

Kessler H., Mathers S.J., Sobisch H.G. (2009). The capture and dissemination of integrated 3D geospatial knowledge at the British Geological Survey using GSI3D software and methodology. *Computers & Geosciences* 35: 1311–1321.

Marchant A.P., Banks V.J., Royse K., Quigley S.P., Wealthall G.P. (2011) An Initial Screening Tool for water resource contamination due to development in the Olympic Park 2012site, London. *Environmental Earth Sciences* 64(2). 483–495.

Maurice L., Barron A.J.M., Lewis M.A., Robins N.S. (2008). The geology and hydrogeology of the Jurassic limestones in the Stroud-Cirencester area with particular reference to the position of the groundwater divide. British Geological Survey Commissioned Report, CR/08/146.

Merritt J.E., Monaghan A.A., Entwisle D.C., Hughes A.G., Campbell S.D.G., Browne, M.A.E. (2007) 3D attributed models for addressing environmental and engineering geoscience problems in areas of urban regeneration: a case study in Glasgow, UK. *First Break* 25: 79–84.

Morgan-Jones M., Eggboro M.D. (1981). The Hydrogeochemistry of the Jurassic Limestones in Gloucestershire, England. *Quarterly Journal of Engineering Geology* 14(1): 25–39.

Neumann I., Brown S., Smedley P., Besien T. (2003). Baseline Report Series: 7. The Great and Inferior Oolite of the Cotswolds District. British Geological Survey report CR/03/202 N.

Parades C., (2012). Hydrogeological control of the river-aquifer interaction in the Cotswolds Limestone Aquifers, UK. MSc Thesis, University of Birmingham, UK.

Robins N.S., Rutter H.K., Dumpleton S., Peach D.W. (2005). The role of 3D visualisation as an analytical tool preparatory to numerical modeling. *Journal of Hydrology* 301(1–4), 287–295.

Royse K., Rutter H., Entwisle D. (2009) Property attribution of 3D geological models in the Thames Gateway, London: new ways of visualising geoscientific information. *Bulletin of Engineering Geology and the Environment* 68(1): 1–16.

Royse K., Kessler H., Robins N., Hughes A., Mathers, S. (2010). The use of 3D geological models in the development of the conceptual groundwater model. *Zeitschrift der Deutschen Gesellschaft für Geowissenschaften, Schweizerbart* 161(2), 237–249.

Rushton K.R., Owen M., Tomblinson L.M. (1992). The Water-Resources of the Great Oolite Aquifer in the Thames Basin, UK. *Journal of Hydrology* 132(1–4): 225–248.

Sumbler M.G, Barron A.J.M., Morigi A.N. (2000). Geology of the Cirencester district. Memoir of the British Geological Survey, Sheet 235 (England and Wales).

Wang L., Tye A., Hughes A. (2010) Riverine floodplain groundwater flow modelling: the case of Shelford (UK). Nottingham, UK, British Geological Survey, 28pp. Report IR/09/043.

Chapter 7

Characterising the spatial distribution of transmissivity in the mountainous region: Results from watersheds in central Taiwan

Po-Yi Chou, Jung-Jun Lin, Shih-Meng Hsu,
Hung-Chieh Lo, Po-Jui Chen, Chien-Chung Ke,
Wong-Ru Lee, Chun-Chieh Huang, Nai-Chin Chen,
Hui-Yu Wen & Feng-Mei Lee
Geotechnical Engineering Research Center,
Sinotech Engineering Consultants Inc., Taipei, Taiwan

ABSTRACT

Complex geological features are present in the mountainous regions of Taiwan. Strongly folded, faulted and highly fractured formations are common, the delineation of groundwater producing zones is, therefore, challenging. However, progress towards characterising the transmissivity in the mountainous region of Taiwan has been made. There are three major watersheds: the Dajia, Wu and Jhuoshuei River. A total of 65 boreholes have been drilled, various types of well logging, rock core inspection and packer tests carried out. Using these data gathered during the past three years, a conceptual model has been developed which delineates the spatial distribution of hydraulically transmissive conduits in the region.

7.1 INTRODUCTION

Groundwater plays a pivotal role in sustainable development. In Taiwan, a great number of studies have been devoted to investigating the groundwater system in the alluvial plains (Hsu, 1998; Ting *et al.*, 1998; Jang & Liu, 2004; Chen *et al.*, 2005; Chou & Ting, 2007; Tu *et al.*, 2011). In total, nine critical groundwater basins have been recognised on the basis of their water production capabilities as well as local water demand. However, along with the ever-growing population and the rising demand for water, recent studies have also shown that the majority of these alluvial groundwater basins are under stress. Interest has been raised over whether the productivity of upstream hard-rock aquifers could provide and meet the needs of downstream water users. Yet, this is a difficult and challenging task as previous studies (Maréchal *et al.*, 2004; Foster, 2012) indicate that groundwater bodies are not universally distributed within hard-rock formations.

The mountainous regions have greater changes in elevation, slope gradient, geological structure and stratigraphic composition than the alluvial areas. The spatial variability and uncertainty of aquifer hydraulic parameters in mountainous regions is higher, and this is one of the main difficulties in modelling such terrain. The earlier investigations concentrated on the alluvium and there remain many gaps in the spatial distribution of transmissivity in the mountainous region. As a result, insufficient base-line data are available to enable reliable analysis and accurate prediction with which to inform policy makers.

Under the co-ordination of the Central Geological Survey (CGS) of Taiwan, a nationally-funded investigation programme was launched in 2010 to improve the understanding of the groundwater resources in the mountainous region. Exploratory drilling and related studies are being carried out by Sinotech Engineering Consultants for, and in collaboration with two technical institutions. The particular aims are (1) to obtain the depth-distribution of geological attributes and geophysical parameters, and (2) to assess the hydrogeological character of the upstream bedrock aquifers. The first phase of the programme (2010–2013) focussed on central Taiwan, which is home to nearly 1 in 4 (5.6 million) of the population of Taiwan. The investigation has been carried out in three major watersheds, including Dajia, Wu and Jhuoshuei River. This chapter provides a brief description of the study area, the methods of interpretation, and the investigation results to date.

7.2   THE STUDY AREA

The study area (Figure 7.1) is situated roughly at latitude and longitude of 24°00'N and 121°00'E and covers an area of 6560square kilometres. The Central Mountain Range (3930 m a.s.l.) runs from north to south and forms the major watershed divide in Taiwan. The catchment watersheds between the Dajia, Wu and Jhuoshuei River are oriented east west with the rivers draining towards the Taiwan Strait. The geographical setting of the study area can be split conceptually into two sub-regions according to the lithological features. Bordering the Central Mountain Range, region one is located in the east of the study area. In this region, the topographic slopes are steep and rugged. The predominant lithologies (ranging from Eocene to Early Miocene) are mainly composed of highly metamorphic argillite, slate and phyllite. Region two is located in the west and is underlain by Miocene to Pleistocene sedimentary rocks. A series of east west fold-and-thrust belts is present in the region. The predominant lithology is massive sedimentary sandstone, siltstone and shale. An apparent lithological transition zone can be recognised in the middle of the area, in which the upper part is composed mainly of hard shale, whereas the lower part is composed of coarse quartz sandstone.

The surficial material comprises unconsolidated weathering materials, *i.e.* the regolith layers, which consist of soil, infills, alluvium, colluvium, and saprolite. The average thickness is 17.5 m. Twenty hydrogeologic units have been recognised based on the age and the rock type, and seventeen of them have been examined over the last three years. The characterisation of these hydrogeologic units is summarised in Table 7.1.

A preliminary map of groundwater potential zones has been prepared to help determine optimum drilling sites. During 2010 to 2012, 65 exploratory boreholes have been drilled to a depth of 100 m, 15 in the Dajia River basin, 16 in the Wu River basin and 34 in the Jhuoshuei River basin using a combination of mechanical and pneumatic drilling rigs. Rock samples are recovered for the initial lithostratigraphic

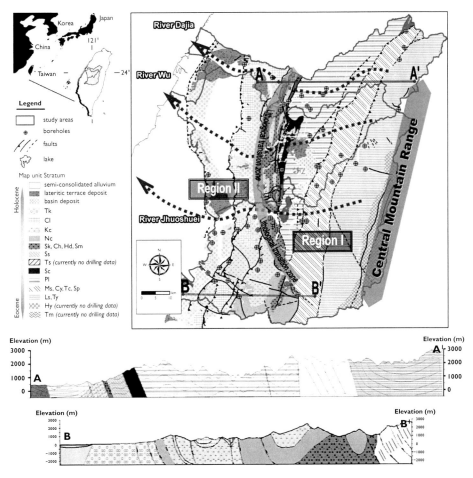

Figure 7.1 Geological setting (up) and two geological cross sections (east-west) (down) of the study area. (*See colour plate section, Plate 22*).

identification and subsequent laboratory analysis. A variety of geophysical logs and flow measurements were carried out.

7.3 FIELD INVESTIGATIONS

The general procedure of field investigations is shown in Figure 7.2. Comparing the driller's log and the geologists' report, a general characterisation can be made for each site of the distribution of fractures and the corresponding permeability of the rock mass. This indicates the depths to which water-bearing fractures are presented, but also any formation breaks within zones of no core recovery. This is as an essential procedure prior to geophysical logging. A number of borehole logging tools are used and aquifer hydraulic testing is carried out in each borehole.

Table 7.1 Geologic formation of hydrogeologic units in the study area.

Geologic formation (code)	Geological epoch	River Basin	Rock type	Number of boreholes
TouKeshan formation (Tk)	Pleistocene	J	S	6
ChoLan formation (Cl)	Pleistocene ~ Pliocene	D, J, W	S	4
KueiChulin formation (Kc)	Pliocene ~ Miocene	D, J	S	7
NanChuang formation (Nc)	Miocene	J, W	S	2
ShenKeng formation (Sk)	Miocene	J, W	S	6
ChangHukeng formation (Ch)	Miocene	J	S	2
ShihSzeku formation (Ss)	Miocene	D, W	S	2
Shihman formation (Sm)	Miocene	D, W	S	2
HouDongkeng (Hd)	Miocene	D, W	S	2
LuShan formaiton (Ls)	Miocene	J, W	M	4
TaYuling formation (Ty)	Miocene	J	M	1
ShuiChanglium formation (Sc)	Oligocene	D, J, W	M	3
MeiShi formation (Ms)	Oligocene	D	M	2
PaiLeng formation (Pl)	Oligocene ~ Eocene	D, J, W	M	12
TaChien formation (Tc)	Oligocene ~ Eocene	J, W	M	3
ChiaYang formation (Cy)	Oligocene ~ Eocene	D, J, W	M	3
ShihPachungchi formation (Sp)	Eocene	J, W	M	2

D: Dajia River basin; **J**: Jhuoshuei River basin; **W**: Wu River basin; **S**: Sedimentary rock; **M**: metamorphic rock

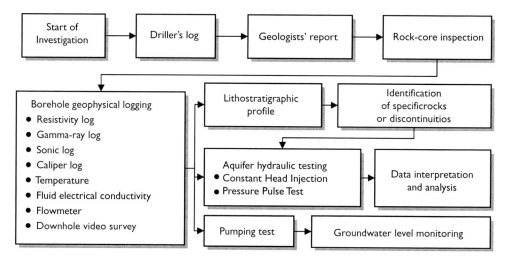

Figure 7.2 Flow-chart of investigation.

7.3.1 Borehole-geophysical logs

-   The **Resistivity log** measures the resistance of the formation fluid to an induced electrical current, and is closely related to the total volume of interconnected pore spaces in the bedrock, namely the effective porosity that contributes to groundwater flow. The range is governed by the electrode spacing, and this is typically set at either 40.6 cm (short) or 162.6 cm (long).
-   The **Gamma-ray log** is used for characterising the radioactivity emanating from the rock surrounding the borehole, and is primarily related to the rock type and the relative proportion of clay minerals in the formation. Generally, a lower gamma-ray response indicates the presence of a coarse-grained formation with a favourable water-transmitting capacity.
-   The **Sonic log** measures the transmission of compressional waves through the formation, and is related to the rock strength, its elastic properties and porosity. The longer the travel time, or the slower the wave propagation velocity, the higher the porosity and water-transmitting capacity.
-   The **Caliper log** provides a continuous profile of the borehole diameter with depth.
-   The **Temperature** and **Fluid electrical conductivity logs** assist in identifying the presence of hydraulically active discontinuities.
-   Two types of **Flowmeter** are used to delineate preferential flow paths and connetivity between fractures. The impeller flowmeter provides a continuous log of the general flow rate in the borehole, while the heat-pulse flowmeter can be used to further delineate the direction of flow.
-   Lastly, the **downhole video survey** provides a clear visual inspection of the borehole walls. Not only can the geologists' report can be confirmed, but the dip-azimuth, aperture and infilling material can also be identified. The high resolution acoustic televiewer (HiRAT) and optical televiewer (OPTV) were used. The HiRAT can be carried out in a mud-filled borehole, while the OPTV is applicable in dry or clean water-filled holes.

Using all the geophysical data a total of 365 segments were identified for in-situ packer tests.

7.3.2 Aquifer hydraulic testing

The packer testing was carried out to determine the transmissivity of the rock mass and of individual fracture-zones that are presumed to be conductive. This standard and reliable procedure that has been applied since the 1950s and is widely employed in groundwater investigation (Mejías *et al.*, 2009; Chou *et al.*, 2012). The basic concept is to isolate a specific interval of the wellbore with inflatable packers, inject water to a set constant pressure, and measure the pressure change over time.

The packer assemblies include discrete rubber packers, pressure transducers, flow regulators, water pumps, nitrogen cylinder, and the hoisting system. All the equipment is calibrated in the laboratory and in the field. A double-packer system is used (Figure 7.3). The sealed-off interval is fixed at 1.5 m. Each test starts by applying the Constant Head Injection (CHI) test, which needs a volume of pressurised water to pass through the test section for a prolonged period of time. The testing procedures

Figure 7.3 Photos taken during packer testing, (A) equipment calibration, (B) rig transfer, (C) 3ata recording and analysis. (*See colour plate section, Plate 23*).

followed the American Society for Testing and Materials (ASTM) methods. Water is injected with a constant pressure of 0.2 MPa in order to not damage to the aquifer. Each CHI test takes at least three hours including calibration (Figure 7.3A), rig transfer (Figure 7.3B), packer inflation, data recording (Figure 7.3C) and pressure recovery. In case no significant water flow rate can be measured, an alternative hydraulic test method, the Pressure Pulse (PP) test, can be applied, i.e. with extremely low transmissivity. A much longer period of testing time is usually needed. The variation of hydraulic head and injected flow rate within the test sections are recorded and stored in a data-logger.

Transmissivity is determined through the interpretation of the injected flow rate by means of the AQTESOLV analysis software. AQTESOLV offers a variety of analytical solutions, which enable both visual and automatic type curve matching of the test data. For all test segments that are capable of being tested by CHI test, the generalised radial flow (GRF) model (Barker, 1988) is applied. The GRF model accounts for the multi-scale fracture network heterogeneity (Le Borgne *et al.*, 2004). The conceptual aquifer geometry in a GRF model is assumed to be a fractal network of conduits, which matches the geological conditions in Central Taiwan. In addition, this model generalises the flow dimension 'to non-integral values. A coefficient 'n' is used, which denotes the flow dimension (the change of the cross-sectional area to the flow) according to the distance from the borehole (Walker and Roberts, 2003), and is set to be 1 for planar flow, 2 for cylindrical flow and 3 for spherical flow. The limitation of the GRF model is that it does not account for any inertial effect (Audouin & Bodin, 2008). The model is written as:

$$S_s \frac{\partial h}{\partial t} = \frac{K}{r^{n-1}} \frac{\partial}{\partial r}\left(r^{n-1} \frac{\partial h}{\partial r} \right) \tag{1}$$

$$T = K \times 1.5^{(3-n)} \tag{2}$$

where S_s represents the specific storage of the aquifer [1/L]; $h(r, t)$ denotes the change in hydraulic head [L] with time, r represents the radial distance from the borehole [L]; K represents the hydraulic conductivity [L/T]; T represents the coefficient of transmissivity [L^2/T]; n denotes the flow dimension according to the distance from the borehole.

For those boreholes that are productive, pumping tests are carried out at the final stage for estimating long-term yield. During the test, water is extracted by using the submersible electric pump (MP1), and a constant pumping rate at a discharge of 70 l/min is maintained for 24 hours. By monitoring the drawdown in every 30 seconds, the transmissivity and storage coefficient of the aquifer are determined using AQTE-SOLV using the analytical solution of Moench (1997). As soon after the intrusive investigations and geophysical surveys are completed, two piezometers were installed at selected depths in each borehole for long-term groundwater level monitoring, one is for observing the groundwater level fluctuation in the regolith, and the other is for that in deep bedrock aquifers.

7.4   RESULTS

7.4.1   Interpretation of hydraulic tests: A statistical perspective

The focus for the packer tests was the 193 segments that were tested by the Constant Head Injection (CHI) method under a steady flow rate. These segments are the primary conductive zone compared to the less productive segments tested by the Pressure Pulse (PP) method. The estimated transmissivity values derived from CHI are plotted (Figure 7.4) on a logarithmic scale against the corresponding depth on a linear scale, with the majority (65%) of estimates lying between 0.1 and 10 m^2/day. A relatively high degree of uncertainty regarding the spatial variability of transmissivity is present. In order to characterise the core data and reduce bias caused by outliers, the geometric mean of the transmissivities across a 10-m depth interval is used, and a slight decrease in transmissivity with depth can be obtained. The higher transmissivities (0.54~0.95 m^2/d) were obtained above 30 m depth, and between 40 to 50 m depth (0.44 m^2/day). The lower depths (fractured bedrock) have transmissivity values in the range 0.22~0.28 m^2/day. A statistical summary of transmissivities is shown in Table 7.2.

Further qualitative characterisation can be made by classifying the magnitude of transmissivity as: (1) high-transmissive zones: $10 > T \geq 1$ m^2/day, (2) medium-transmissive zones: $1 > T \geq 0.1$ m^2/day, (3) low-transmissive zones: $0.1 > T \geq 0.01$ m^2/day. It is apparent that several significantly high transmissive zones are present at depths between 30 to 60 m. and these may be the predominant pathway for groundwater flow. This suggests that assigning depth-dependent-transmissivity in multiple layers is not sensible within a conceptual hydrogeological model for the mountain region and that it is more appropriate to specify the representative transmissivity values (bounded by the arithmetic and the harmonic means) for different depth domains.

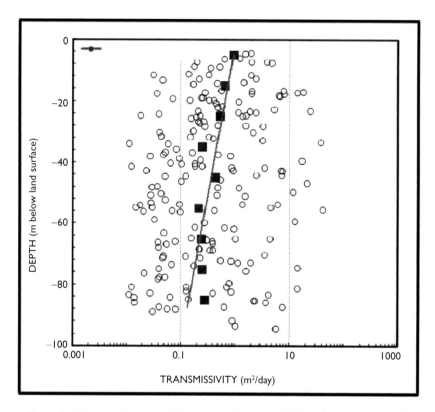

Figure 7.4 Estimated transmissivity derived from the CHI method against depth.

Table 7.2 Statistical summary of transmissivities (m²/day) across a 10-m depth interval.

Depth (m)	Geometric mean	Arithmetic mean	Standard deviation	Harmonic mean
<10	0.95	1.57	1.53	0.55
10–20	0.66	2.64	4.47	0.19
20–30	0.54	1.90	4.84	0.18
30–40	0.25	3.16	8.97	0.07
40–50	0.44	2.74	5.04	0.09
50–60	0.22	3.30	9.22	0.05
60–70	0.24	0.55	1.02	0.13
70–80	0.25	1.35	3.28	0.08
80–90	0.28	1.74	3.23	0.05
Total	**0.37**	**2.18**	**5.38**	**0.09**
highest				
lowest				

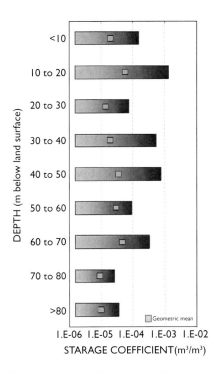

Figure 7.5 Estimated values of storage coefficient against depth.

The estimated distribution (1st–3rd quartile and geometric means) of storage coefficient for each depth interval is shown in Figure 7.5. The higher storage coefficients occur above a depth of 20 m, coinciding with the average thickness of regolith. This indicates that the transition zone between regolith and bedrock forms the upper productive water-bearing channel. The second high zone occurs at depths between 30 and 50 m. By comparing the arithmetic mean of transmissivity values (Table 7.2) for the same depth range, the second productive water-bearing layer is identified coinciding with dense fracture networks. In comparison with the first layer, this layer is a relatively narrow conduit for groundwater flow. According to the definition given by Lin (2010), both layers can be viewed as the Critical Zone (CZ), in which continuous exchange of water, energy and mass take place.

In the deeper zones the water storage capacity of the bedrock is limited. The slightly higher storage capacity layers are at depths from 60 to 70 m, but the corresponding transmissivity values are low. An explanation is that only a small portion of the open fractures above this depth range are hydraulically active, and the majority are sealed by quartz, calcite, chlorite, or filled with mud or clay. This is confirmed by the geophysical logs and rock core inspection.

A conceptual hydrogeologic model can be established, which provides a general but practical characterisation of groundwater flow in the mountain region. The water-bearing capacity of each horizon is shown in Figure 7.6 by different grey scales. The

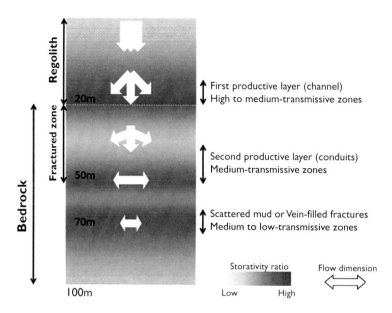

Figure 7.6 A conceptual hydrogeological model for groundwater flow in the mountainous region (not to scale).

darker-grey area corresponds to higher storativity ratio. The principal flow dimension in each horizon is also given, the longer arrow represents a potential for higher transmissivity values. There are two principal productive layers in this conceptual hydrogeologic model. The first is likely to appear at the transition zone from the regolith to bedrock. The second is the hydraulically active fractured zone within the uppermost portion of bedrock, which, as described by Sukhjia *et al.*, (2006), needs to connect to a distant recharge source beyond the edge of the overlying weathered zone. Dewandel *et al.*, (2006) have also illustrated that, in granite-type rocks, this layer may assume most of the transmissive capacity of the aquifer.

7.4.2   Vertical weathering profile in different geological formations: An integrated *view*

Certain zones within the regolith and the underlying bedrock have sufficient permeability to yield water to wells (Chilton & Foster, 1995). It is important to characterise the transmissivity at different zones. Generally, three distinct zones in terms of the degree of weathering can be identified above the bedrock: (1) the soil or infill deposits (R-s/b), (2) the highly weathered alluvium- or colluvium zone (R-a/c), and (3) the less weathered saprolite- or saprock zone (R-sl/sr). By integrating data obtained from drill core and aquifer hydraulic testing, the conceptual weathering profile of sedimentary and metamorphic formations (Figure 7.7), as well as the water-transmitting capacity of each weathering zones can be presented.

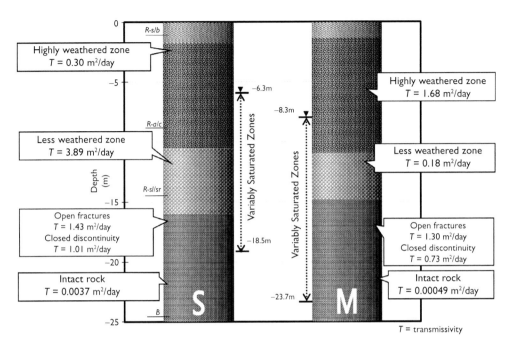

Figure 7.7 Vertical weathering profile in different lithologic settings (S for sedimentary, M for metamorphic rock).

Based on the data obtained from the HiRAT and OPTV, it is possible to distinguish the open fractures (with aperture ≥ 1 mm) from the closed ones (such as hairline cracks) within the bedrock. Therefore, the corresponding water-transmitting capacity possessed by the open fractures and the closed discontinuities is also presented in Figure 7.7. The open fractures identified in both types of formations are generally the high-transmissive zones; their water-transmitting capacities are 1.3~1.8 times greater than the estimates obtained from the closed discontinuities.

The zone of soil and backfill in the mountainous region is thin, generally less than 2 m, while the thickness of R-a/c zone and R-sl/sr zone are much greater. The arithmetic mean of transmissivities at each weathering zones suggests that in the sedimentary formations, the R-a/c zone possesses a lower transmissive capacity (T = 0.30 m²/day) than the R-sl/sr zone (T = 3.89 m²/day). The R-sl/sr zone is the first productive layer in the sedimentary formations (Figure 7.6), and the hydraulically active fractured zone in the uppermost portion of bedrock (T = 1.01~1.43 m²/day) is the second one. Comparatively, in metamorphic formations, the R-a/c zone is found to have a higher transmission capacity (T = 1.68 m²/day) than the R-sl/sr zone (T = 0.18 m²/day). As such, the R-a/c zone ought to be the first productive layer in the fractured bedrock, while the second one is the hydraulically active fractured zone at the uppermost portion of bedrock (T = 0.73~1.30 m²/day).

Since groundwater-level data are collected at each borehole, the relative groundwater elevations with respect to the weathering profile are also displayed in Figure 7.7.

In the sedimentary formations, the groundwater level lies approximately 6.3 to 18.5 m below ground level. In fractured bedrock, the groundwater level generally lies at deeper depths from 8.3 to 23.7 m. This shows the spatial variability of the vadose zone and the actual range of the Critical Zone in the two different settings. The thicker saturated weathered zone and its shallow water table offers a more favourable groundwater potential than fractured bedrock.

7.5   CONCLUSIONS

Interest in the yield of aquifers in the mountainous headwater basins is driven by increasing demand on the alluvial plains in Taiwan. It is difficult to assess the groundwater potential in such upland fractured bedrock and regolith terrain, which, as Klemeš (1988) argues, is the "blackest black box" in the hydrological cycle. Information from field investigations including lithostratigraphic identification, borehole-geophysical logs, aquifer hydraulic tests, enables a preliminary understanding of the distribution of storage and transmissivity in the Dajia, Wu and Jhuoshuei River catchments.

Two principal productive layers can be identified. The first is the transition zone between the regolith and the bedrock at a depth above 20 m, and the second is the hydraulically active fractured zone existing within the range from 30 to 50 m below the surface in the fractured bedrock. A conceptual hydrogeologic model is proposed that accounts for the spatial distribution of groundwater potential zones in this mountainous region. In the sedimentary formations and regolith, the higher transmissivity zone is found corresponding to the saprolite – or saprock, while in the metamorphic formations, the alluvium – or colluvium zone near the ground surface is possesses the best groundwater potential.

ACKNOWLEDGMENTS

The authors are grateful to our colleagues from the geotechnical testing group, Sinotech, in assisting the field work, and to the financial support of the Central Geological Survey, Ministry of Economic Affairs (MOEA) of Taiwan. We are also grateful to Prof. Dr. Stephen Foster and Irena Krusic-Hrustanpasic for their critical review and useful comments.

REFERENCES

Audouin O., Bodin J. (2008) Cross-borehole slug test analysis in a fractured limestone aquifer. *Journal of Hydrology* 348, 510–523.

Barker J.A. (1988) A generalised radial flow model for hydraulic tests in fractured rock. *Water Resources Research* 24, 1796–1804.

Chen J.F., Lee C.H., Yeh T.C., Yu J.L. (2005) A water budget model for the Yun-Lin Plain, Taiwan. *Water Resources Management* 19, 483–504.

Chilton P.J., Foster S.S.D. (1995) Hydrogeological characterization and water-supply potential of basement aquifers in Tropical Africa. *Hydrogeology Journal* 3, 36–49.

Chou P.Y., Lo H.C., Hsu S.M., Lin Y.T., Huang C.C. (2012) Prediction of hydraulically transmissive fractures using geological and geophysical attributes: a case history from the mid Jhuoshuei River basin, Taiwan. *Hydrogeology Journal* 20, 1101–1116.

Chou P.Y., Ting C.S. (2007) Feasible groundwater allocation scenarios for land subsidence area of Pingtung Plain, Taiwan. *Water Resources* 34, 259–267.

Dewandel B., Lachassagne P., Wyns R., Maréchal J.C., Krishnamurthy N.S. (2006) A generalised 3-D geological and hydrogeological conceptual model of granite aquifers controlled by single or multiphase weathering. *Journal of Hydrology* 330, 260–284.

Foster S., (2012) Hard-rock aquifers in tropical regions: using science to inform development and management policy. *Hydrogeology Journal* 20, 659–672.

Hsu S.K. (1998) Plan for a groundwater monitoring network in Taiwan. *Hydrogeology Journal* 6, 405–415.

Jang C.S., Liu C.W. (2004) Geostatistical analysis and conditional simulation for estimating the spatial variability of hydraulic conductivity in the Choushui River alluvial fan, Taiwan. *Hydrological Processes* 18, 1333–1350.

Klemeš V. (1988) Foreword. In: L. Molnâr L. (Eds.) *Hydrology of mountainous areas*. IAHS Publication No. 90.

Le Borgne T., Bour O., de Dreuzy J.R., Davy P., Touchard F. (2004) Equivalent mean flow models for fractured aquifers: Insights from a pumping tests scaling interpretation. *Water Resources Research* 40, W03512.

Lin H. (2010) Earth's Critical Zone and hydropedology: concepts, characteristics, and advances. Hydrology and Earth System Science 14, 25–45.

Maréchal J.C., Dewandel B., Subrahmanyam K. (2004) Use of hydraulic tests at different scales to characterize fracture network properties in the weathered-fractured layer of a hard rock aquifer, Water Resour Res 40, W11508.

Mejías M., Renard P., Glenz D. (2009) Hydraulic testing of low permeability formations: a case study in the granite of Cadalso de los Vidrios, Spain. *Engineering Geology* 107, 88–97.

Moench A.F. (1997) Flow to a well of finite diameter in a homogeneous, anisotropic water table aquifer. *Water Resource Research* 33, 1397–1407.

Sukhija B.S., Reddy D.V., Nagabhushanam P., Bhattacharya S.K., Jani R.A., Kumar D. (2005) Characterisation of recharge processes and groundwater flow mechanisms in weathered-fractured granites of Hyderabad (India) using isotopes. *Hydrogeology Journal* 14, 663–674.

Ting C.S., Zhou Y., de Vries J.J., Simmers I. (1998) Development of a preliminary ground water flow model for water resources management in the Pingtung Plain, Taiwan. *Ground Water* 35, 20–36.

Tu Y.C., Ting C.S., Tsai H.T., Chen J.W., Lee C.H. (2011) Dynamic analysis of the infiltration rate of artificial recharge of groundwater: a case study of Wanglong Lake, Pingtung, Taiwan. *Environmental Earth Science* 63, 77–85.

Walker D.D., Roberts R.M. (2003) Flow dimensions corresponding to hydrogeologic conditions. Water Resources Research 39: 1349–1357.

Chapter 8

Spring discharge and groundwater flow systems in sedimentary and ophiolitic hard rock aquifers: Experiences from Northern Apennines (Italy)

Alessandro Gargini[1], Maria Teresa De Nardo[2], Leonardo Piccinini[3], Stefano Segadelli[4] & Valentina Vincenzi[5]

[1]*Department of Biological, Geological and Environmental Sciences, Alma Mater Studiorum – University of Bologna, Italy*
[2]*Geologic, Seismic and Soil Survey, Emilia Romagna Region, Bologna, Italy*
[3]*Department of Geosciences, University of Padua, Italy*
[4]*Department of Earth Sciences and Physics "Macedonio Melloni", University of Parma, Italy*
[5]*Geotema s.r.l., Ferrara, Italy*

ABSTRACT

This chapter deals with hydrogeological characterization and springs/upreach streams monitoring activity on different test sites in Northern Apennines (Italy) where turbiditic and ophiolitic hard rock aquifers occur. The work aimed to define the investigation tools able to parameterize the aquifer and build-up a methodology for hydrogeological mapping. Hydrological analysis of base flow, tracer testing with fluorescent dyes and numerical flow modelling by the equivalent porous medium approach were used for the evaluation of direct recharge, quantification of flow and transport parameters and implementation of numerical models, calibrated by continuous discharge monitoring, finalized to define spring flow paths and protection areas. The ratio between direct recharge and total precipitation resulted in the range 0.13–0.17 for turbidites and 0.35–0.40 for igneous ophiolites. Groundwater average effective velocity ranges from 3.6 m/d in the thin-bedded pelitic turbidites to 39 m/d in the calcareous turbidites.

8.1 INTRODUCTION

This chapter is a review of the research activities, along with the main results, conducted in the hard rock aquifers of Northern Apennines, at the boundary between Tuscany and Emilia-Romagna regions in Northern Italy (Figure 8.1). Specific tools for hydrogeological characterization in such an area, such as: analysis of base flow hydrological recession, equivalent porous medium modelling approach calibrated against continuous monitoring of discharge and stream-tunnel tracer tests are applied. The main aim is to address the knowledge gaps in groundwater flow systems in low to medium bulk permeability fractured aquifers.

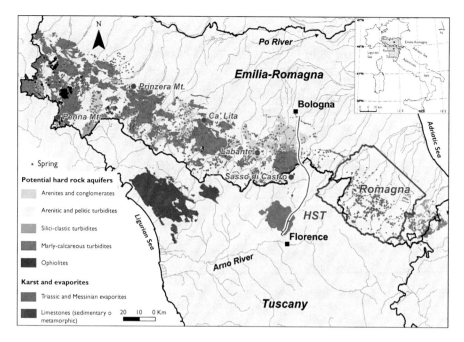

Figure 8.1 Aquifers of Northern Apennines along with location of test-sites. HST: High Speed Train tunnels path. (*See colour plate section, Plate 24*).

The dominant lithology of the aquifers in the area is turbidites and arenites, either siliciclastic or marly calcareous. Scattered outcrops of ophiolites occur, both effusive basalts or intrusive peridotites. These geological units constitute the most valuable regional groundwater reservoir in contrast with the overexploited and locally contaminated and salinized alluvial and coastal aquifers of nearby Po plain.

The term hard rock aquifer (or discontinuous aquifer), originally related to igneous and metamorphic rocks (Neuman, 2005; Dewandel *et al.*, 2006; Shakeel *et al.*, 2008; Dewandel *et al.*, 2011; Lachassagne *et al.*, 2011), is well suited also for sedimentary formations in the Northern Apennines for the following reasons (Gargini *et al.*, 2006, 2008; Piccinini *et al.*, 2012):

1   primary porosity is often fully obliterated by precipitation of calcite or, if present, has practically no relevance for groundwater flow;
2   alteration of arenitic and pelitic layers, along with the complex structural setting (locally post-orogenic extensional tectonics has induced the formation of regional fault-zone aquifers with very high horizontal hydraulic transmissivity), makes the groundwater flow systems compartmentalized and heterogeneous in terms of permeability distribution;
3   the tectonic history of the chain, with the erosion of upper nappes originally overthrusted above turbiditic units, has determined the formation of an upper stress-release induced detensioned mantle (deep down to 150 m below ground surface) where fractures opening and connectivity is higher.

The interest toward a better knowledge of these aquifers arose at the end of the 1990s due to the hydrogeological effects, in terms of groundwater drainage and impacts against Groundwater Dependent Ecosystems (GDE), of the boring of a sequence of tunnels for the high speed railway connection between Bologna and Florence across the northern Apenninic chain (Gargini *et al.*, 2008; Canuti *et al.*, 2009; Vincenzi *et al.*, 2013).

8.2 THE GEOLOGY OF AQUIFERS OF THE NORTHERN APENNINES

The Northern Apennines are a fold-and-thrust chain within the Alpine System, formed largely by the deformation of the Adriatic plate (Adria micro-plate) as a result of the collision between the African and Eurasian plates (Marroni & Treves, 1998; Castellarin, 2001). As a result, the structure of the Apennines presents a series of tectonic units thrusted over each other; the outcropping units are formed by sedimentary rocks, with very few exceptions.

The sketch map of potential hard rock aquifers in the Northern Apennines (Figure 8.1) summarizes the geological data produced by the Geological Surveys of Emilia-Romagna and Tuscany Regions and shows the location of the test sites. Spring data derive mainly from information available on tapped springs; further information was derived from a specific dataset related to springs documented in historical topographic maps. The map, relative to the Emilia-Romagna region, was drawn up as part of measures for implementation of European Union (EU) regulations (Water Framework Directive 2000/60/EC); in this sense aquifers identified in Figure 8.1 constitute groundwater bodies, mapped inside the Geological Survey of Emilia-Romagna region according the definition given by the Directive. Potential hard rock aquifers include:

1   Siliciclastic turbidites, marly calcareous turbidites and arenites (Late Cretaceous, Eocene and Miocene ages). These units generally overlay argillites which behave like aquitards, forming mesa-like reliefs (plates) with spring discharges at the base (*Labante* test site in Figure 8.1);
2   Pelitic turbidites (Oligo-Miocene and Miocene), e.g. Marnoso-Arenacea Formation (Miocene; Figure 8.2). Springs are concentrated within facies with a high Arenite/Pelite (A/P) layers ratio and/or in areas with strong tectonic deformations (particularly along post-orogenic extensional faults), as in *Romagna* and *High Speed Train (HST)* test sites in Figure 8.1;
3   Ophiolites, the unique case of non-sedimentary rocks, already identified as main groundwater reservoirs (Dewandel *et al.*, 2003, 2004). In Northern Apennines Jurassic peridotites or basalts often form isolated blocks at the top of mountains, surrounded by shaley aquitards giving rise to permeability thresholds (Figure 8.3; *Penna*, *Prinzera* and *Sasso di Castro* test sites in Figure 8.1). In such a setting an important favourable condition for groundwater flow is gravitational slope deformation, inducing detensioning of the rock-mass (Figure 8.4).

Figure 8.2 Outcrop of turbiditic hard rock aquifer: Premilcuore member of Marnoso-Arenacea Formation (Upper Bidente valley, Romagna). (*See colour plate section, Plate 25*).

Figure 8.3 Ophiolitic hard rock aquifer: peridotite block of Ragola Mountain (near Prinzera Mountain in Figure 8.1). (*See colour plate section, Plate 26*).

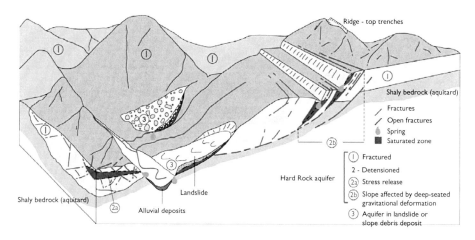

Figure 8.4 Hydrogeological sketch map of groundwater flow systems (saturated zone) in an ophiolitic block involved by fracturing and gravitational processes. (*See colour plate section, Plate 27*).

8.3 MAIN HYDROGEOLOGICAL INVESTIGATION TOOLS IN NORTHERN APENNINES

Monitoring and quantification of stream base flow during low flow conditions (from June to the end of October, on the average) is a main prerequisite to evaluate groundwater resources occurrence (Welch & Allen, in press). The base flow of a stream, during summer, represents the least influenced portion of the hydrograph where the noise of surface flow or interflow is fully removed (Dewandel *et al.*, 2003; Tetzlaff & Soulsby, 2008). In such conditions the stream flow produced by a mountainous watershed is actually the integral contribution of all up-gradient discharge points (linear, point-type or diffuse; Appleby, 1970; Dewandel *et al.*, 2004) and is an indirect expression of the groundwater recharge at the watershed scale assuming that hydrographic and hydrogeological boundaries overlap, as they often do in hard rock aquifer (Korkmaz, 1990; Foster, 1998; Scanlon *et al.*, 2002). Stream and spring mean summer discharge and the recession coefficient according Maillet (1905) represent the key factors to classify groundwater flow systems in turbiditic aquifers (Gargini *et al.*, 2008), whereas this is not common in dual or multi slope recession curve karst systems (Tallaksen, 1995; Amit *et al.*, 2002).

The rugged stream bed morphology of creeks, together with the very low flow values to be detected (in the order of 1–10 L/s on average), need precise and accurate monitoring devices, such as frictionless flow meters, electromagnetic induction flow sensors (Hofmann, 2003), the acoustic doppler velocimeter (Rehmel, 2007) or the salt dilution method (Käss, 1998). Multiple flow measurements, along a transect longitudinal to the channel route provide evidence of groundwater-surface water relationships or the effects of water abstractions for agricultural uses.

When the goal of the investigation is the definition of the zone of contribution of a spring (Paradis *et al.*, 2007) or of the spring-head protection areas, in terms of travel time (EPA, 1994; Paradis *et al.*, 2007), numerical modelling can provide valuable support by the equivalent porous medium approach (Long *et al.*, 1982; Rayne *et al.*, 2001; Cook, 2003; Kanit *et al.*, 2003; Scanlon *et al.*, 2003; Mun & Uchrin, 2004) developed, in the presented case-studies, by finite difference numerical modelling. The reliability of the approach is enhanced by detailed geological mapping and by the availability of continuous spring discharge record dataset, in different stages of the hydrograph (replenishment and recession), for model calibration purposes (boreholes and hydraulic head record dataset are often not available in mountainous settings); tracer test can provide, in specific settings direct data about groundwater flowpaths and hydrodynamic parameters.

8.4 ANALYSIS OF BASE FLOW AND INDIRECT ESTIMATION OF RECHARGE

The discharge of base flow from groundwater flow systems could be spring-focused or stream-focused. In both cases local or intermediate flow systems, according the pattern proposed by Tóth (1963), could be involved.

Exploiting the huge dataset of the hydrogeological monitoring of springs and streams for the High Speed Train (HST) tunnel borings between Florence and Bologna, through siliciclastic and marly-calcareous turbidites (Figure 8.1), Gargini *et al.* (2008) conducted a detailed analysis on the base flow regime defining either a ranking method of Apenninic springs or indirect tools to estimate direct recharge.

In the ranking method, *S* (Slope) type and *T* (Trans-watershed) type springs have been differentiated according to the following parameters: differential elevation of the spring above the local base level, recession coefficient α and average base flow discharge (average summer discharge in the Mediterranean climate).

Scattered springs are connected to shallow circuits developed along slopes inside the regolith, the detensioned upper mantle of the rock mass or inside landslides and debris unconsolidated deposits (*S*-type springs, Slope Springs). Mean spring flow rate for *S*-type springs is low (VII-VI class of Meinzer, 1923), consequently their actual or potential use is local (domestic, livestock watering). The final receptor of these circuits could be the stream flowing along the valley downslope, through the interflow or stream-bed intersecting fractures. Typical examples are shown in Figure 8.4.

Fewer major springs represent the discharge of intermediate flow systems (*T*-type springs, with *T* standing for Trans-watershed) related to the main extensional post-orogenic fault zones, cutting a series of watersheds, in some instances occurring in plateaus made by a permeable lithology completely surrounded and underlain by clayey and shaley low permeability complexes. In proper conditions, *T*-type springs maintain discharges of 10–20 L/s during low flow period (dry summer, typical of subtropical Mediterranean climate) and, in such a case, they sustain perennial headwater streams along with the riparian and fresh-water habitats (Bertrand *et al.*, 2012), constituting the typical GDE of Northern Apennines. Some of them sustain water supplies for local communities. If the stream-bed crosses a fault zone the surface waters are the final receptor of such a system.

At the watershed scale, the base flow index or base flow yield, equal to the ratio between average base flow discharge, during the recession period, and the surface area of the reach upstream to the measuring section, puts in evidence those sub-watersheds where localised groundwater discharge or sink sectors toward other watersheds occur. A base flow occurrence map is shown in Figure 8.5, in relationship to the Upper Montone river watershed in Romagna Apennines (pelitic-arenaceous turbidites). Dark-grey sub-watersheds display a gaining behaviour (as shown by the arrows), constituting the final receptors of intermediate flow systems.

The recharge estimation method is based upon an empirical relationship between Q_a (average annual discharge) and Q_s (average summer discharge) of the final receptor (spring or stream section). The dataset includes 82 springs and 11 stream sections, monitored on a monthly basis for 11 hydrological years (with stream discharge measurements taken always at least 5 days after rainfall events). The relationship is linear, on the log scale (see the graph in Gargini *et al.*, 2008), and varies according to the value of the recession coefficient (α Maillet's coefficient). Starting from few discharge measurements during low flow season, the value of α and Q_s are determined and, applying the relationship, the Q_a value is estimated as a proxy of the effective infiltration (direct recharge) inside the area of contribution (in the case of a spring) or the whole up-gradient watershed (in the case of a stream). The expected accuracy is in the range ±15% respect to the actual value.

As an example of the application of the method (see Figure 8.6) a discontinuous monitoring hydrograph for a spring discharging from a pelitic turbidite is shown

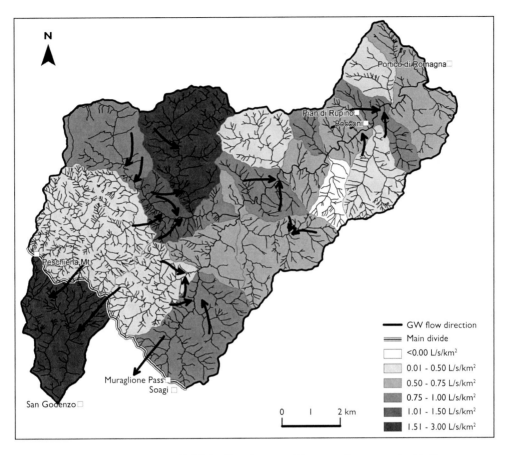

Figure 8.5 Analysis of base flow yield (L/s/km²) in the upper Montone river watershed in *Romagna* area test-site. The main divide is between the Adriatic Sea and the Ligurian Sea.

along with the analysis of the recession: α value is 5.4×10^{-3} d^{-1} and this produces, starting from a Q_s value of 50.4 L/s, an indirect estimation of Q_a equal to 66.3 L/s by the application of the proper exponential relationship ($Q_a = 1.3 * Q_s^{0.93}$, for the α range between 3.0×10^{-3} and 6.0×10^{-3} d^{-1}). This value has an error of -14% with respect to the actual average experimental value of Q_a as determined according the monitored data (Figure 8.6).

The method has been applied not only to other analogous hydrogeological settings (not involved in tunnel drainage), characterized by outcropping pelitic turbidites and arenites (*Romagna* test sites in Figure 8.1), but also to ophiolitic aquifers as Penna Mountain, Sasso di Castro and Prinzera Mountain (Figure 8.1), three blocks formed, respectively, by basalts (the first two) and peridotites (Figure 8.4). It was possible to estimate, for different lithologies, the infiltration coefficient C, intended as the ratio between direct recharge R (expressed by Q_a) and total rainfall P over the contribution area (Table 8.1). Similar C values were obtained, for Marnoso-Arenacea Formation pelitic turbidites (*Romagna* test site) and Pantano Formation arenites (*Labante*

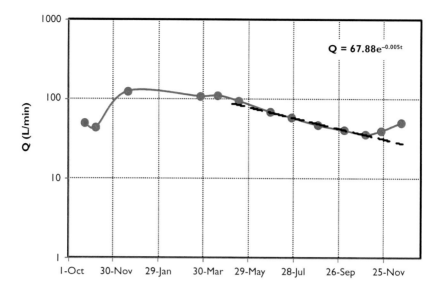

$$Q = 67.88e^{-0.005t}$$

Figure 8.6 Hydrograph of *Gorgogliosa* spring (Romagna area) along the hydrological year 2002–2003 with the analysis of recession according to Maillet's equation.

Table 8.1 Infiltration coefficient values (*C*) for different lithologies in Northern Apennines. *P* = mean annual precipitation over the area of contribution (values relative to meteo stations most representative of the area and to the hydrological year of the analysis); *R* = estimation of direct recharge (mm). The values indicated by * was derived by numerical modelling.

Test-site (Fig. 8.1)	Aquifer	P (mm)	C = R/P
HST; Romagna	Pelitic turbidites, Marnoso Arenacea Formation	992–1387	0.13–0.15; 0.16*
Labante	Arenites, Pantano Formation	1216	0.17*
Prinzera Mt.	Peridotites	862	0.36
Penna Mt.	Basalts	2378	0.38
Sasso di Castro Mt.	Basalts	1519	0.34

test site), by numerical modelling calibrated versus continuous monitoring of spring discharge. This similarity confirms the validity of the approach of recharge indirect estimation.

Streams are the final receptor of the major proportion of base flow discharge, the effects of tunnel drainage or of intensive pumping (man induced environmental impacts) are addressed particularly against surface waters, as demonstrated by Vincenzi *et al.* (2009, 2013). Stream-tunnel tracer tests, performed inside a turbiditic aquifer, proved effective either to demonstrate the severity and extent of the impacts or to define parameters useful for the numerical modelling application to analogous settings.

8.5 TRACER TESTING IN TURBIDITIC AQUIFERS

Groundwater tracing with fluorescent dyes is a methodology commonly applied to karst systems, but rarely to fractured aquifers (Käss 1998). In the last ten years it has been tested on different sites representative of hard rock groundwater flow systems in the Northern Apennines.

During the years 2001–2007 multi-tracer tests with uranine, sulforhodamine-G and tinopal have been tested on four different test sites (coded as *S/T* in Table 8.2; watersheds located nearby *HST* trace in Figure 8.1) in order to evaluate the connection between mountain streams and draining tunnels on the High Speed Train Line between Bologna and Florence. The injections in streams required high quantities of tracers, because only a small fraction infiltrates into the aquifer, while most was lost in streams waters and by dilution-dispersion inside the aquifer; this implied low recovery rates at the sampling points and required precise analytical work with the scanning technique of spectrofluorimetry. The methodology and the results at two of the test sites are detailed in Vincenzi *et al.* (2009). The results of all the test sites represent unprecedented data on groundwater flow and transport in turbiditic aquifers and made it possible to better characterise the differential impacts of drainage along a tunnel.

The impact radius, assessed through hydraulic data and hydrogeological monitoring of streams and springs, increases with the rock mass permeability and is directly related to the linear flow velocities assessed by means of tracer tests: they range from

Table 8.2 Summary of main data related to tracer tests. *S/T* = Stream/Tunnel tracer test (*HST* test-site); * length of monitored tunnel sector, considering the tunnel perpendicular to the main flow direction; ° width and °° length of the landslide involved by tracer test. Explanation of symbols: Thickness of arenitic layers (VH: very high; H: high; M: medium); Tracers (U: Uranine; S: Sulfo-G; T: Tinopal; E: Eosine; A: Amino G_Acid); L: size range of traced pathways (m); t_i: time of first detection (days after injection); v_{max}: maximum groundwater velocity (first detection); v_p: modal groundwater velocity (peak velocity); n.d.: not determined.

	S/T 1	S/T 2	S/T 3	S/T 4	Ca' Lita	Corniolo
Turbidite type	Pelitic M	Pelitic H-M	Pelitic VH-H	Marly-calcareous	Pelitic and calcareous	Pelitic M
Site size parallel to flow (m)	900	1,000	1,500	900	600°°	800°°
Site size perpendicular to flow (m)	3,400*	7,000*	2,900*	4,000*	400°	700°
Tracers	5 kg U 2 kg S 8 kg T	10 kg U 8 kg S	5 kg U 1 kg S	5 kg U 25 kg LiCl	3 kg U	2 kg U 2 kg E 2 kg A
# sampling points	34	95	52	22	13	25
# sampling surveys	22	24	32	29	31	18
months of monitoring	10	4	10	7	8	8
L (m)	127–307	91–468	1,107–1,390	469–615	No results	244–570
t_i (d)	14–25	4–121	18–74	5–107	No results	36–225
v_{max} (m/d)	9.1–12.3	2.1–32.4	18.8–63.3	5.5–93.8	No results	n.d.
v_p (m/d)	2.4–5.8	3.8–14.7	16.1–20.9	39.1	No results	1.46–6.78

3.6 m/d in the thin-bedded pelitic turbidites (where the impact radius is 200 m) to 39 m/d in the calcareous turbidites and thick-bedded arenitic turbidites (where it reaches respectively 2.3 and 4.0 km).

At several places, tracer tests allowed discrete fault zones as main hydraulic pathways between impacted streams and draining tunnels to be identified.

Recently, the methodology has been tested on two landslides affecting slopes inside turbiditic outcroppings ("S-type" GFS according Gargini *et al.*, 2008). At Ca' Lita landslide, a roto-translational rock slide in pelitic and calcareous turbidites of Paleocenic Monghidoro Formation (Ca'Lita test-site; Figure 8.1), a tracer test was performed in order to investigate the groundwater exchange between the bedrock and the landslide body. 3 kg of uranine were injected inside a piezometer located up-slope of the crown zone; sampling was carried out for one year, daily in the first period and weekly afterwards, in 12 control points located in the landslide body, represented by piezometers, drainage wells and sub-horizontal drains. The monitoring procedure required both the collection of water samples and a cumulative sampling with charcoal bags (passive samplers for adsorbing fluorescent dyes). Samples were analysed by scanning spectrofluorimetry. However, one year after the injection no evidence of the tracer was found in the landslide. Different interpretations are possible (no connection, high dispersion and/or very low travel time due to low permeability) and are discussed in Ronchetti *et al.* (2009).

At Corniolo landslide, a translational rock slide in pelitic turbidites of Marnoso-Arenacea Formation (located inside the Romagna test-site area; Figure 8.1), three fluorescent dyes (uranine, eosine and amino-G-acid) have been injected in the upper-part of the landslide body in two piezometers, bored inside the fractured rock mass, or an underground tank located inside the shallow debris above the rock mass (Vincenzi *et al.*, 2012). The monitoring and analysing procedure was the same as for the Ca' Lita landslide; monitoring points were represented by scattered springs and seepages down the slope. Two groundwater flowpaths have been characterized: one inside the shallow debris, with a velocity of about 50 m/d; the other one inside the turbiditic rock mass with lower velocities (1.46–6.78 m/d) and very low recovery rates due to the high dispersivity.

Some general conclusions can be inferred (Table 8.2). The medium-low permeability of turbidites requires high injection quantities of tracers and cannot assure positive results for the tests; at these sites salt tracers have also been tested, but are not applicable due to the higher detection limits; uranine always showed the most conservative behaviour; the behaviour of isotopic tracers (Deuterium, Tritium) should be tested, but they are more expensive. Groundwater tracing under natural gradient conditions has a lower probability of success, while artificial gradient, generated by groundwater drainage or pumping, can represent a good solution, by increasing flow velocities and so lowering recovery times and tracer dispersion.

The tracer tests proved effective, the results helped the verification of groundwater flow and transport numerical modelling to finalize the definition of spring-head protection areas.

8.6 NUMERICAL MODELLING OF SPRING FLOW

Numerical modelling was applied in two case studies in order to locate the contribution areas and springhead protection areas for springs discharging from sedimentary hard rock aquifers. Fully 3D numerical models were constructed to simulate continuous discharge measurements for the Brenziga spring (2001–2005 years), discharging out at 650 m a.s.l. from the Premilcuore member of Marnoso-Arenacea Formation (Miocenic pelitic turbidite; Figure 8.2), in the Romagna test-site, and the Labante spring (2001–2011 years), discharging out at 612 m a.s.l. from the Pantano formation (Oligocenic arenite; Labante test-site in Figure 8.1). The equivalent porous media approach was applied, respectively, by the finite difference codes MODFLOW-2000 (Harbaugh *et al.*, 2000) for Brenziga spring (Piccinini *et al.*, 2012) and MODFLOW-2005 (Harbaugh, 2005) for Labante spring.

The domain of the Brenziga model is a rectangle of 3500 × 5000 m, subdivided into cells of 25 × 25 m on the horizontal plane. The depth, from a plane at 300 m a.s.l. up to the surface topography, has been divided into five layers of variable thickness. The main axis of the domain is oriented north-south, according to the presumed regional groundwater flow direction, and is delimited by the main Appenninic divide to the north, by the outcropping of aquitard members of the Marnoso Arenacea Formation (pelitic turbidites with very low A/P ratio value) to the east and west and by a main stream to the south (Figure 8.7). The domain of the Labante model is a rectangle of 4350 × 3250 m, subdivided into cells of 10 × 10 m on the horizontal plane. The depth, from a plane at 250 m a.s.l. to the surface topography, has been divided into five layers. The domain is delimited by the outcropping of aquitard formations (argillites) to the north, east and south and by hydrographic network to the west (Figure 8.8). Both model domains were defined on geological basis, in relationship to distribution of lithologies and thicknesses of different units, and according to the conceptual hydrogeological model of the groundwater flow system whereas the assignment of minimum cell size has been based upon the Representative Elementary Volume (Long *et al.*, 1982; Kanit *et al.*, 2003) according the fracture density obtained from geo-mechanical field surveys.

The hydraulic conductivity values (K) were always assigned isotropically and were differentiated according to lithology, fracture density and fault occurrences, as located by geological survey (Table 8.2), while storage parameters (Specific storage and specific yield; Table 8.3) were derived from the calibration process, taking also into account the results of tracer test performed in analogous lithology.

The surface water-groundwater interactions were always simulated by means of a third type boundary condition, i.e. the Streamflow-Routing Package of MODFLOW (SFR1, Prudic *et al.*, 2004). SFR1 simulates the surface water flow inside streams through the propagation of flow rates from cell to cell, contemporarily to their interaction with groundwater (controlled by the head difference between the stream and the aquifer and by the hydraulic conductance of stream-bed medium).

Both springs were simulated by a third type boundary condition, i.e. the Drain Package of MODFLOW (DRN; Harbaugh *et al.*, 2000). DRN removes groundwater from the cells in which it is applied as a function of head difference between the aquifer and the drain elevation. Drain elevation values were fixed equal to spring elevation while a high drain conductance was used (100 000 m^2/d) to allow unrestricted discharge of spring water (Scanlon *et al.*, 2003). A DRN with a high conductance value corresponds to a seepage face condition, where a hydraulic head equal to the nodal elevation (first type boundary) is combined with a constraint condition allowing only outflow.

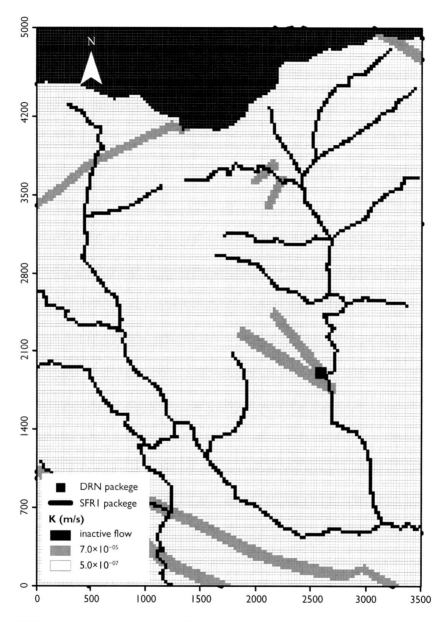

Figure 8.7 Hydraulic conductivity values (K) and boundary conditions assigned to the layer 1 in the *Brenziga* model.

Recharge to aquifer has been simulated as a second type boundary condition (Recharge in MODFLOW) and applied to all the cells of the first layer, distinguishing between high and low A/P ratio turbidites. Recharge values were optimised during the calibration process according to rainfall and spring discharge. The assignment of such a boundary condition follows the conceptual model of the groundwater flow system

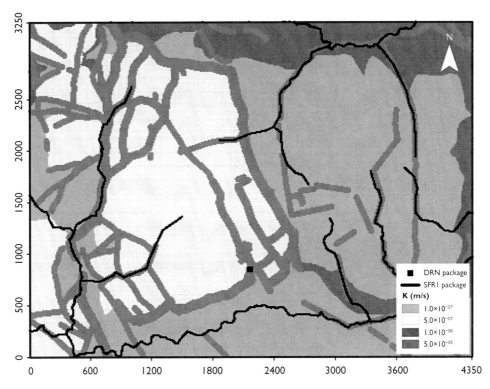

Figure 8.8 Hydraulic conductivity values (K) and boundary conditions assigned to the layer 1 in the *Labante* model.

Table 8.3 Simulated hydrodynamic parameters, after calibration, for *Brenziga* and *Labante* spring models, relative to units located inside the model grid. K, hydraulic conductivity; S_s, specific storage; S_y, specific yield; n_e, effective porosity.

Hydrogeological Unit	Brenziga			Labante		
	K (m/s)	S_s (1/m)	S_y/n_e ( )	K (m/s)	S_s (1/m)	S_y/n_e ( )
Normally fractured aquifer	5.0×10^{-7}	1.0×10^{-7}	0.01	5.0×10^{-7}	1.0×10^{-5}	0.0025
Main faults and fractures	7.0×10^{-5}	1.0×10^{-5}	0.02	5.0×10^{-5}	1.0×10^{-5}	0.005
Aquitard (marls)	–	–	–	1.0×10^{-7} 1.0×10^{-8}	1.0×10^{-4}	0.001

in sedimentary hard rock aquifers where groundwater rainfed recharge is the primary input and groundwater discharge is focused on springs or streams.

The infiltration coefficient (C) has been derived from a three stage calibration process. First, the recession hydrographs were calibrated, at transient state, in order to optimise the hydrodynamic properties (Figures 8.9A and 8.9B). Second, the high flows, expressed by average daily discharge, were calibrated, at transient state, in

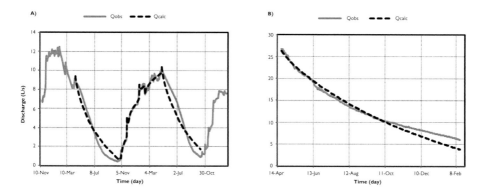

Figure 8.9 Calibration of the models: A) observed versus simulated daily flow rate for *Brenziga* spring during 2003–2004; B) observed vs. simulated average daily flow rate for *Labante* spring in the recession 2001–2002.

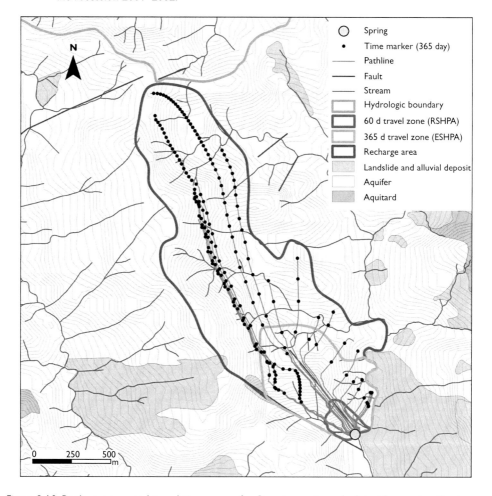

Figure 8.10 Recharge area and travel time zones for *Brenziga* spring calculated from advective transport simulation. (See *colour plate section, Plate 28*).

order to estimate recharge values (Figure 8.9A). Third, the recharge values were compared with rainfall data on a yearly basis. Finally, the main flow paths and travel times were calculated by steady state advective groundwater simulations with MOD-PATH (Pollock, 1994) in order to define the contribution area. The effective porosity (n_e) used in transport simulation has been fixed equal to S_y (Table 8.3).

The contribution (or recharge) areas obtained from both simulations are the result of derived C values compared to spring discharge and take also into account morphological and geological constraints. The Brenziga or Labante spring are T-type, according the ranking methodology, and the model confirms that the springs are receiving water that was recharged either inside or outside of their up-gradient surface watershed. In the Brenziga model, a recharge area of 1.51 km^2 produces a spring mean annual flow rate of 6.05 L/s (Figure 8.10) whereas in the Labante model an area of 2.61 km^2 produces 12.65 L/s. Springhead protection areas were defined by the envelop of the 60 day and 365 day travel times. Their extensions are, respectively, 2.6% and 23.9% of the contribution area in the *Brenziga* model (Figure 8.10) and 1.6% and 11.4% in the Labante model. Finally, C above the contribution area resulted equal to 0.16 for the Brenziga spring and to 0.17 for the Labante spring (Table 8.1). The infiltration coefficient for Brenziga spring resulted very close to the values calculated, by the indirect estimation recharge method, for the same turbiditic lithology of the aquifer.

8.7 CONCLUSIONS

Groundwater resources in the Northern Apennines are located in hard rock aquifers although their genesis is mostly sedimentary. The structural complexity of the geological setting, together with the high heterogeneity of the units in terms of hydraulic conductivity distribution and of the specific climatic conditions of the area, emphasises the role of base flow analysis to the study of the groundwater flow system. The major proportion of the base flow discharge is stream-focused, so monitoring of stream base flow, with appropriate and accurate tools, becomes the starting point of any hydrogeological quantitative research in the area.

Due to the huge dataset of hydrogeological monitoring data from the boring of the high speed railway tunnels between Bologna and Florence, a methodology, based upon springs and stream base-flow during recession season (summer), has been defined in order either to differentiate local and intermediate (trans-watershed) systems or to estimate direct recharge. The stream-tunnel tracer tests, performed in different settings, revealed the impact on groundwater dependent ecosystems (perennial headwater streams) and allowed a parameterisation of flow and transport inside the turbiditic aquifers.

Similar investigations have been tested at other test-sites, in analogous hydrogeological settings, with the aim of estimating recharge, using tracer tests in proper hydrogeological conditions and applying numerical groundwater flow modelling for the definition of spring contribution and protection areas. The developed recharge estimation method was also applied to ophiolitic aquifers. The main general purpose is to define a methodology of hydrogeological mapping for Northern Apennines hard rock aquifers, potentially valid also for similar geological and climatic settings in other countries.

Among the main results are:

1 the coefficient of infiltration, intended as the ratio between direct recharge and total rainfall, and derived either by base flow hydrological analysis or numerical modelling, varies between 0.13–0.17 for turbidites and arenites and 0.36–0.38 for igneous ophiolites;

2 the best tracer to be used in turbiditic aquifers is uranine, requiring: a geological setting with the occurrence of fracture enlarged by detensioning of the rockmass; high injection quantities; an enhancing, if possible, of hydraulic gradient by pumping or tunnel draining in order to increase flow velocities and so lower recovery times and tracer dispersion;

3 groundwater flow numerical modelling by the equivalent porous medium approach is a useful tool to delineate realistic contribution and spring-head protection areas through a calibration based on continuous spring discharge monitoring and with constraints, to the model grid implementation, relying upon the parallel estimation of coefficient of infiltration and the detailed geological mapping of spring upstream area.

ACKNOWLEDGEMENTS

We acknowledge Daniele Bonaposta and Annalisa Parisi for the contribution to the geological setting description. We acknowledge two anonymous reviewers for the precious and detailed recommendations and corrections.

REFERENCES

Amit H., Lyakhovsky V., Katz A., Starinsky A., Burg A. (2002) Interpretation of spring recession curves. *Ground Water* 40, 543–551.

Appleby V. (1970) Recession flow and the baseflow problem. *Water Resources Research* 6, 1398–1403.

Bertrand G., Goldscheider N., Gobat J.M., Hunkeler D. (2012) Review: From multi-scale conceptualization to a classification system for inland groundwater-dependent ecosystems. *Hydrogeology Journal* 20, 5–25.

Canuti P., Ermini L., Gargini A., Martelli L., Piccinini L., Vincenzi V. (2009) Le gallerie TAV attraverso l'Appennino Toscano: impatto idrogeologico ed opere di mitigazione (In Italian; transl.: High Speed Train tunnels across Tuscan Apennine: hydrogeological impact and mitigation measures). Ermini L. Ed., EDIFIR edizioni, Florence (Italy). ISBN 978-88-7970-411-3.

Castellarin A. (2001) Alps-Apennines and Po Plain-frontal Apennines relations. In: Anatomy of an orogen, the Apennines and adjacent Mediterranean basin. Kluwer, Dordrecht, The Netherlands, pp 177–195.

Cook P.G. (2003) A guide to regional groundwater flow in fractured rock aquifers. CSIRO, Land and Water, Glen Osmond, SA, Australia, pp 115.

Dewandel B., Lachassagne P., Bakalowicz M., Weng P., Al-Malki A. (2003) Evaluation of aquifer thickness by analysing recession hydrographs. Application to the Oman ophiolite hardrock aquifer. *Journal of Hydrology* 274, 248–269.

Dewandel B., Lachassagne P., Qatan A. (2004) Spatial measurements of stream baseflow a relevant methodology for aquifer characterization and permeability evaluation. Application to a hard-rock aquifer, the Oman ophiolite. *Hydrological Processes* 18, 3391–3400.

Dewandel B., Lachassagne P., Wyns R., Maréchal J.C., Krishnamurthy R.S. (2006) A generalized 3-D geological and hydrogeological conceptual model of granite aquifers controlled by single or multiphase weathering. *Journal of Hydrology* 330(1), 260–284.

Dewandel B., Lachassagne P., Zaidi F.K., Chandra S. (2011) A conceptual hydrodynamic model of a geological discontinuity in hard rock aquifers: example of a quartz reef in granitic terrain in South India. *Journal of Hydrology* 405, 474–487.

EPA (1994) Ground water and wellhead protection handbook. EPA-625/R-94-001. EPA (US-Environmental Protection Agency), Office of Water, Washington DC (USA).

Foster S.S.D. (1998) Groundwater recharge and pollution vulnerability of British aquifers: a critical overview. In: Robins N.S. Editor, Groundwater pollution, aquifer recharge and vulnerability, Geological Society, London, Special Publication 130, 7–22.

Gargini A., Piccinini L., Martelli L., Rosselli S., Bencini A., Messina A., Canuti P. (2006) Idrogeologia delle unità torbiditiche: un modello concettuale derivato dal rilevamento geologico dell'Appennino Tosco-Emiliano e dal monitoraggio ambientale per il tunnel alta velocità ferroviaria Firenze-Bologna (In Italian; transl: Hydrogeology of turbidites: a conceptual model derived by the geological survey of Tuscan-Emilian Apennines and the environmental monitoring for the high speed railway tunnel connection between Florence and Bologna). *Bollettino della Società Geologica Italiana* 125, 293–327.

Gargini A., Vincenzi V., Piccinini L., Zuppi G.M., Canuti P. (2008) Groundwater flow systems in turbidites of the Northern Apennines (Italy): natural discharge and high speed railway tunnel drainage. *Hydrogeology Journal* 16(8), 1577–1599.

Harbaugh A.W. (2005) MODFLOW-2005, the U.S. Geological Survey modular ground-water model – the Ground-Water Flow Process. U.S. Geological Survey Techniques and Methods 6-A16.

Harbaugh A.W., Banta E.R., Hill M.C., Mcdonald G. (2000) MODFLOW-2000, The U.S. Geological Survey modular ground-water model – User Guide to modularization concepts and the ground-water flow process. U.S. Geological Survey, Open-File Report 00-92.

Hofmann F. (2003) *Fundamental principles of Electromagnetic Flow Measurement*. KROHNE Messtechnik GmbH & Co. KG, Duisburg.

Kanit T., Forest S., Galliet I., Mounoury V., Jeulin D. (2003) Determination of the size of the representative volume element for random composites: statistical and numerical approach. *International Journal of Solids and Structures* 40, 3647–3679.

Käss W. (1998) *Tracing Technique in Geohydrology*. Balkema, Rotterdam.

Korkmaz N. (1990) The estimation of groundwater recharge from spring hydrographs. *Hydrological Sciences Journal* 37, 247–261.

Lachassagne P., Wyns R., Dewandel B. (2011) The fracture permeability of hard rock aquifers is due neither to tectonics, nor to unloading, but to weathering processes. *Terra Nova* 23(3), 145–161.

Long J.C.S., Remer J.S., Wilson C.R., Witherspoon P.A. (1982) Porous Media Equivalent for Networks of Discontinuous Fractures. *Water Resources Research* 18(3), 645–658.

Maillet E. (1905) Essai d'hydraulique souterraine et fluviale. Herman, Paris.

Marroni M., Treves B. (1998) Hidden terranes in the Northern Apennines, Italy: a record of late Cretaceous-Oligocene transgressional tectonics. *Journal of Geology* 106, 149–162.

Meinzer O.F. (1923) The occurrence of groundwater in the United States. Government Printing Office, USGS, Water Supply Paper 489, Washington D.C. (USA).

Mun Y., Uchrin C.G. (2004) Development and Application of a MODFLOW Preprocessor Using Percolation Theory for Fractured Media. *Journal of the American Water Resources Association* 40(1), 229–239.

Neuman S.P. (2005) Trends, prospects and challenges in quantifying flow and transport through fractured rocks. *Hydrogeology Journal* 13(1), 124–147.

Paradis D., Martel R., Karanta G., Lefebvre R., Michaud Y. (2007) Comparative study of methods for WHPA delineation. *Ground Water* 45(2), 158–167.

Piccinini L., Gargini A., Martelli L., Vincenzi V., De Nardo M.T. (2012) Upper catchment of Montone river. Hydrogeological map of Northern Apennines. Emilia-Romagna Region and University of Bologna, Infocartografica S.n.c., Piacenza (Italy), 119 pp (in italian).

Pollock D.W. (1994) User's Guide for MODPATH/MODPATH-PLOT version 3: A particle tracking post-processing package for MODFLOW, the USGS finite-difference groundwater flow model. U.S. Geological Survey Open-File Report 94-464.

Prudic D.E., Konikow L.F., Banta E.R. (2004) A new streamflow-routing (SFR1) package to simulate streamaquifer interaction with MODFLOW-2000. U.S. Geological Survey Open-File Report 2004-1042, 95 pp.

Rayne T.W., Bradbury K.R., Muldoon M.A. (2001) Delineation of Capture Zones for Municipal Wells in Fractured Dolomite, Sturgen Bay, Wisconsin, USA. *Hydrogeology Journal* 9, 432–450.

Rehmel M. (2007) Application of Acoustic Doppler Velocimeters for Streamflow Measurements. *Journal of Hydraulic Engineering ASCE*, 133(12), 1433–1438.

Ronchetti F., Borgatti L., Cervi F., Gorgoni C., Piccinini L., Vincenzi V., Corsini A. (2009) Groundwater processes in a complex landslide, Northern Apennines, Italy. *Natural Hazard Earth Systems Science* 9, 895–904.

Scanlon B.R., Healy R.W., Cook P.G. (2002) Choosing appropriate techniques for quantifying groundwater recharge. *Hydrogeology Journal* 10, 18–39.

Scanlon B.R., Mace R.E., Barrett M.E., Smith B. (2003) Can we simulate regional groundwater flow in a karst system using equivalent porous media models? Case study, Barton Springs Edwards aquifer, USA. *Journal of Hydrology* 276, 137–158.

Shakeel A., Ramaswamy J., Abdin S. (2008) *Groundwater dynamics in hard rock aquifers.* Springer (New York).

Tallaksen L.M. (1995) A review of baseflow recession analysis. *Journal of Hydrology* 165, 349–370.

Tetzlaff D., Soulsby C. (2008) Sources of baseflow in larger catchments-Using tracers to develop a holistic understanding of runoff generation. *Journal of Hydrology* 359, 287–302.

Tóth J. (1963) A theoretical analysis of groundwater flow in small drainage basins. *Journal of Geophysical Research* 68, 4785–4812.

Vincenzi V., Gargini A., Goldscheider N. (2009) Using tracer tests and hydrological observations to evaluate effects of tunnel drainage in the Northern Apennines (Italy). *Hydrogeology Journal* 17(1), 135–150.

Vincenzi V., Benini A., Gargini A. (2012) Hydrogeological study of Corniolo landslide (high Bidente Valley, Italy) In: 7th EUREGEO – EUropean congress on REgional GEOscientific cartography and Information systems, Bologna, Italy, 2012. Proceedings: Volume 1 + Volume 2-837 pages. Emilia-Romagna Region – Geological Seismic and soil Survey, 2012.

Vincenzi V., Gargini A., Goldscheider N., Piccinini L. (2014) Differential hydrogeological effects of draining tunnels through the Northern Apennines, Italy. *Rock Mechanics and Rock Engineering*, 47(3), 947–965.

Welch L.A., Allen D.M. (in press) Consistency of groundwater flow patterns in mountainous topography: implications for valley bottom water replenishment and for defining groundwater flow boundaries, *Water Resources Research* 48 W05526 doi:10.1029/2011WR010901.

Chapter 9

Fracture transmissivity estimation using natural gradient flow measurements in sparsely fractured rock

Andrew Frampton
Department of Physical Geography and Quaternary Geology, Stockholm University, Stockholm, Sweden

ABSTRACT

Numerical simulations are conducted to evaluate connectivity and estimate fracture transmissivity, based on field measured borehole flows in sparsely fractured crystalline rock. The data set considered consists of directional fracture-specific flows obtained without pumping, corresponding to natural flow conditions where topography is the main driving force for flow. A method for conditioning transmissivity against the flow measurements is developed and applied for a semi-generic discrete fracture network representation of the bedrock. The model is conditioned against depth-aggregated flows and is able to provide a description of the general features of the subsurface flow system. Results indicate that direct flow-conditioned fracture transmissivities for natural flow conditions can be up to an order of magnitude smaller than transmissivities obtained from flow measurements under pumped conditions and inferred using simplifying homogenisation assumptions. The flow-conditioned transmissivities are shown to be more representative for the regions of bedrock local to the borehole vicinity and consistent with transmissivities obtained from traditional transient pressure response tests. The effect of open boreholes is notable in the field data and its applicability in estimation of upper bounds of the effective vertical transmissivity of rock is significant.

9.1 INTRODUCTION AND BACKGROUND

Hydrogeological field characterisation of subsurface flow and transport plays an important role in evaluating the performance of subsurface repositories. In the Finnish and Swedish programmes for permanent storage of spent nuclear fuel, the main geological environment under consideration is crystalline rock. These environments are considered advantageous since they are sparsely fractured systems in a relatively intact and geologically stable environment. The main flow conduits are fractures, which may be only moderately connected and typically have very low transmissive properties. However, understanding and characterising flow and transport in fractured rocks is a formidable challenge (Neuman, 2005), largely due to the high degree of heterogeneity and variation in structural properties typically encountered in fractured media. Field characterisation in combination with conceptual and numerical modelling is necessary for predictive analysis. To understand transport behaviour and to make predictions on contaminant spreading in sparsely fractured rock, a robust understanding of the behaviour of subsurface flow is required (Frampton & Cvetkovic, 2007a, 2007b), since flow and transport pathways in bedrock, and in particular sparsely fractured rock, are strongly impacted by flow and structural heterogeneity (Cvetkovic *et al.*, 2004; Frampton & Cvetkovic, 2011; Selroos & Painter, 2012).

Multiple field measurement techniques are often employed to characterise hydrogeological systems. In particular, borehole core logs and hydraulic pump tests may provide information on subsurface geological and hydraulic properties, where pump tests can provide information on the hydraulic head field and effective hydraulic conductivities or transmissivities of a network of fractures. However, interpretation of hydraulic testing in fractured rock can be challenging, due to unknowns in heterogeneity, anisotropy, connectivity and flow geometry (National Research Council, 1996). A description of the hydraulic head field combined with transmissive and structural properties of the subsurface would, in principle, suffice to describe subsurface flow. However, in situ flow measurements may be used to directly characterise flow rates, flow patterns and flow behaviour in the subsurface, which may be seen as favourable (Neuman, 2005) since this is a measure more closely related to transport pathways.

Several in-situ borehole measurement techniques have been developed for determination of flow rates and flow patterns in the subsurface (Hess, 1986; Tsang *et al.*, 1990; Paillet & Pedler, 1996; Tsang & Doughty, 2003; Hatfield *et al.*, 2004; Doughty *et al.*, 2005; Klammler *et al.*, 2007). The Posiva Flow Log (PFL) is a thermal-based instrument which is particularly useful for fractured media, since it can measure a wide range of flow magnitudes with a small support scale (Öhberg & Rouhiainen, 2000; Ludvigson *et al.*, 2002). Another advantage of the PFL device is that it can determine the flow rate entering or leaving a borehole. This is in significant contrast to hydraulic head information which is a scalar field quantity. The PFL device has been extensively applied and used in the Finnish and Swedish site investigation and characterisation programmes for high-level radioactive waste (Follin *et al.*, 2006; Hartley *et al.*, 2006; Follin *et al.*, 2007).

There are, however, difficulties in interpretation of fracture-specific flow measurements which concern the validity of assumptions used for relating flow to transmissivity. For example, observations by Niemi *et al.* (2000) have indicated that continuum-based analysis of field measured conductivities from one-dimensional borehole tests may underestimate conductive characteristics of discontinuous fractured media. A stochastic modelling approach for conditioning of fracture transmissivity distributions against flow distributions obtained from PFL measurements has been developed and evaluated by Frampton & Cvetkovic (2010); it was observed that flow-conditioned transmissivity distributions could produce somewhat higher transmissivities than those obtained from traditional pump test analysis where homogenisation assumptions are required. Furthermore, there is a need to reconcile transmissivity obtained from flow meters with transmissivity obtained from classical evaluation of transient responses from hydraulic tests. Thus it is necessary to improve the understanding and ways of implementing fracture transmissivity for systems of sparsely fractured rock, where different field techniques may produce different measures of hydraulic properties.

In this chapter, in situ borehole flow measurements obtained using the Posiva Flow Log (PFL) device are analysed and applied in numerical discrete fracture network flow models, where the main aim is to obtain a general understanding of the subsurface flow behaviour of a region of sparsely fractured rock under natural (non-pumped) flow conditions. A specific objective is to develop a method in which these non-forced fracture-specific flow measurements, which include both directional information as well as flow magnitudes, can be used to characterise and better understand the subsurface flow system for sparsely fractured media. Flow measurements are thereby used to constrain and condition the hydraulic properties of a discrete fracture network model. The resulting fracture transmissivities are evaluated and consolidated against those obtained from homogenisation assumptions using steady-state flow logging and transient hydraulic response tests.

9.2 SITE DESCRIPTION AND FIELD DATA

The data originate from the Finnish site characterisation programme for final disposal of spent nuclear fuel conducted at the island of Olkiluoto in the municipality of Eurajoki, Finland. The island is situated off the west coast of Finland on the Gulf of Bothnia in the Baltic Sea at approximate location 61.24° N, 21.48° E (Figure 9.1). It is about 9 km² and has a relatively flat topography, with mean elevation of about 5 m above sea level, and has relatively low hydraulic gradients even at shallow depths. The

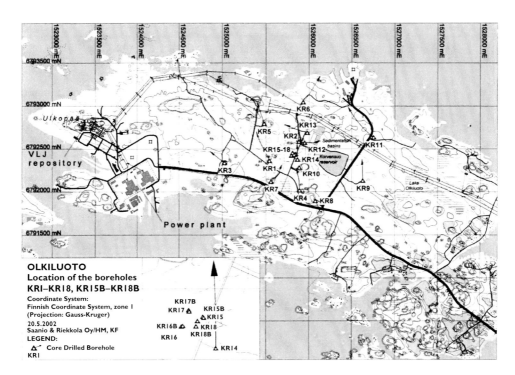

Figure 9.1 Olkiluoto Island, located approximately at 61.24° N, 21.48° E in the WGS84 coordinate system. The grid used in the figure is the Finnish coordinate system. The locations of several deep (KR15, KR16, etc.) and shallow (KR15B, KR16B, etc.) core drilled boreholes are shown. The data originate from the deep KR14 – KR18 boreholes located roughly in the centre of the island. (*Image courtesy of Posiva Oy*). (*See colour plate section, Plate 29*).

island hosts the underground rock characterisation and research facility ONKALO, currently in the final stages of construction, which is planned to be converted into a final disposal facility (Posiva Oy, 2003).

The site characterisation programme is an extensive campaign covering multiple scientific and engineering disciplines which commenced over two decades ago. The geological investigations are summarised by Aaltonen *et al.* (2010). The hydrogeological borehole investigations include hydraulic tests and geological and geophysical investigations using numerous techniques for both single-hole and multiple cross-hole examinations with associated interpretation and conceptual model design (Vaittinen *et al.*, 2009). Numerical modelling investigations and analyses are also conducted both in applying and interpreting data, and constructing suitable site-specific hydrogeological models (Hartley *et al.*, 2009) for current and future projections of water flow and solute transport (Löfman *et al.*, 2009). Boreholes are used for geological core-log analysis, in situ borehole TV imaging, electric conductivity measurements, and pressure and flow measurements under transient and steady-state pumping as well as under non-pumped conditions. Flow measurements and associated fracture transmissivities are obtained from over 40 deep (100 to 1000 m) and 16 shallow (10 to 100 m) core-drilled boreholes throughout the island, as well as additional underground tunnel boreholes associated with the ONKALO facility. The cumulative length of all boreholes currently amounts to nearly 20 km. A comprehensive overview of available hydrogeological data and analysis is presented by Tammisto *et al.* (2009).

In this study, a small selection of available field data are used, mainly related to PFL flow measurements obtained from the five deep boreholes OL-KR14 through OL-KR18, which are situated at the approximate centre of the island (Figure 9.1, inset). (Hereafter the prefix OL for boreholes is omitted.) The primary data for these boreholes is available (Pöllänen & Rouhiainen 2002a, 2002b), which were obtained prior to excavation of the ONKALO facility. The PFL data undergo a quality-assurance procedure and is partially processed, mainly by relating the location of PFL logged flow measurements with observed fracture locations obtained from geological core logs and borehole imaging. Flow values are typically associated with individually observed fracture features. The data selections used in this study are consistent with the database provided by Tammisto *et al.* (2009).

The PFL device measures flow entering or leaving a small section of a borehole by thermal dilution. For the data used in this study, the effective support scale is a length of 0.1 m along the axis of the borehole. In most cases for sparsely fractured rock, this is sufficient to associate flow values with individual or a few geologically observed fractures. The typical range of measurable flows is approximately 2 to 5000 ml/min and the lower limit of measureable transmissivities is typically in the range 10^{-10} to 10^{-9} m^2/s. A technical description of the PFL device is provided by Öhberg & Rouhiainen (2000).

The PFL-observed flowing features only constitute a small subset of the total number of geologically identified fractures. Even though the system may have very few PFL flowing features which are usually associated with fractures, there is typically a significantly greater number of geologically identified fractures, without measurable flow. In addition to PFL flow logging, transient pressure response tests are conducted with the Hydraulic Testing Unit (HTU) system for selected boreholes, which is a conventional constant-head double packer testing system (Hämäläinen, 2005). The HTU system has a coarser support scale and requires more time to log a borehole, but is able to detect lower transmissive features than the PFL device. Also, since the HTU is based on short-term pressure response tests, and the PFL device is based on steady-state pumping, the effective radius and regions of fractured rock tested are expected to be significantly different. Hence, transmissivity values are likely to be representative for different portions of the bedrock.

A combination of geological and geophysical investigations, as well as hydraulic response tests, has been carried out to identify and characterise major fracture deformation zones (Vaittinen *et al.*, 2009). Of particular interest for the region (encompassing the KR14 – KR18 set of boreholes) are two deformation zones denoted as HZ19 A and HZ19C; they are sub-horizontally inclined and intersect the boreholes at depths of about 10–20 m and about 40–50 m respectively (Karvonen, 2011), and have a relatively high transmissivity in the range 10^{-5} to 10^{-6} m²/s. Conventional hydraulic head measurements are also carried out where density dependence due to salinity is accounted for when residual pressures and hydraulic gradients are calculated (Ahokas *et al.*, 2008).

9.3 APPROACH

A discrete fracture network (DFN) approach is used to model the hydrogeological flow system. With a DFN approach, flow only occurs in fractures embedded in an impermeable rock matrix. The approach represents fractures by rectangular conduits with assigned transmissivity. This can be seen as flow through a thin slab of porous media with an assigned effective hydraulic conductivity. The steady-state, constant density groundwater flow equation is used to obtain fluid pressure at intersections of the fracture network, as

$$(K \nabla h) = 0 \qquad (1)$$

Where h is hydraulic head [L] and K is the hydraulic conductivity tensor [L T^{-1}]. Then volumetric flow Q [L³ T^{-1}] through each fracture is calculated from the spatial gradient of the head field ∇h [-] between intersections by assuming a constant transmissivity T [L² T^{-1}] (or hydraulic conductivity K with effective aperture $2b$ [L]) within each fracture and between intersections using Darcy's law, as

$$Q = w\, T\, \nabla h \qquad (2)$$

for a fracture plane of width w [L]. In this study, each fracture is assumed to be homogeneous and is assigned a constant transmissivity, but the transmissivity of fractures can vary. Furthermore, boreholes are represented as hydraulically open conduits in the DFN, where a well model is used to correct for the pressure in the borehole relative to the pressure in the fracture plane. The industry-standard numerical code ConnectFlow (AMEC, 2012a) is used with the NAPSAC module (AMEC, 2012b) to create the DFN and to solve for the pressure field and water flow.

Fracture transmissivity can be inferred from PFL flow measurements using steady-state, constant drawdown single-hole pump tests and by adopting the Thiem (or Dupuit) homogenisation assumptions. These include assuming fractures are isolated and confined features connected to a boundary of constant head at a given radial distance from the borehole, i.e., effectively eliminating a network of fractures. Then, fracture transmissivity can be inferred using the Thiem or Dupuit equation (Bear, 1979),

$$T = \frac{Q}{2\pi\Delta h}\log\frac{r_1}{r_2} \tag{3}$$

where r_1 [L] is the radius of the borehole, r_2 [L] is the radial distance to a cylindrical boundary with constant head, Δh [L] is the constant drawdown in the borehole due to extraction pumping, and Q [$L^3\,T^{-1}$] is the fracture-specific volumetric flow entering the measurement section. Since r_2 is generally not known typically an assumed value of 20 m is used; nonetheless, since it appears as a term in the logarithm it is not a very sensitive parameter (Bear, 1979). The measured hydraulic head values are converted to corresponding freshwater head values as part of the data processing procedure. The PFL device can efficiently log the full depth of most boreholes in a matter of hours, but requires a stable drawdown and steady-state pumping. The main limitations are expected to lie in the inconsistency of homogenisation of a fracture network as single thin slabs of porous media and the requirement of a constant head boundary at a fixed radial distance (Frampton & Cvetkovic, 2010).

In addition to PFL flow logging, transient pressure response tests are conducted with the Hydraulic Testing Unit (HTU) system, a conventional constant-head double packer testing system. An overpressure is induced in the packed-off section for approximately 15 minutes followed by a fall-off period of about 10 minutes. The stationary phase is used to obtain transmissivity estimates based on the Moye interpretation (Moye, 1967), and the transient phase is used to obtain transmissivity estimated based on the Horner and inverse flow interpretations (Hämäläinen, 2005). The lower limit of measurable transmissivities with the HTU system is approximately 10^{-11} m²/s and its support length is 2 m. Hence, the lower measurement limit is smaller than for PFL-inferred transmissivities, however, the support scale is coarser.

9.4   APPLICATION OF FLOW MEASUREMENTS TO SIMULATION DESIGN

The relative depths and surface locations of the five boreholes KR14 – KR18 are shown in Figure 9.2, and the PFL measured flows under non-pumped conditions for each of the five boreholes are shown in Figure 9.3 (solid blue arrows). Boreholes KR14 and KR15 are the deepest, reaching depths of approximately 500 m. Borehole KR14 is furthest away and contains a notably greater number of PFL measured flows than the other boreholes (Table 9.1). A greater density of flowing features are also observed in the upper 200 m section of KR14 (Figure 9.3, solid blue arrows), with a few additional flowing features to a depth of approximately 480 m. Interestingly,

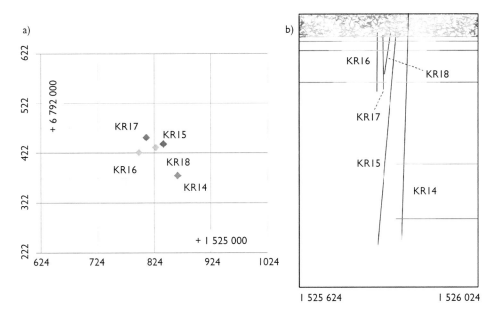

Figure 9.2 (a) Surface view of the model domain in the Finnish coordinate system (cf. Figure 9.1) and location of boreholes. The positive vertical axis is metres Northing and positive horizontal axis is metres Easting. (b) Vertical cross-section of the model domain with boreholes and semi-synthetic fracture network. The vertical axis is depth and the positive horizontal axis is Easting. Fractures are coloured by hydraulic head obtained from simulation case C (colour legend provided in Figure 9.4). The surface of the domain is 400 m by 400 m and it is 600 m deep.

in borehole KR15 there are no observed flowing features below approximately 70 m depth for these non-pumped conditions, even though it extends to approximately 500 m depth. Boreholes KR16, KR17 and KR18 are drilled to depths of approximately 170 m, 160 m, and 130 m respectively and contain only a few flowing features. Also, the upper portions of boreholes KR15 – KR18 are cased to 40 m depth and KR14 to 10 m depth.

Since the measured flows are caused by a natural, non-pumped flow field, the main driving force is likely to be due to the local island topography (Löfman, 1999). Also, since the bedrock system is sparsely fractured, the effect of open boreholes is notable. Positive flow values are defined as flow entering the borehole from fractures in the bedrock and negative values as flow leaving the borehole and entering fractures. Hence, flow enters boreholes in the top regions, approximately between depths of 10 m to 40–60 m, depending on the borehole, and leaves the borehole at depths below approximately 60 m (solid blue arrows, Figure 9.3). Thus the boreholes are effectively acting as vertical drainage conduits.

A semi-synthetic discrete fracture network model is designed with the aim of capturing the general flow behaviour and main features in this system. The numerical model domain is a cuboid with surface area of 400 m by 400 m and depth of 600 m (Figure 9.4). The size of the domain is intended to represent a compromise between

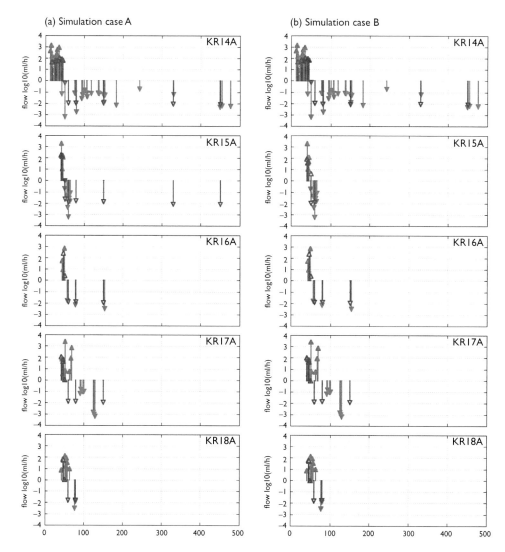

Figure 9.3 Fracture-specific flows in boreholes KR14 through KR18 obtained from field PFL measurements (blue solid arrows) and numerical simulations (red open arrows) from (a) case A and (b) case B. Positive values indicate flow entering the borehole from the fracture and negative values indicate flow leaving the borehole. Note the vertical axes show flow using a logarithmic scale.

distance to the hydraulic conductors in contact with the sea (a few km) and effects of neglecting fracture network heterogeneity in the bedrock (here only a simple DFN is considered). The surface of the model is centred at the approximate midpoint of the five boreholes, and the depth extends beyond the deepest boreholes (500 m for KR14 and KR15). A boundary condition of zero head is assigned to the lateral sides of the numerical model domain. The surface boundary condition is such that head

Table 9.1 Sum of measured flows Q entering/leaving boreholes during non-pumped conditions and estimated transmissivities T (m²/s) of corresponding hypothetical vertical fractures extending through the full extent of the simulation domain..

	KR14	KR15	KR16	KR17	KR18
Nr flow observations (non-pumped)	34	10	6	16	8
Sum flow entering borehole (ml/h)	6302	3207	1235	5279	446
Sum flow leaving borehole (ml/h)	4780	3227	777	5001	588
Mean borehole flow (ml/h)	5541	3217	1006	5140	517
Mean borehole flow (m³/s)	9.2E-05	5.4E-05	1.7E-05	8.6E-05	8.6E-06
Corresponding fracture T (m²/s)	1.4E-05	8.0E-06	2.5E-06	1.3E-05	1.3E-06

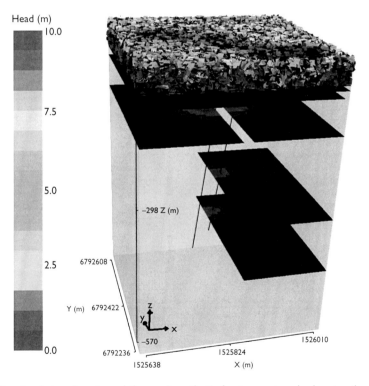

Figure 9.4 The simulation domain and the semi-synthetic fracture network, showing the case B configuration with eight horizontal fractures at depth as well as a near-surface random network. The random set of fractures is used to represent hydraulic effects of unconsolidated rock and high-conductive fracture zones observed between approximately 10 m to 50 m depth. Also, the random network connects the top surface boundary condition with the boreholes and horizontal fractures at depth. The top surface boundary condition is such that head is set equal to surface elevation, and the lateral sides have head set to zero. The surface of the domain is 400 m by 400 m and it is 600 m deep. The (x, y) coordinates refer to the Finnish coordinate system (cf. Figure 9.1).

is equal to topographical elevation, that is $h(x,y) = z(x,y)$ along the top surface. This corresponds to a simplifying assumption that the system is fully saturated and that the water table coincides with the ground surface. On an annual basis the water table is typically 1–2 m below the surface (Voipio *et al.*, 2004), hence the effect of this top surface boundary condition is to slightly raise the actual water table. This combination of boundary conditions for this sub-region of the domain is consistent with a simplification of results obtained from regional modelling work (Frampton *et al.*, 2009; Blessent *et al.*, 2011).

In order to represent the near-surface regime consisting mainly of unconsolidated rock, a generic random DFN network with high fracture density is generated for the depth range 0–50 m. Here fractures are generated with uniformly distributed orientations (strike and dip angles), constant size of 10 m by 10 m, and are assigned a transmissivity T of 10^{-6} m²/s. The volumetric intensity, that is the surface area of fractures per unit volume of rock mass, is set to 0.5 m²/m³. Approximately 7000 fractures are generated in the 400 m by 400 m by 50 m region. The main purpose of this near-surface random fracture set is to represent unconsolidated rock and it also serves as an aggregated representation of the hydraulic effects of the two major fracture zones (HZ19 A and HZ19C) observed at about 10 m and 40 m depth, which have effective transmissivities in the range 10^{-5} to 10^{-6} m²/s (Vaittinen *et al.*, 2009; Karvonen, 2011). In addition, it serves to capture effects of the spatial variation in head from the top surface boundary condition and enables water to be transmitted into the open boreholes and the set of discrete fractures at depth.

Horizontal discrete fractures are included at depth to attempt to capture the main flow of the system by aggregating flows into effective depth compartments. The locations of the fractures are determined by representing the overall/major flow behaviour as measured by the PFL device for the depth range 50 to 500 m. Six locations are determined and placed at the depths 50 m, 60 m, 80 m, 150 m, 330 m, and 450 m based on the main flowing features in the boreholes (solid blue arrows, Figure 9.3). The fracture at 50 m depth intersects the random network mentioned previously. This ensures flow is transmitted from the top boundary to all boreholes, regardless of position and casing depth. Since these fractures are horizontal all vertical flow below depths of 50 m occurs solely through the boreholes. Thus, flow from the top surface boundary occurs first via the random fracture network which intersects the fracture at 50 m depth which in turn intersects all boreholes. Also, since KR15 through KR18 are cased down to about 40 m depth, only the top-most approximate 10 m sections of these boreholes are available for intersections with the random network. KR14 is only cased down to 10 m depth and a greater extent of about 40 m of the top of this borehole is available for intersections with the random network.

In the initial model configuration (case A) the six deterministic and horizontal fractures are assigned a transmissivity T of 10^{-8} m/s and extend throughout the entire domain, so that they have a size of 400 m by 400 m and connect to the lateral sides of the domain boundary, where head is set to zero. This creates fracture flows not observed with the PFL device; note the simulated connections between KR14 and KR15 at depths 80 m, 150 m, 330 m and 450 m are not observed with the PFL instrument in KR15 (Figure 9.3a, second panel, open red arrows).

KR16 experiences field measured flows near both 80 m and 150 m depth, and KR17 experiences field measured flows down to depths of about 130 m depth (Figure 9.3a, third and fourth panels respectively). The fractures intersecting KR17 may either be inclined and intersecting the borehole at a slightly different depth or they may be different fractures. Even though these boreholes do not extend to the same depths as KR14 and KR15, it indicates that there are superfluous connections at depth in the simulated system between KR14 and KR15, and that the fracture system observed at depths in KR14 may not be connected to KR15 – KR18 group of boreholes. There are flows in boreholes KR16 – KR18 that are deeper than those observed in KR15. The deepest flow in KR15 occurs at 65 m depth, whereas KR16, KR17 and KR18 experience flows down to 150 m, 130 m and 80 m depth respectively. Thus this may indicate there are connections between the KR16 – KR18 group which are not connected to KR15.

A second simulation configuration is designed to disconnect KR14 from KR15 – KR18 at depths below 65 m in order to be more consistent with field measured flows. This is achieved by removing parts of the fracture surface for the four fractures below 65 m (i.e. fractures at 450 m, 330 m, 150 m, and 80 m depth) such that they only intersect KR14 and connect to the domain boundaries. This causes the observed flows in KR16 – KR18 at depths below 65 m to be removed. Therefore, two new fractures are included at 150 m and 80 m which connect KR16 – KR18 to the domain boundaries without connecting to KR15 and KR14, which results in a configuration of eight horizontal fractures (case B, cf. Figure 9.4).

The flows obtained from the second simulation configuration (case B) are shown in Figure 9.3b. In this setup the main connections are captured, even though the field measurements clearly indicate a greater number of fractures spread over multiple depth ranges. Again the main objective of the semi-synthetic network is to capture the overall flow behaviour. Note also the directions in terms of flow leaving or entering the borehole correspond to the field measured flow directions for all boreholes except KR17 at 60 to 70 m depth (Figure 9.3b, fourth panel) and KR18 at depths around 60 m (Figure 9.3b, fifth panel). There is a greater mismatch in intersection location for the simulated fractures in KR17.

The case B structure is used to condition simulated flow against field measured flow. Since this structure is at best a coarse representation of the flow system, only the cumulative sum of flows entering and leaving each borehole are considered. Fracture transmissivities are adjusted until the simulated flows approximately correspond to the aggregated field-measured flows (case C). The resulting transmissivities are typically a factor of 5 or 10 greater than the initial assignment of 10^{-8} m^2/s, with the exception of the fracture at 330 m depth which is a factor of 10 lower (Table 9.2, case B versus case C). The simulated results in terms of cumulative flow along boreholes are shown in Table 9.3 (case C); the ratio of PFL measured and simulated cumulative flows are relatively close to unity for most boreholes and are notably improved when compared to cases A and B.

Another simulation based on the case B structure but using field measured transmissivities has been carried out (case D). Single-hole pumping and simultaneous PFL flow measurements can be used to infer fracture transmissivity based on flow homogenisation assumptions. In order to adopt those values to the case B fracture structure, each borehole is partitioned into depth sections and the sum of transmissivities for each section is calculated (Table 9.4). The depth intervals are designed to be consistent with the location of the eight simulated fracture depth locations. The geometric

Table 9.2 Fracture configuration, transmissivities and borehole connections used in simulation cases B–E.

| Fracture ID | Fracture Depth (m) | Transmissivities (m²/s) | | | | Borehole connections |
| | | Case B | Case C | Case D | Case E | |
		Initial assignment	Modified by flow comparison	Geometric mean field values	Geometric mean x 0.1	
#1	−50	1.00E-08	1.0E-07	6.2E-06	6.2E-07	All (KR14-18)
#2	−60	1.00E-08	1.0E-07	9.0E-07	9.0E-08	All (KR14-18)
#3a	−80	1.00E-08	5.0E-08	2.3E-06	2.3E-07	KR14
#3b	−80	1.00E-08	1.0E-07	2.2E-07	2.2E-08	KR16, KR17, KR18
#4a	−150	1.00E-08	5.0E-08	3.7E-07	3.7E-08	KR14
#4b	−150	1.00E-08	1.0E-07	4.9E-07	4.9E-08	KR16, KR17, KR18
#5	−330	1.00E-08	1.0E-09	6.3E-09	6.3E-10	All (KR14-18)
#6	−450	1.00E-08	5.0E-08	3.2E-08	3.2E-09	All (KR14-18)

Table 9.3 Comparison of PFL measured effective flow (ml/h) along boreholes under non-pumped conditions with corresponding simulation results.

| Flows in ml/h | Borehole | | | | |
	KR14	KR15	KR16	KR17	KR18
PFL measured					
Flow along borehole	5541	3217	1006	5140	517
Simulation case A	Six deterministic fractures intersecting entire domain				
Flow along borehole	800	762	388	355	193
Ratio PFL/sim	6.9	4.2	2.6	14.5	2.7
Simulation case B	Setup with 6+2 deterministic fractures				
Flow along borehole	851	269	381	332	177
Ratio PFL/sim	6.5	12.0	2.6	15.5	2.9
Simulation case C	Case B with transmissivity adjusted against flow				
Flow along borehole	4030	1210	3260	3150	1770
Ratio PFL/sim	1.4	2.7	0.3	1.6	0.3
Simulation case D	Case B but adopting field measured transmissivities				
Flow along borehole	40800	9030	16500	15100	9430
Ratio PFL/sim	0.1	0.4	0.1	0.3	0.1
Simulation case E	Case D but transmissivities reduced by a factor of 10				
Flow along borehole	5360	1070	2080	1880	1120
Ratio PFL/sim	1.0	3.0	0.5	2.7	0.5

Table 9.4 Application of field measured transmissivity data for simulation case D.

Selected depth interval (m)		Sum of measured transmissivities T (m²/s) in boreholes within selected depth interval					Geometric mean T (m²/s)	Assigned to fracture	Fracture depth (m)
		KR14	KR15	KR16	KR17	KR18			
0	−55	2.2E-05	2.2E-06	6.9E-06	4.3E-06	6.3E-06	6.2E-06	#1	−50
−55	−70	1.3E-08	7.7E-06	8.1E-07	6.8E-06	1.1E-06	9.0E-07	#2	−60
−70	−115	2.3E-06					2.3E-06	#3a	−80
−70	−115			4.6E-07	2.7E-08	9.1E-07	2.2E-07	#3b	−80
−115	−240	3.7E-07					3.7E-07	#4a	−150
−115	−240			2.2E-07	1.1E-06		4.9E-07	#4b	−150
−240	−390	6.3E-09	6.3E-09				6.3E-09	#5	−330
−390	−520	2.5E-08	4.0E-08				3.2E-08	#6	−450
Sum T (m²/s)		2.4E-05	1.0E-05	8.4E-06	1.2E-05	8.4E-06			

mean of the summed transmissivities over all boreholes is assigned to its respective simulated fracture, so that the simulated fracture represents the effective transmissivity. However, the simulated flows obtained with these transmissivities are significantly greater than the flows measured by PFL (Table 9.3, case D).

If these transmissivities are decreased by an order of magnitude then the simulated flow along boreholes better correspond to field measured borehole flow (Table 9.3, case E). Note also that these transmissivities are in close correspondence with the previous flow-conditioned transmissivities (Table 9.2, cases C versus E). Thus the field measured transmissivities are required to be reduced by about an order of magnitude to better agree with flow measurements under non-pumped conditions based on this simplified simulation configuration.

9.5   ANALYSIS OF FLOW-CONDITIONED TRANSMISSIVITIES

The eight-fracture network, although simple in design, is capable of capturing the overall flow behaviour. The configuration also allows for adopting a process of direct conditioning of transmissivities against flow, rather than direct assignment of field-inferred transmissivities. Here the main objective is to obtain reasonable estimates and refine the structure of the model design. As a next step, it would, in principle, be possible to adopt formal calibration approaches. The process is demonstrated first by comparison against field-measured flows without making use of field-inferred transmissivity information (case C), and then compared these against field-inferred transmissivity information (case D). It is shown that the field-inferred transmissivities may be up to approximately a factor of 10 greater (case E) than that obtained by direct flow conditioning (case C).

Although the main flows can be captured this way, the fracture system has limitations in its capability of projecting details of the flow system. For example, if other data is included, such as flow observations in additional boreholes, the configuration would need to be extended and further refined. The implementation of the boundary conditions is somewhat simplified; the top surface water table is exaggerated and transient effects neglected, and the distance to the lateral boundaries, which are limited by the model domain size, are expected to be larger. However, the lateral boundaries are more than 20 m from any borehole, which is the radial distance to a constant head assumed when inferring transmissivity based on pumping with homogenisation assumptions (cf. Section 3). Despite the uncertainties and the limitations of this model, the value of the approach is that it provides a reference case model for further constructing and restraining more elaborate and complex DFN models.

Conditioning against flow rather than directly adopting transmissivity information has previously been shown to be useful in the context of stochastic DFN modelling, as it can avoid homogenisation assumptions (Frampton & Cvetkovic, 2010), which are commonly assumed in field applications (e.g. Pöllänen & Rouhiainen, 2002a,b; Rouhiainen & Sokolnicki, 2005). The approach adopted in Frampton & Cvetkovic (2010) makes use of DFN simulations with pumping in the boreholes, which is consistent to the field approach used to obtain PFL-inferred transmissivities. Therefore, it is interesting to note that those stochastically flow-conditioned transmissivities obtained are about a factor of 2 greater than PFL-inferred median transmissivities, which is in contrast to the analysis conducted here, where flow conditioned transmissivities are up to ten times smaller than PFL-inferred transmissivities.

The simulated transmissivities obtained by conditioning against natural flows using this semi-synthetic DFN representation are smaller than transmissivities measured by adopting Thiem-Dupuit homogenisation assumptions on PFL flow measurements under forced (pumped) conditions. The latter also means that the flow system is evaluated under relatively long-term pumping such that steady-state conditions prevail. Thus, the flows and hence inferred transmissivities obtained are due to a relatively large radius of influence, and essentially represent effective transmissivity for a large portion of bedrock. They may consist of additionally connected fractures in the network. For this reason, PFL inferred transmissivities are generally considered to be accurate to within an order-of-magnitude of the values obtained (Rouhiainen & Sokolnicki, 2005).

Other hydraulic investigation techniques are also employed, such as short-term overpressure responses in packered-off sections of boreholes with the HTU system. A comparison of transmissivities obtained with the PFL and HTU devices from the KR14 – KR18 boreholes is shown in Figure 9.5 as the cumulative distribution of values aggregated for all boreholes. The minimum transmissivity value is expected to be lower for HTU transmissivities due to its lower measurement limit; similarly, the density of small transmissivity values is expected to be greater for the HTU data for the same reason. However, since the hydraulic responses are evaluated for relatively short time periods with the HTU system, the transmissivities obtained are considered to be more representative for the hydraulic conditions close to the borehole vicinity. Thus, whereas the PFL and HTU transmissivities are obtained from the same boreholes, the region of influence of bedrock which is tested and hence the transmissivity values obtained are expected to be different. Thus the difference in the distributions should reflect differences in local versus larger-scaled fractured rock mass. This is perhaps more notable by comparing the median or mean transmissivity values, where the median is almost one order of magnitude smaller for HTU measured values than

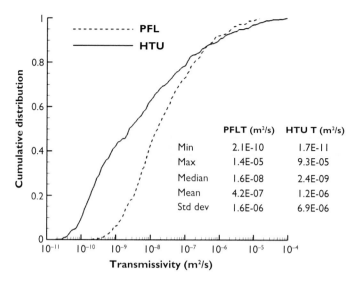

Figure 9.5 Comparison of the cumulative sample distribution of fracture transmissivities obtained from PFL flow logging and HTU response tests as well as sample statistics. The HTU device has a greater range of measurable transmissivities than that obtained from PFL flow logging, however the effective radius of influence of the tested section of bedrock is expected to be significantly smaller, hence the approximate one order of magnitude difference in median and mean values of transmissivity. The spread in the respective distributions is somewhat greater for the HTU measured values, mainly due to its lower measurement limit, also seen by the difference in minimum values.

for PFL measured values (cf. tabulated statistics inset in Figure 9.5). The standard deviations are all relatively similar.

The transmissivities obtained by flow conditioning of natural flows are likely to be representative for the regions of the fractures local to the boreholes. There is coherence between PFL flow measurements conducted under natural, non-forced flow conditions and their interpretation of transmissivity by flow-conditioned modelling, with transmissivities obtained from short-term transient response tests, as both are likely impacting only the close vicinity of the rock system near the borehole location. This serves as a possible explanation for the difference in transmissivities obtained between flow conditioning against non-forced PFL flows, as analysed in this study, and flow conditioning against forced PFL flows, as conducted in Frampton & Cvetkovic (2010).

9.6   ESTIMATION OF VERTICAL TRANSMISSIVITY

A significant uncertainty is the degree to which the existence of open boreholes impacts the natural undisturbed flow system. The magnitude of the vertical flow in the boreholes under non-pumped conditions is an indicator of the vertical transmissivity of the system, albeit with the presence of open boreholes. Since the fracture system is generally sparse with low transmissivity, the open boreholes are likely to enhance the vertical transmissivity component of the bedrock.

In a steady-state flow system, borehole inflow should equal borehole outflow, however, the PFL measured inflows and outflows are not in precise equilibrium. One reason for this could be that the natural flow system may be in a slight transient state due to effects of surface recharge during the measurement period. However, since the measurement period is rather short (a few hours), transient effects are assumed to be minor. Another reason may be related to measurement errors, whereby flows below the lower measurement limit are not detected and hence may contribute to an under-estimated sum. This is likely to be more prevalent at depth since typically there are fewer fractures and with lower transmissive properties at depth. Also, measurement errors can be caused by flow bypassing the device if the support section is not completely sealed, e.g. due to irregularities in the borehole (Ludvigson *et al.*, 2002). This may be more prevalent in unconsolidated rock in the upper sections of the boreholes.

The effective PFL measured vertical flow is estimated by taking the arithmetic mean value of the cumulative inflows and outflows from the boreholes (Table 9.1). Based on these estimates of effective vertical borehole flow, Darcy's law is used to obtain estimates of effective transmissivity (Table 9.1, last row) for hypothetical/generic vertical fractures replacing each corresponding borehole, based on assuming fracture sizes which intersect the full extent of the model domain. Also, an assumed hydraulic gradient is based on a head loss obtained from the difference between the typical maximum topographical elevation (10 m) and the head value assigned on the boundaries of the domain (0 m) over the length scale of the domain depth (600 m); hence the gradient used is 1/60. The vertical transmissivities obtained in this way should at best be seen as reference values. Nonetheless the resulting values range between about 10^{-5} m^2/s and 10^{-6} m^2/s and are consistent with the sum of transmissivities measured in each of the five boreholes (Table 9.4, last row). These values are also consistent with the effective transmissivities of the sub-horizontal fracture zones which transect the upper 50 m sections of this region of bedrock. The vertical transmissivity of the bedrock without the presence of open boreholes must be lower, if not significantly lower, especially in the deeper sections of the bedrock, and these estimates may be seen as upper bounds of vertical transmissivity.

9.7   CONCLUSIONS

An analysis of field measured flows under non-pumped conditions was conducted combined with numerical modelling of a semi-synthetic discrete fracture network representation of the bedrock. The following conclusions are obtained.

–   It is determined that the fracture system observed at depth in borehole KR14 is likely to be not strongly connected to borehole KR15 through KR18, even though the distance between them is only between approximately 70 to 90 m. Also, KR15 seems to be only partially connected to KR16, KR17, and KR18 at depths down to about 80 m.

- A method for direct conditioning of transmissivity based on *in situ* directional borehole flow measurements in sparsely fractured rock is developed and demonstrated for a semi-synthetic discrete fracture network model.
- Fracture transmissivities conditioned against aggregated measured flows result in values up to an order of magnitude lower than transmissivities obtained from steady-state pump tests. However, the flow-conditioned transmissivities are in agreement with transmissivities obtained from pressure response tests.
- The flow-conditioned transmissivities based on natural gradients are more representative of the regions of fractured rock in the vicinity of the borehole. Transmissivities obtained from steady-state pump tests are likely to be representative of a significantly larger portion of the fractured system.

The effective vertical transmissivity of the system was estimated based on the cumulative flow magnitude observed in boreholes, and values obtained are in agreement with typical transmissivity values observed for major sub-horizontal fracture zones of the region. Since boreholes are open, the inferred vertical transmissivities of the system correspond to a disturbed system, where the presence of open boreholes significantly enhances the actual vertical transmissivity. Therefore, the inferred values at best represent upper bounds.

Details of the flow system cannot be captured with the fracture network design. More complex models are necessary to consolidate the flow system incorporating additional data, such as flow measurements obtained from pumped conditions or additional boreholes. Nevertheless, the approach is still readily applicable and could be used and further refined with extended borehole information. A larger domain which extends to the actual vertical hydraulic zones in contact with the Baltic Sea would provide more realistic lateral boundaries. This would also require effects of fractures beyond the local network to be considered. Transient effects of the water table are expected to be less significant. However, the depth to the average water table would reduce the simulated hydraulic gradients slightly, and thereby reduce the simulated flow rates. In the flow-conditioned simulation case, most aggregated borehole flows are above the measured flows and an adjustment is expected to improve the agreement with field measured flows.

Despite its inherent limitations, the model can be constrained to be in agreement with the main flow observations of the non-pumped system. It can be used for direct conditioning against directional flow measurements obtained from deep boreholes in crystalline rock, which is achievable mainly due to the sparseness of the fractured system as well as the relatively few observed flowing features under non-pumping conditions. The conditioning is applied by aggregating *in situ* borehole flow measurements according to depth partitions which approximate the major flowing features.

ACKNOWLEDGEMENTS

This work has been supported by the Swedish Nuclear Fuel and Waste Management Company (SKB). Site data has kindly been provided by Posiva Oy through the Äspö Task Force on Modelling of Groundwater Flow and Transport of Solutes.

REFERENCES

Aaltonen I., Lahti M., Engström J., Mattila J., Paananen M., Paulamäki S., Gehör S., Kärki A., Ahokas T., Torvela T., Front K. (2010) Geological Model of the Olkiluoto Site, Version 2.0 (Working Report). Posiva Oy, Eurajoki, Finland.

Ahokas H., Tammisto E., Lehtimäki T. (2008) Baseline Head in Olkiluoto (Working Report No. 2008-69). Posiva Oy, Eurajoki, Finland.

AMEC (2012a) ConnectFlow Technical Summary Release 10.4. Safety and Risk Consultants (UK) Limited, part of AMEC plc., Harwell Oxford, United Kingdom.

AMEC (2012b) NAPSAC Technical Summary Release 10.4. Safety and Risk Consultants (UK) Limited, part of AMEC plc., Harwell Oxford, United Kingdom.

Bear, J. (1979) *Hydraulics of Groundwater*. McGraw-Hill, New York.

Blessent D., Therrien R., Gable C.W. (2011) Large-scale numerical simulation of groundwater flow and solute transport in discretely-fractured crystalline bedrock. *Advances in Water Resources* 34, 1539–1552.

Cvetkovic V., Painter S., Outters N., Selroos J.O. (2004) Stochastic simulation of radionuclide migration in discretely fractured rock near the Äspö Hard Rock Laboratory. Water resources research 40, W02404.

Doughty C., Takeuchi S., Amano K., Shimo M., Tsang C.F. (2005) Application of multirate flowing fluid electric conductivity logging method to well DH-2, Tono Site, Japan. *Water Resources Research* 41, W10401.

Follin S., Stigsson M., Svensson U. (2006) Hydrogeological DFN modelling using structural and hydraulic data from KLX04. Preliminary site description, Laxemar subarea – version 1.2. (Report No. R-06-24). Swedish Nuclear Fuel and Waste Management Co (SKB), Stockholm, Sweden.

Follin S., Levén J., Hartley L., Jackson P., Joyce S., Roberts D., Swift B. (2007) Hydrogeological characterisation and modelling of deformation zones and fracture domains, Forsmark modelling stage 2.2 (Report No. R-07-48). Swedish Nuclear Fuel and Waste Management Co (SKB), Stockholm, Sweden.

Frampton A., Cvetkovic V., 2007a. Upscaling particle transport in discrete fracture networks: 1. Nonreactive tracers. *Water resources research* 43, W10428.

Frampton A., Cvetkovic V., 2007b. Upscaling particle transport in discrete fracture networks: 2. Reactive tracers. *Water Resources Research* 43, W10429.

Frampton A., Cvetkovic V., and Holton D. (2009) Äspö Task Force on modelling of groundwater flow and transport of solutes – Task 7A. Task 7A1 and 7A2: Reduction of performance assessment uncertainty through modelling of hydraulic tests at Olkiluoto, Finland. International Technical Document ITD-09-05. Svensk Kärnbränslehantering AB, Stockholm, Sweden.

Frampton A., Cvetkovic V. (2010) Inference of field-scale fracture transmissivities in crystalline rock using flow log measurements. *Water Resources Research* 46, W11502.

Frampton A., Cvetkovic V. (2011) Numerical and analytical modeling of advective travel times in realistic three-dimensional fracture networks. *Water Resources Research* 47, W02506.

Hämäläinen H. (2005) Hydraulic Conductivity Measurements with HTU at Eurajoki Olkiluoto, Borehole OL-KR14 (Working Report No. 2005-42). Posiva Oy, Eurajoki, Finland.

Hartley L., Hunter F., Jackson P., McCarthy R. (2006) Regional hydrogeological simulations using ConnectFlow. Preliminary site description Laxemar subarea – version 1.2 (Report No. R-06-23). Swedish Nuclear Fuel and Waste Management Co (SKB), Stockholm, Sweden.

Hartley L., Hoek J., Swan D., Roberts D., Joyce S., Follin S. (2009) Development of a Hydrogeological Discrete Fracture Network Model for the Olkiluoto Site Descriptive Model 2008 (Working Report No. 2009-61). Posiva Oy, Eurajoki, Finland.

Hatfield K., Annable M., Cho J., Rao P.S.C., Klammler H. (2004) A direct passive method for measuring water and contaminant fluxes in porous media. *Journal of Contaminant Hydrology* 75, 155–181.

Hess A.E. (1986) Identifying hydraulically conductive fractures with a slow-velocity borehole flowmeter. *Canadian Geotechnical Journal* 23, 69–78.

Karvonen T. (2011) Olkiluoto Surface and Near-Surface Hydrological Modelling in 2010 (Working Report No. 2011-50). Posiva Oy, Eurajoki, Finland.

Klammler H., Hatfield K., Annable M.D. (2007) Concepts for measuring horizontal groundwater flow directions using the passive flux meter. *Advances in Water Resources* 30, 984–997.

Löfman J., 1999. Site Scale Groundwater Flow in Olkiluoto (Report No. 99-03). Posiva Oy, Eurajoki, Finland.

Löfman J., Mészáros F., Keto V., Pitkänen P., Ahokas H. (2009) Modelling of Groundwater Flow and Solute Transport in Olkiluoto – Update 2008 (Working Report No. 2009-78). Posiva Oy, Eurajoki, Finland.

Ludvigson J.-E., Hansson K., Rouhiainen P. (2002) Methodology study of Posiva difference flow meter in borehole KLX02 at Laxemar (Report No. R-01-52). Swedish Nuclear Fuel and Waste Management Co (SKB), Stockholm, Sweden.

Moye D.G., 1967. Diamond drilling for foundation exploration. *Civil Engineering Transactions* 9(1), 95–100.

National Research Council (1996) *Rock Fractures and Fluid Flow: Contemporary Understanding and Applications*. The National Academies Press, Washington, D.C.

Neuman S.P. (2005) Trends, prospects and challenges in quantifying flow and transport through fractured rocks. *Hydrogeology Journal* 13, 124–147.

Niemi A., Kontio K., Kuusela-Lahtinen A., Poteri A. (2000) Hydraulic characterization and upscaling of fracture networks based on multiple-scale well test data. *Water Resources Research* 36, 3481–3497.

Öhberg A., Rouhiainen P. (2000) Posiva Groundwater Flow Measuring Techniques (Working Report No. 2000-12). Posiva Oy, Eurajoki, Finland.

Paillet F.L., Pedler W.H. (1996) Integrated borehole logging methods for wellhead protection applications. *Engineering Geology* 42, 155–165.

Pöllänen J., Rouhiainen P. (2002a) Difference flow and electric conductivity measurements at the Olkiluoto site in Eurajoki, boreholes KR13 and KR14 (Working Report No. 2001-42). Posiva Oy, Eurajoki, Finland.

Pöllänen J., Rouhiainen P. (2002b) Difference flow and electric conductivity measurements at the Olkiluoto site in Eurajoki, boreholes KR15-KR18 and KR15B-KR18B (Working Report No. 2002-29). Posiva Oy, Eurajoki, Finland.

Posiva Oy 2003. ONKALO Underground Characterisation and Research Programme (UCRP) (Report No. 2003-03). Posiva Oy, Eurajoki, Finland.

Rouhiainen P., Sokolnicki M. (2005) Oskarshamn site investigation, Difference flow logging of borehole KLX04, Subarea Laxemar (Report No. P-05-68). Swedish Nuclear Fuel and Waste Management Co (SKB), Stockholm, Sweden.

Selroos J.O., Painter S.L., 2012. Effect of transport-pathway simplifications on projected releases of radionuclides from a nuclear waste repository (Sweden). *Hydrogeology Journal* 1–15.

Tammisto E., Palmén J., Ahokas H. (2009) Database for Hydraulically Conductive Fractures (Working Report No. 2009-30). Posiva Oy, Eurajoki, Finland.

Tsang C.F., Hufschmied P., Hale F.V. (1990) Determination of fracture inflow parameters with a borehole fluid conductivity logging method. *Water Resources Research* 26, 561–578.

Tsang, C.F., Doughty C. (2003) Multirate flowing fluid electric conductivity logging method. *Water Resources Research* 39, 1354.

Vaittinen T., Ahokas H., Nummela J. (2009) Hydrogeological structure model of the Olkiluoto site – Update in 2008 (Working Report No. 2009-15). Posiva Oy, Eurajoki, Finland.

Chapter 10

Prediction of fracture roughness and other hydraulic properties: Is upscaling possible?

John M. Sharp, Jr.[1], *Mishal M. Al-Johar*[2],
Donald T. Slottke[3] *& Richard A. Ketcham*[1]

[1]*Department of Geological Sciences, Jackson School of Geosciences, The University of Texas, Texas, USA*
[2]*Arcadis, Portland. Oregon, USA*
[3]*Schlumberger, Inc., Houston, Texas, USA*

ABSTRACT

Where present, open, connected fractures dominate fluid flow and solute transport. A challenge in predicting flow and transport in fractured media is describing representative physical characteristics appropriate to modelling. Fracture aperture, roughness, and channeling characteristics are important to predict flow and transport in hard rock terrains. Upscaling from hand or core sample properties would be highly beneficial but must assume a scale invariant or smoothly transformable relationship between fracture morphology and discharge. We analyze results of flow tests and flow modelling through natural fractures imaged by computed tomography. Using an areal roughness metric, statistics of roughnesses are plotted against sample size. For fracture specimens measuring up to 725 cm^2, a 10 cm^2 sample yields a representative roughness, but apertures cannot be scaled from this size. Our data indicate that hydraulic aperture cannot be predicted by single aperture measurements or averaging along scanlines.

Keywords: fractured rocks, fluid flow, transport, computed tomography, roughness, aperture, upscaling

10.1 INTRODUCTION

Characterisation of fluid flow in fractured rock systems has been the subject of a number of symposia in the fields of groundwater (e.g., Krásný & Sharp, 2007a; Shakeel *et al.* 2007; Banks & Banks, 1993) and petroleum geology (e.g., Petford & McCaffrey, 2003). These systems produce water and hydrocarbons, but their characterisation is difficult. Intensive drilling and characterisation programs for either resource in these settings are rare. Consequently, it would be valuable if data from studies at a small (e.g., hand sample or thin section) scale could be upscaled to well-field or regional systems. Scaling of permeability has been examined in karstic systems; Kiraly (1975) and Halihan *et al.* (2000) show that regional flow is dominated by conduits, flow to wells by fractures, and flow at the hand sample scale by the carbonate matrix. In fractured crystalline rocks, it is commonly assumed that this also holds as more permeable fractures should be encountered with increasing scale. However, Robins (1993) and Clauser (1991) suggest this need not always be the case. Indeed, Clauser stated that permeability should upscale but at very large scales would approach a constant. On the other hand, Krásný (1996), and Krásný and Sharp (2007b) suggest that geomorphic and tectonic features need to be carefully evaluated. Because of data limitations, a variety of mathematical upscaling approaches have been proposed to scale permeability in a variety of media, including power law distributions (Blum *et al.*, 1997), rock constitutive laws (Exadaktylos and Stavroploulou, 1997), multiple subregion models (Gong *et al.*, 2008), and percolation probability (Masihi and King, 2008). However, in order to upscale the hydraulic properties of fractures, the properties must possess a relationship between fracture morphology and discharge that is either scale invariant or smoothly transformable. We examine these relationships using flow tests, high resolution X-ray computed tomography (HRXCT) of natural fracture surfaces, and statistics.

10.2   METHODS

We selected natural fractures in indurated rocks (granites, tuffs, and sandstones) for laboratory analysis. Table 10.1 lists specimens used for flow testing and HRXCT analysis of aperture and roughness. Collecting intact natural fractures is a key as both fracture walls must be obtained and the *in situ* position between the fracture walls maintained and registered. More often, sampling involves only one fracture wall or analysis across scanlines. We show below these are, at best, of questionable value for ascertaining fracture hydraulic properties.

10.2.1   Aperture

The natural fractures are kept in their field position to the maximum possible extent and sealed along the lateral edges except for 2 specially designed end plates

Table 10.1 Specimens used for S/S_0 roughness scaling and flow testing (after Slottke, 2010). CC02-2 is the sample shown in Figures 10.1 and 10.7.

Specimen	Location	Description	Size [cm^2]
E11	Elberton, GA	Medium grained granite with iron oxide and clay coating.	160
Fr-Wr	Fredericksburg, TX	Medium grained granite with 1 cm weathering rind.	154
Fr-MnO	Fredericksburg, TX	Medium grained granite with pyrolusite coating.	266
Oatman Creek	Llano, TX	Fine grained unweathered granite.	52
CC01-1	Big Bend State Park, TX	Welded rhyolitic tuff, unweathered.	29
CC01-2	Big Bend State Park, TX	Welded rhyolitic tuff, weathered.	54
CC01-3	Big Bend State Park, TX	Impact fracture in welded rhyolitic tuff (2 surfaces), unweathered.	134
CC02-1	Big Bend State Park, TX	Semi-welded rhyolitic tuff (2 surfaces), clay coating.	120
CC02-2	Big Bend State Park, TX	Semi-welded rhyolitic tuff (2 surfaces), clay coating.	142

(Figure 10.1). The fracture is then placed in a constant permeameter (Thompson, 2005; Slottke, 2010) and both the discharge and the hydraulic gradient measured. From the cubic law, a hydraulic aperture (b) can be calculated by:

$$b = \left(\frac{12\mu}{\rho_w} \frac{Q}{\nabla h \cdot w} \right)^{1/3} \tag{1}$$

where ρ_w and μ are, respectively, the density the viscosity of the fluid (water), ∇h is the hydraulic gradient and Q the discharge over a fracture of width (w).

The entire fracture is then placed in The University of Texas High Resolution X-Ray Computed Tomography (HRXCT) facility for imaging the fracture aperture and the roughness of the fracture surfaces. The resolution of the scan is on the order of 0.25 mm. Surface locations obtained are conservatively accurate to 30 μm. Thus, over a single fracture surface of 100 cm² approximately 160 000 pixels of location data for each fracture surface and the equivalent number of determinations of fracture aperture are obtained. Details of these procedures and the HRXCT facility are described in Ketcham and Carlson (2001), Thompson (2005), Slottke (2010), Al-Johar (2010), and Ketcham et al. (2010).

These data are used to calculate apertures at each pixel and surface roughness. The mechanical or arithmetic mean (b_a), geometric mean (b_g), and harmonic mean (b_h) apertures are calculated, respectively, by

$$b_a = \frac{1}{n} \sum_{i=1}^{n} b_i \tag{2}$$

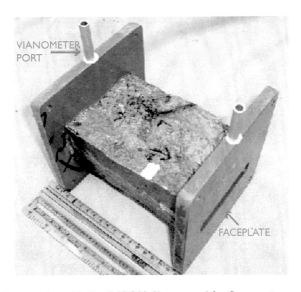

Figure 10.1 Fracture in welded tuff (CC02-2) prepared for flow testing and HRXRCT.

$$b_g = \left(\prod_{i=1}^{n} b_i \right)^{1/n} \tag{3}$$

$$b_h = n \Bigg/ \sum_{i=1}^{n} 1/b_i \tag{4}$$

where b_i refers to each individual estimate/measurement of aperture and n is the number of measurements.

In these estimates, we used the apparent aperture – the difference between the heights of the top and bottom surfaces measured on a coordinate system applied to the entire fracture. There are other estimates of aperture. The Mourzenko (Mourzenko *et al.*, 1995) aperture is local aperture estimated by the diameter of a sphere that is tangent to the top and bottom (adjacent) fracture surfaces. Other geometrical aperture estimates are suggested be Ge (1997) and Oron & Berkowitz (1998), but these are less applicable to 3-D fracture settings.

10.2.2   Roughness

We used several methods to estimate surface roughness. The departure from planarity was the primary method. Other methods were used to compare the roughness statistics on mated fracture surfaces.

The departure from planarity (surface area to footprint ratio) is calculated by

$$S/S_0 = \frac{\iint\limits_{R} \sqrt{\left(\frac{\partial z}{\partial x} \right)^2 + \left(\frac{\partial z}{\partial y} \right)^2 + 1}\, dA}{LW} \tag{5}$$

where $z = f(x, y)$ is defined over a region of surface elevations with the planar trend removed, L and W are the linear dimensions of the region, and dA is an area element. For discrete data sets as processed from the computed tomography data, the integral is simplified by solving for the area between every 4 adjoining points by dividing each subregion into 2 triangles with no loss or addition of data (Figure 10.2) and using Heron's formula,

$$A = \sqrt{s(s-a)(s-b)(s-c)} \tag{6}$$

where A is the area of a triangle, a, b and c are the lengths of the sides of the triangle, and s is the semi-perimeter or $1/2(a + b + c)$. A perfect plane has $S/S_0 = 1.0$.

S/S_0 is not the only means of assessing surface roughness. For instance, Stout (2000) lists 14 surface roughness parameters. In comparing surface roughness on mated surfaces, Al-Johar (2010) calculated mean-square (RMS) deviation, RMS slope, skewness, and kurtosis of the fracture surfaces and analyzed fracture surface. S/S_0, RMS deviation, and RMS slope data are shown in Table 10.2.

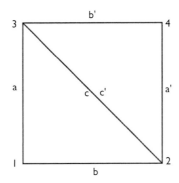

Figure 10.2 Two triangles defined by each set of 4 points derived from HRXCT allow calculation of surface area without loss or addition of data.

Table 10.2 Mated (top and bottom) surface roughness metrics (after Al-Johar, 2010). CC02-2 is the sample shown in Figure 10.1 and 10.7.

Specimen	Description	Surface area to footprint ratio	RMS roughness	RMS slope
Hick top	Cambrian Hickory Sandstone, Burnet Co. TX, medium-grained sandstone	1.136	1.19	0.57
Hick bot		1.133	1.19	0.57
El 1 top	Elberton granite, see Table 10.1	1.058	1.46	0.36
El 1 bot		1.048	1.62	0.32
El 2 top	Elberton granite, see Table 10.1	1.071	2.70	0.42
El 2 bot		1.049	1.19	0.33
Oatman top	Oatman granite, see Table 10.1	1.298	1.10	1.28
Oatman bot		1.320	1.08	1.36
Pack top	Precambrian Packsaddle Schist, Llano, TX	1.368	1.91	1.87
Pack bot		1.361	1.82	1.85
CC02-1 top	Santana Tuff, see Table 10.1	1.077	2.12	0.43
CC02-1 bot		1.078	2.14	0.44
CC02-2 top	Santana Tuff, see Table 10.1	1.120	2.14	0.56
CC02-2 bot		1.122	2.20	0.56

Fracture surface fractal dimensions were compared using the roughness-length, first return probability, and power spectral density (PSD) methods (Table 10.3). The roughness-length method (Malinverno, 1990) calculates RMS roughness at different scales and the Hurst exponent from the power law relationship between RMS roughness and profile length scale. The first return probability method (Schmittbuhl *et al.*, 1995) calculates the fractal dimension of a surface profile by measuring the horizontal distance required to intersect or return the vertical height of each point along a traverse. The power spectral density (PSD) method (Brown and Scholz, 1985) determines the fractal dimensions of self-affine fractal surfaces by applying the fast

Table 10.3 Mated (top and bottom) fractal dimensions (after Al-Johar, 2010). Sample descriptions are shown in Tables 10.1 and 10.2.

Specimen	Roughness-length		First return probability		PSD	
	x	y	x	y	x	y
Hick top	0.822	0.593		0.661	1.082	0.963
Hick bot	0.610	0.610		0.672	1.086	0.985
El 1 top	0.594	0.665	0.270	0.804	0.764	0.849
El 1 bot	0.626	0.650	0.448	0.663	0.820	0.851
El 2 top	0.658	0.745	0.571	0.726	0.859	0.933
El 2 bot	0.619	0.546	0.447	0.664	0.814	0.814
Oatman top	0.714	0.553	0.081	0.414	0.955	0.887
Oatman bot	0.698	0.555	0.237	0.420	0.964	0.917
Pack top	0.536	0.817	0.585		0.610	0.816
Pack bot	0.538	0.835	0.511		0.635	0.817
CC02-1 top	0.834	0.770		0.830	1.117	1.120
CC02-1 bot	0.842	0.777		0.823	1.165	1.111
CC02-2 top	0.829	0.777	0.688	0.769	1.208	1.178
CC02-2 bot	0.815	0.771	0.760	0.756	1.29	1.154

Fourier transform and plotting the squared amplitude against the frequency of each sine and cosine function.

10.3   RESULTS AND DISCUSSION

Three results are discussed below: 1) a comparison of aperture means with Q as a function of ∇h in the flow test; 2) an illustration of how well hydraulic apertures are estimated by measurements of aperture along scanlines; and 3) an analysis of scalability of fracture roughness and hydraulic aperture.

10.3.1   Hydraulic aperture estimates

Figure 10.3 shows the predicted discharge from using the arithmetic, geometric, and harmonic means and the actual discharge for sample CC02-2 (Table 10.1). The means are calculated on 0.25 mm intervals over the entire 10 cm by 15 cm fracture. The cubic law is valid over a considerable range of hydraulic gradients for this sample. Non-laminar conditions were apparent only at hydraulic gradients greater than 0.4. For this sample, Reynolds Numbers greater than 100 occur at hydraulic gradients of 0.4 and non-laminar conditions are expected above these gradients. The cubic law using a hydraulic gradient is, therefore, a robust estimator of flow in fractures at hydraulic gradients expected in nature.

It is apparent that the arithmetic mean consistently over-estimates discharge in this laminar flow regime. At gradients normally expected in the field, the geometric mean provides the best estimator of the hydraulic aperture. At higher gradients (>0.15), the

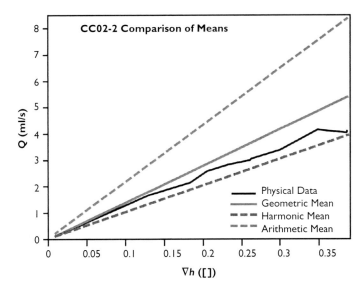

Figure 10.3 Comparison of means as a function of hydraulic gradient (∇h) for flow tests on sample CC02-2 (shown in Fig. 10.1). Hydraulic apertures are between the geometric and harmonic means. The geometric mean is the best approximation at low gradients. (*See colour plate section, Plate 30*).

hydraulic aperture is between the geometric and harmonic mean apertures. The latter under-estimates discharge at all points in the laminar flow regime.

10.3.2   Estimating hydraulic apertures by point or scanline estimates

Figure 10.4 shows the hydraulic aperture, calculated from equation (1) using data from the flow test (Fig. 10.3). It also shows the arithmetic and geometric means for individual scanlines normal to flow. These are what might be observed on an outcrop or thin section.

These data show that neither single measurement of aperture on this fracture nor an arithmetic mean of many measurements along a fracture profile provide confidence of estimating the actual hydraulic behavior. An areal estimator is required. For typical groundwater gradients, the geometric mean of all aperture values most closely predicts the hydraulic aperture.

10.3.3   Scalability of roughness and aperture

Figure 10.5 shows the mean and median of surface roughness (S/S_0) as a function of increasing scan area of the same sample documented in Figures 10.3 and 10.4. Also shown is the variance of roughness as a function of increasing scan areas. Similar values were obtained for scans of other fracture surfaces (Tables and Fig. 10.6). Mean roughness of 64×64 data point samples nearly matches that of the whole

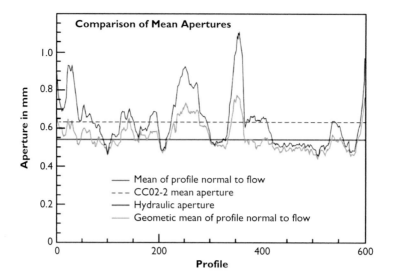

Figure 10.4 Arithmetic and geometric means of scanlines normal to flow through sample CC02-2 with actual hydraulic aperture and total fracture arithmetic mean aperture. (*See colour plate section, Plate 31*).

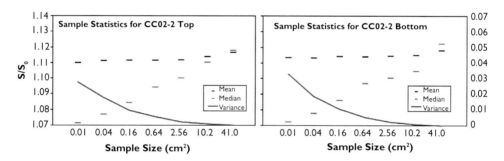

Figure 10.5 Roughness mean, median, and variance as a function of area scanned for the top and bottom surfaces of sample CC02-2. (*See colour plate section, Plate 32*).

surface for all cases analyzed (Fig. 10.6a), but the range of median roughnesses at this sample size (Fig. 10.6c) shows the effect of significantly rough outliers resulting in considerable variance in roughness among samples (Fig. 10.6b). In all cases, the variance approached zero and the means and medians became nearly equal at areas above 10 cm². This indicates that an area of 10 cm² is sufficient to characterize surface roughness of a given fracture.

Table 10.2 lists roughnesses and Table 10.3 lists some fractal dimensions calculated from the HRXRCT data for the mated surfaces of 7 natural fractures. Commonly only one surface of a natural fracture is available for measuring roughness, so if the mated surfaces have similar properties, this would be very convenient. The data from our analyses indicate that the surfaces that we measured have, in general, similar

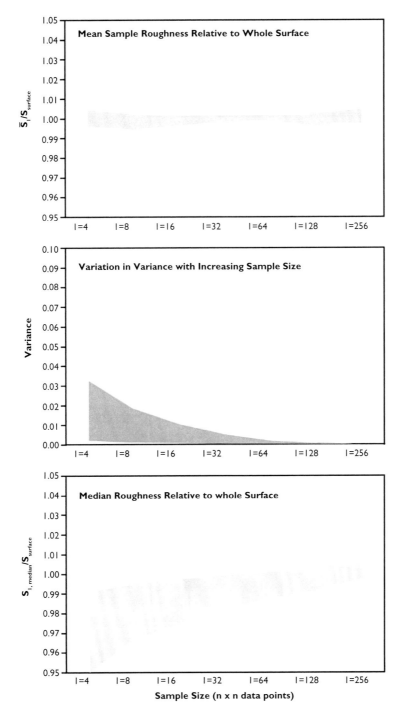

Figure 10.6 Range of S/S_0 statistics for surfaces (Table 10.1) as a function of sample size. (a) Arithmetic mean; (b) Variance of sample means; (c) Median value of means. (*See colour plate section, Plate 33*).

roughnesses and similar fractal dimensions. Thus, if the corresponding fracture surfaces are not available, similar properties can be assumed.

We caution, however, that site-specific conditions must be evaluated. For instance, near surface subhorizontal fractures, such as sheeting joints, could have one surface more altered than the other; visual inspection of the fracture surfaces is important.

Our study, as indicated, compared apparent local aperture distributions with measured fluid flow. The HRXCT data allow the selection of subsections of the larger fracture. Figure 10.7 shows sample CC02-02 and the subdivision into 20 smaller fractures. Each sub-section was numerically adjusted so that 3 points were in contact and a flow rate and accompanying hydraulic aperture estimated using MODFLOW and HRXCT apertures data. Slottke (2010) develops a program for this that adjusts the digital surfaces to 3 points of contact.

Figure 10.8 shows the flow calculated using MODFLOW for the 20 subsamples compared with the actual flow test measurement. Each of the 20 subsamples was adjusted numerically to have 3 points of contact as would occur if we have been able to test them in the flow apparatus. It is apparent that that there is no consistent method of upscaling hydraulic aperture from the small ~10 cm^2 to the larger ~140 cm^2 scale. This is apparently due to channeling even at such a small scale. If a representative elemental volume exists, it must be at a scale larger than this fracture. Similar results were inferred from even our largest sample (Paintbrush Tuff of Nevada) which was on the order of 3000 cm^2.

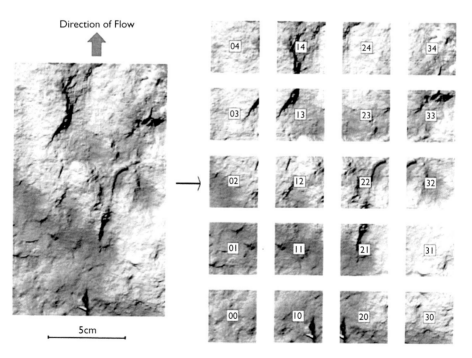

Figure 10.7 Sample CC02-2 transformed into 20 discrete subsamples (Slottke, 2010). See Plate 3 for a color coded relative elevation of one of the surfaces. (*See colour plate section, Plate 34*).

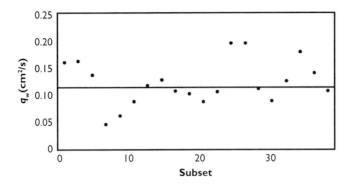

Figure 10.8 Calculated discharge per unit width (q_w) for the 20 subsamples of CC02-2 discharge per unit width. Horizontal line is the overall sample discharge per unit width. Subset numbers correspond to Figure 10.7.

10.4   CONCLUSIONS

We analyzed natural fractures of indurated rocks, welded tuffs and granites in particular, using flow cells and HRXCT data on fracture aperture and surface roughness. We find that:

- the cubic law using the hydraulic aperture is valid for the range of gradients expected in the field;
- the hydraulic aperture is best estimated using the geometric mean fracture aperture;
- neither single measures of fracture aperture nor estimates over a single scanline are adequate for estimating hydraulic aperture;
- mated fracture surfaces have very similar roughnesses and fractal dimensions so that the properties of one surface can be inferred from the other;
- fracture surface roughness can be adequately estimated from samples of 10 cm² for a variety of fracture surfaces; and
- upscaling of hydraulic aperture was not feasible from this same representative size due to difficulty in replicating *in situ* spacing of small samples in the laboratory and the effects of channeling and pinch out at small scales.

From these results, we infer that upscaling of surface roughness is feasible, but upscaling of hydraulic aperture is questionable.

ACKNOWLEDGEMENTS

This material is based upon work partially supported by the National Science Foundation under Grant No. EAR-04986 and the Geology Foundation of The University of Texas.

REFERENCES

Al-Johar M.M. (2010) Constraining fracture permeability by characterizing fracture surface roughness: MS thesis, The University of Texas, Austin, TX, USA.

Banks D., Banks S. (eds.) (1993) *Hydrogeology of hard rocks*. Memoires of the 24th Congress. International Association of Hydrogeologists, Oslo, Norway.

Blum P., Mackay R., Riley M., Knight J.L. (1997) Performance assessment of a nuclear repository, upscaling coupled hydro-mechanical properties for far-field transport analysis. *International Journal of Rock Mechanics and Mining Science* 42, 781–792.

Brown SR Scholz CH (1985) Broad bandwidth study of the topography of natural surfaces. *Journal of geophysical research*: Solid Earth 90(B14), 12575–12582.

Clauser C. (1991) Permeability of crystalline rocks. *EOS* 73, 233–238.

Exadaktylos G., Stavropoulo, M. (1997) A specific upscaling theory of rock mass parameters exhibiting spatial variability: analytical relations and computational scheme. *International Journal of Rock Mechanics and Mining Science* 45, 1102–1125.

Ge S. (1997) A governing equation for fluid flow in rough fractures. *Water Resources Research* 33(1), 53–61.

Gong B., Karimi-Ford M,, Durlofsky L.F. (2008) Upscaling discrete fracture characteristics to dual porosity, dual permeability models for efficient simulation of flow with strong gravitational effects. *SPE Journal* 13: 58–67.

Halihan T., Mace R.E., Sharp J.M., Jr. (2000) Flow in the San Antonio segment of the Edwards aquifer: matrix, fractures, or conduits? In: Sasowsky I.D. Wicks C.M. (eds.) *Groundwater flow and contaminant transport in carbonate aquifers*, Balkema, Rotterdam, 129–146.

Ketcham R.A., Carlson W. (2001) Acquisition, optimization and interpretation of X-ray computed tomographic imagery: Applications to the geosciences. *Computers and Geosciences* 27:381–400.

Ketcham R.A., Slottke D.T., Sharp J.M., Jr. (2010) Three-dimensional measurement of fractures in heterogeneous materials using high-resolution X-ray CT. *Geosphere* 6, 499–514.

Király L. (1975) Rapport sur l'état actuel des connaissances dans le domaine des caractères physiques des roches karstiques. *Hydrogeology of karstic terrains*, International Association of Hydrogeologists, Paris, 53–67.

Krásný J. (1996) Hydrogeological environment in hard rocks: an attempt at its schematizing and terminological considerations. In: Krásný J., Mls J. (eds.) *First workshop on hardrock hydrogeology of the Bohemian Massif*, Acta Universitatis Carolinae Geologica 40, 115–122.

Krásný J., Sharp J.M., Jr. (eds.) (2007a) *Groundwater in fractured rocks*: Selected Papers 9, International Association of Hydrogeologists, Taylor & Francis, London.

Krásný J., Sharp J.M., Jr. (2007b) Hydrogeology of fractured rocks from particular fractures to regional approaches: State-of-the-art and future challenges. In: Krásný J., Sharp J.M., Jr. (eds) *Groundwater in fractured rocks*, Selected Papers, International Association of Hydrogeologists, Taylor & Francis, London 9, 1–30.

Malinverno A. (1990) A simple method to estimate the fractal dimension of a self-affine series. *Geophysical Research Letters* 17, 1953–1956.

Masihi, M., King, P.R. (2008) Connectivity prediction in fractured reservoirs with variable fracture size: analysis and validation. *SPE Journal*, 13, 88–98.

Mourzenko V., Thovert J., Adler P. (1995) Permeability of a single fracture: validity of the Reynolds equation. *Journal de Physique II*, 5(3).

Oron A.P., Berkowitz B. (1998) Flow in rock fractures: The local cubic law assumption reexamined. *Water Resources Research* 34, 2811–2826.

Petford N., McCaffrey K.J.W. (eds.) (2003) *Hydrocarbons in crystalline rocks*. Special Publication 214, Geological Society, London.

Robins N.S. (1993) Reconnaissance survey to determine the optimum groundwater potential of the Island of Jersey. In: Banks D., Banks S. (eds.) *Hydrogeology of hard rocks*: Memoires of the 24th Congress, International Association of Hydrogeologists 24(1), 327–337.

Schmittbuhl J., Vilotte J-P., Stephane R. (1995) Reliability of self-affine measurements. *Physical Review* E1 (1), 131.

Shakeel A., Jayakumar R., Salih A. (eds.) (2007) *Groundwater dynamics in hard rock aquifers.* Capital Publishing Company, New Delhi.

Slottke D.T. (2010) Surface roughness of natural rock fractures: Implications for prediction of fluid flow. PhD dissertation, The University of Texas, Austin, TX, USA.

Stout K.J. (ed.). (2000) Development of methods for the characterisation of roughness in three dimensions. Penton Press, London.

Thompson C. (2005) Investigation of surface roughness of natural rock fractures using high-resolution X-ray computed tomography and laboratory flow test measurements. MS thesis, The University of Texas, Austin, TX, USA.

Chapter 11

Scale dependent hydraulic investigations of faulted crystalline rocks – examples from the Eastern Alps, Austria

Gerfried Winkler[1] *& Peter Reichl*[2]
[1]*Institute for Earth Sciences, University of Graz, Graz, Austria*
[2]*Resources – Institute for Water, Energy and Sustainability, Joanneum Research Forschungsges, Graz, Austria*

ABSTRACT

Fault permeabilities can range over several orders of magnitude due to complex spatio-temporal heterogeneities and anisotropy. Investigation methods at field scale (packer tests) and laboratory scale (triaxial cell tests) were performed to characterise faulted crystalline rocks in the Semmering area, Eastern Alps (Austria). The analyses of more than 180 packer tests performed in boreholes with depths down to 700 m below ground surface allow a clear differentiation of the hydraulic properties of the four crystalline tectonic units present, even though they are composed of similar lithologies. In addition these tectonic units show a general decrease in hydraulic conductivity with depth following the correlation $\log_{10} K = -2,49 \log_{10}(z) - 2,59$ where (z) is the depth in metres. Two tectonic units show an increase of the hydraulic conductivity above the threshold of fault rock content of 15% within the test interval. Additional investigations of fault core domains at outcrop with a thickness of 10 to 15 m and of drill cores were performed at laboratory scale. Core samples (length of 15 cm, diameter 10 cm) were taken perpendicular and parallel to the fault foliation within a kinematic coordinate system. The samples were analysed hydraulically by triaxial permeability cell testing at a flow pressures of 36 kPa. The analyses show clear hydraulic anisotropies with an up to two orders of magnitude higher hydraulic conductivity parallel to the fault foliation than perpendicular within the core zone domains.

11.1 INTRODUCTION

Collisional belts such as the European Alps comprise crystalline and metamorphic sedimentary rocks affected by multiple tectonic events. This results in the formation of heterogeneous and complex rock masses with a range of petrophysical properties. The hydraulic properties of fractured rocks and the hydrogeologic assessment of rock masses are influenced by the occurrence of faults and fault zones. Faults can involve distinct discontinuities with localised shear (shear fractures) (National Research Counil, 1996) or can be fully developed as fault zones including a damage zone and a fault core or any development in between of these two extremes. The fault domains (damage zones and/or core zones) are characterised by diverging hydraulic properties, and aquifer types, such as fractured aquifers in damage zones or porous aquifers in core zones with incohesive and loose rock masses. Permeability structures of fault zones vary within different zonation configurations (Figure 11.1).

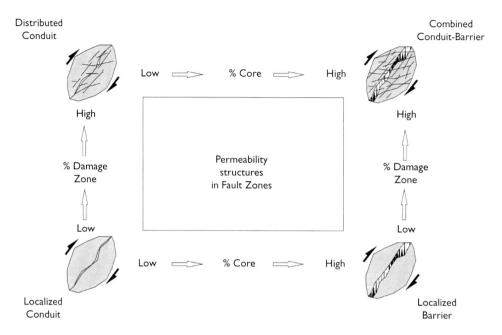

Figure 11.1 Permeability structures in fault zones after Caine *et al.* (1996).

Fault zones can be barriers or conduits for water flow or a combination of both (e.g. Maclay & Small, 1987; Antonellini & Aydin, 1994; Caine *et al.*, 1996; Evans *et al.*, 1997; Storti *et al.*, 2003). Fault permeabilities can range over several orders of magnitude due to complex spatio-temporal heterogeneities and anisotropy (e.g. Hesthammer & Fossen; 2001, Flodin & Aydin, 2004a; Flodin & Aydin, 2004b; Jourde *et al.*, 2002; Shipton *et al.*, 2002; Odling *et al.*, 2004; Winkler *et al.*, 2010).

Prediction of water flow in faults and fault zones is a critical aspect in the hydrogeological assessment concerning for example (nuclear) wastes disposals, infrastructural sites (e.g. prognosis work in tunnelling, tunnel design, the long-term stability of dams and underground powerhouses), and spring catchments (recharge areas, protection zones). Groundwater resources management in fractured aquifers requires knowledge of field-scale hydraulic properties, flow dynamics and transport properties of potential contaminants (e.g. National Researh Council, 1996, Lee *et al.*, 2001). In fractured crystalline aquifers it is generally assumed that the matrix conductivity and intergranular porosity is negligible, and that flow occurs mainly within the fracture system and the fault zones (e.g. Singhal & Gupta, 1999; Lachassagne *et al.*, 2001; Neumann, 2005). The challenges related to faults include their substantial heterogeneity with regard to internal structure and the evolution of the fault in space and time. The evolution of a fault is bound on the tectonic stress regime and the petrophysical and mechanical properties of the protolith (e.g. National Researh Council, 1996; Faulkner *et al.*, 2003; Woodcock *et al.*, 2007).

The heterogeneities of complex rock masses at different scales require adequate investigation methods. Hydraulic packer tests in open boreholes are proper field scale investigation methods to quantify the hydraulic properties of lithological/tectonic units at discrete borehole sections in situ and are generally used for exploration in applied hydrogeology (e.g. Freeze & Cherry, 1979; Frohlich *et al.*, 1996; Bühler & Thut, 1999; Reichl *et al.*, 2006). In addition packer tests provide information about the flow dynamics, flow model and the limitations of the system (e.g. Stober, 1986; Stober & Bucher, 2007; Horne, 1995). Assuming a dominant flow path relative to the testing equivalent fracture apertures can be calculated from packer tests in fractured aquifers (Halihan *et al.*, 1999, 2000). But packer tests are strongly limited in characterising small scale heterogeneities or anisotropy effects in cohesionless fault core zones. The limitations comprise the minimal size of the test interval especially for test intervals in deep boreholes. The limitation is bound by the flow direction parallel to the fault being in most cases the hydraulically dominant direction and on the fact that boreholes generally intersect one or more fault domains. Thus, it is not possible to determine the hydraulic anisotropic behaviour of fault domains/fault zones.

The purpose of this chapter is the hydraulic characterisation of the intensively faulted crystalline units of the Eastern Alps at different scales. The hydraulic properties of the entire lithological/geological units are presented including their correlation to depth. A method is applied to determine small scale heterogeneities and anisotropies of fault zones at exposure scale and the application to drill core samples. The combination of these data enables a scale dependent characterisation of the lithological/geological units providing important information to understand complex hydrogeological systems.

11.2   TEST SITE

The test site is the Semmering region in the Eastern Alps (Austria) (Figure 11.2). The region comprises crystalline rocks of the Graywacke zone belonging to the Upper Austro-Alpine and faulted metamorphic carbonate and crystalline rocks of the Semmering-Wechsel nappe complex (Lower Austro-Alpine) and is affected by a polyphased complex tectonic evolution. The lithological units show a contrasting deformational behaviour at the upper crust (brittle clastic) conditions, with various types of cataclastic fault structures and fault rocks. The Semmering-Wechsel nappe complex comprises the pre-Alpine basement rocks of the units Semmering Crystalline (SCR) and Wechsel Crystalline (WCR) and cover sequences of the Central Alpine Permomesozoic (CAPM) consisting of metamorphic carbonate and siliciclastic rocks (Figure 11.2). In the investigation area the Salzach-Ennstal-Mariazell-Puchberg (SEMP) fault system forms the boundary between the Semmering-Wechsel nappe system (Lower Austro-Alpine), in the south, and the Upper Austro-Alpine nappe system, comprising the Graywacke zone and the Northern Calcareous Alps, in the north.

Assembly of Austro-Alpine nappes took place during the Early Cretaceous. The Lower Austro-Alpine nappes and part of the Graywacke zone were subsequently affected by greenschist facies metamorphism during Late Cretaceous times (for summary, see Schmid *et al.*, 2004; Schuster & Kurz, 2005). The main thrusts and lithological boundaries, however, were re-activated and overprinted during Oligocene to Miocene strike-slip faulting. The latest dominated fault system is the SEMP (Ratschbacher *et al.*, 1989, 1991; Decker *et al.*, 1993; Decker & Peresson, 1996; Wang & Neubauer, 1998; Frisch *et al.*, 2000). The west south west–east north east oriented SEMP fault extends for

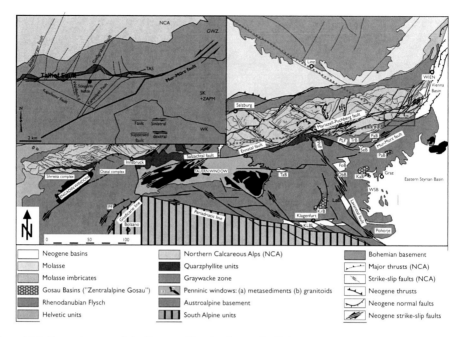

Figure 11.2 Tectonic map of the Eastern Alps displaying the Palaeogene to Neogene fault system and the location of the test site (after Linzer *et al.*, 2002). The site is located along the Talhof segment of the Salzach-Ennstal-Mariazell-Puchberg fault (SEMP). GöB: Göriach basin, FoB: Fohnsdorf basin, KLB: Klagenfurter basin, LaB: Lavanttal basin, ObB: Obdach basin, PaB: Parschlug basin, PF: Peijo fault, PeF: Pernitz fault, PLF: Palten-Liesing fault, PöF: Pöls fault, SeT: Seegraben basin, TaB: Tamsweg basin, WSB: Weststyrian basin.

400 km along the Eastern Alps. The maximum left lateral displacement along the SEMP is 70 km (Linzer *et al.*, 2002). The central and eastern segments of the SEMP show brittle deformation distributed over a broad segmented shear zone with a thickness of up to several hundred metres (Frisch *et al.*, 2000; Linzer *et al.*, 2002). The fault trends sub – parallel to the regional orogenic strike from the western part of the Tauern Window to the Vienna Basin (Figure 11.2), and crosses all Austro-Alpine tectono-stratigraphic units. Thus, there exist a lot of other fault zones striking parallel or acute-angled to the SEMP as the sub-vertical, approximately east to west striking Talhof fault segment with a left-lateral displacement of approximately 2–3 km. It is cut by the north south trending left-lateral Altenberg and Mur-Mürz faults in the west and in the east, respectively.

11.3   METHODS AND DATA BASE

11.3.1   Packer tests

The hydraulic properties of fractured and/or faulted crystalline aquifers are preferably determined by hydraulic tests in boreholes such as packer tests (Freeze & Cherry, 1979; National Research Council, 1996). The hydraulic conductivity K (m/s) is the proportionality constant in the Darcy flow law and can be derived from packer tests to provide the transmissivity T (m²/s). T is the product of K over the whole aquifer thickness b (m) and is given by $K \times b$. The hydraulic conductivity is a value of a combined system of rock matrix, interconnected fracture network and fault domains dominated by the highest permeable zone within a test interval. Packer tests enable the hydraulic characterisation at specific depths within a borehole as limited by the location of the packer in relation to borehole stability especially in faulted rock masses. Thus, in practice test intervals intersect more than one fault domain (protolith, damage zone or core zone) and the results reflect the most permeable intersected zone. In the investigation area more than 180 packer tests (mainly single packer tests) were performed in more than 80 wells to characterise the lithologic/geologic units of GWZ, SCR, WCR and CAPM the tectonic units. The packer tests performed are summarised in the Table 11.1.

The packer tests were performed in boreholes down to 700 m below ground surface (b.g.s.) with interval lengths between 4 m and 253 m. The tests and their analyses were conducted by Golder Associates using a test sequence with an initial, diagnostic, main and end phase resulting in combinations of injection and production tests with constant head or constant flow rate or slug and pulse tests. The data analyses was done with the software FlowDim v2.14 applying transient solutions based on Gringarten *et al.* (1979); Bourdet *et al.* (1984); Peres *et al.* (1989); Chakrabarty & Enachescu (1997). In addition the basic flow model of each test interval was determined with respect to the hydraulic conditions of the test intervals. Additional borehole tests as tracer fluid logging and flow meter logging have also been carried out in some boreholes. The fracture network was determined in situ by acoustic borehole imager logging (ABI) and by drill core analyses such as RQD (Priest, 1993), spacing and degree of separation. The drill core analyses included the lithological profile and the determination and quantification of fault rock percentage related to the test intervals. The macroscopic description of the fault rocks is based on the fault rock classification of Riedmüller *et al.* (2001).

11.3.2 Triaxial permeability tests – fault rock sampling

The tectonic units are characterised by intensively faulted lithologies where the fault core zones mainly could be classified as cohesionless (soil like material) cataclasites

Table 11.1 Number and interval length variability of the packer tests and the variability of the mean test depth for each tectonic unit. GWZ: Graywacke zone, WCR: Wechsel Crystalline, SCR: Semmering Crystalline, CAPM: Central Alpine Permomesozoic, min: lowest value, max: highest value, b.g.s.: below ground surface.

Tectonic units	Number of tests	Depth [m b.g.s.]		Interval length [m]	
		min	max	min	max
GWZ	31	15	578	5	93
WCR (basement)	42	18	645	4	253
SCR (basement	76	27	687	7	137
CAPM (crystalline cover series of SCR and WCR)	38	37	693	5	242

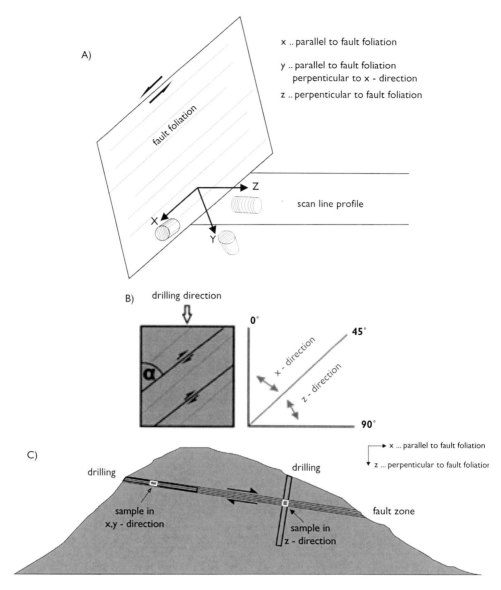

Figure 11.3 A) sample orientation in relation to the kinematic coordinate system of the exposed fault zone along scan line profiles. B) drill core sampling relating to the drilling direction and C) with the assignment to the kinematic coordinate system of the fault.

with varying grain size distributions (Riedmüller *et al.*, 2001). Thus, it was possible to sample the fault rock material for small scale heterogeneities and hydraulic anisotropy at laboratory scale. At exposure scale steel pipe samples ($\varnothing = 10$ cm, $h = 15$ cm) were taken and additionally drill core samples out of fault domains in drillings with the same sample sizes. The hydraulic properties of the samples were determined by triaxial permeability cell tests based on the standard ÖNORM B 4422-1 (1992) in the geotechnical laboratory of the Institute of Soil Mechanics and Foundation Engineering at the University of Technology Graz. Triaxial permeability cells are modified permeameter tests with a filter plate at the bottom and the top of the sample including water in – and outlet constructions. The sample is surrounded by a rubber diaphragm and hold in shape by the hydrostatic pressure in the triaxial cell which is slightly higher than the flow pressure in the sample. The tests were performed as constant head tests with a standard pressure of 36 kPa. The hydraulic anisotropies and heterogeneities of fault domains were investigated at a well exposed fault core zone of 10–15 m thickness which is built up by cohesionless cataclasites and fault gouges (Riedmüller et al., 2001) with a planar fabric parallel to the fault zone boundary. The architecture and the hydraulic properties of the core zone were investigated along two scan-lines. More than 50 steel pipe samples were taken in three orientations referring to a kinematic coordinate system (Figure 11.3A).

Outcrops at the surface are often small and/or they are exposed to alteration and covered by vegetation and thus are limited by scale, quantity and quality. Exploration boreholes provide the investigation of fault zones with as high data quality as possible at specific depths. Thus, in addition to the exposure samples fault domains encountered within packer test intervals of eight boreholes were analysed using 20 drill core samples. To compare the core samples with the samples of the outcrop, the reference kinematic coordinate system of the outcrop was applied to define the drill core sample with respect to the fault orientation (Figures 11.3B and 11.3C). At a first step the spatial orientation of the sampled fault domain was determined using borehole measurements (acoustic or optical borehole imager, ABI/OBI) and drill core analyses. In a second step the angle between the drilling direction representing the sample orientation and the fault orientation was determined. If the angle between fault foliation and drill direction (α) is below 45° the sample was assigned to be in x or y direction to the cinematic coordinate system of the fault. If α has an angle higher than 45°, the z direction was assigned.

11.4   RESULTS AND DISCUSSION

The packer tests result a similar hydraulic properties variability of the tectonic units GWZ and WCR with medians of the hydraulic conductivity of 1.2E-09 m/s and 2.0E-09 m/s. SCR shows even lower minimum values but is generally half an order of magnitude higher permeable than the units WCR and GWZ (Table 11.2). Compared to the basement units the cover series (CAPM) are characterised by the highest hydraulic conductivities (Table 11.2; Figure 11.4).

Additionally a correlation between hydraulic conductivity and depth could be observed (Figure 11.4). Based on the means of 100 m intervals of all data the correlation to depth follows the equation

$$\log_{10}K = -2,49 \log_{10}(z) - 2,59 \tag{1}$$

where (z) is the depth in metres. Stober & Bucher (2007) published similar results of gneiss dominated units in the Black Forest area. In their paper they gave an overview of world-wide hydraulic property data of crystalline basement down to depth of more

Table 11.2 Overview of the packer test intervals. Hydraulic conductivity and cataclasite content ranges of the test intervals according to the tectonic unit; cataclasite content: percentage of fault rock material related to the total packer test interval.

Tectonic units	Hydraulic conductivity [m/s]			Cataclasite content [%]		
	min	*max*	*median*	*min*	*max*	*mean*
GWZ	4.1E-11	2.3E-07	1.2E-09	0	78	24
WCR (basement)	5.2E-11	2.3E-06	2.0E-09	0	100	08
SCR (basement	2.9E-11	1.6E-05	7.4E-09	0	76	13
CAPM (crystalline cover series of SCR and WCR)	1.3E-11	6.5E-04	4.7E-08	0	98	21

than 4 km. In Table 11.3 selected investigation areas with gneissic rocks are listed compared to the Semmering data.

Compared to the Black Forest data (down to 1.2 km) the Semmering data generally show lower hydraulic conductivities which can be explained by the different lithologies. The Semmering units are mica schists and phyllites dominated while the Black Forest data are bound on gneisses and granite types. In addition it has to be considered that the units in the Semmering area are intensively faulted rock masses, where some packer test intervals consist of 100% cohesionless fault rock.

Analyses considering the cataclasite content within the test intervals indicate a threshold of about 15% for the tectonic units SCR and CAPM. Packer test intervals with more than 15% cataclasite content show significantly higher hydraulic conductivities than the test intervals below the threshold (Figure 11.5A). Additional analyses of the fracture density and spacing in relation to hydraulic properties within the test intervals showed no correlation. The fracture network analyses were based on acoustic and optical borehole imager data (ABI, OBI) and drill core analyses including the parameters RQD, fracture spacing and density (Priest, 1993). Preliminary results with respect to fracture equivalent apertures (Halihan *et al.*, 1999) calculated for all packer test intervals performed in the crystalline units range over nearly up to three orders of magnitude (0.2 mm to 100 mm; Figure 11.5B). The packer tests results are hydraulically affected by fault domains and open fractures not combined with the fault domains. The differentiation between these two potential flow paths has to be investigated in prospective research work.

Fault zone heterogeneity analyses were applied to one well exposed fault core zone and on fault domains intersected by exploration boreholes of the Semmering Basetunnel project. Preliminary analyses at the exposed fault core zone indicate specific zones of deformation with varying clay content (Winkler *et al.*, 2010). Similar observations are presented by Crawford *et al.* (2002) who divided the fault rocks broadly between two end member types: (i) cataclastic fault rocks and (ii) shale shears. They observed permeability ranges over six orders of magnitude depending on clay content and shearing characteristics. Thus, the main zones of deformation strongly influence the hydraulic properties parallel and perpendicular to the fault foliation.

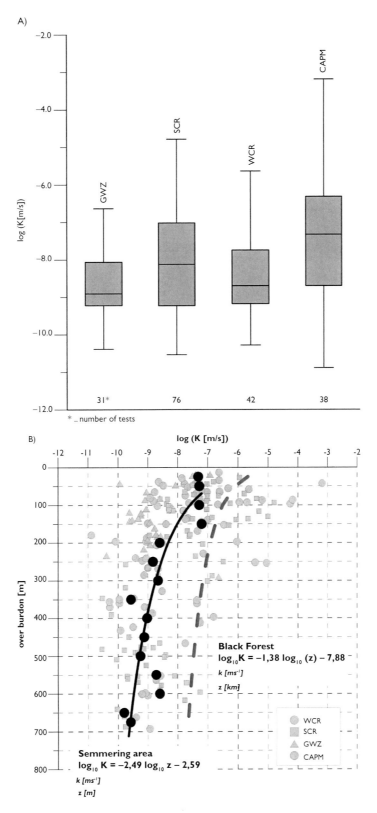

Figure 11.4 Hydraulic conductivities (K) variability of the tectonic units and depth dependency. Solid line: fit curve through Semmering data based on means of 100 m intervals (black dots), dashed line: depth correlation of Black Forest data (Stober & Bucher, 2007).

Table 11.3 Semmering data with selected investigation areas and their hydraulic characterisation (Stober & Bucher, 2007); DP: depth dependence, rep: representative value of K, ddp: decreasing depth dependence, idp: increasing depth dependence, ndp: no depth dependency.

Investigation area	Depth [m]	Geology/Lithology	Hydraulic conductivity K [m/s]			DP
			min	max	rep	
GWZ, Austria	<578	Phyllites, schists	4.1E-11	2.3E-07	1.2E-09	ddp
WCR (basement), Austria	<645	Phyllites, schists, gneisses	5.2E-11	2.3E-06	2.0E-09	ddp
SCR (basement), Austria	<687	Mica-rich schists, gneisses	2.9E-11	1.6E-05	7.4E-09	ddp
CAPM (crystalline cover series of SCR and WCR), Austria	<693	Quartzites, schists, phyllites	1.3E-11	6.5E-04	4.7E-08	ddp
Olkiluoto, Finland	<1,000	Mica-rich gneisses, minor granite	1.0E-09	1.0E-08		ndp
NAGRA deep wells, N-Switzerland	<2,500	Granite, gneiss	1.0E-13	4.0E-04	1.0E-09	ndp
Urach 3, S-Germany	<4,400	Gneisses			1.1E-09	
Cajon Pass, Califronia, USA	<2,077	Gneisses	3.8E-12	1.4E-09	1.0E-11	idp
Black Forest all data, S-Germany	<1,000	Granite, gneisses	5.0E-10	1.0E-04	7.0E-06	ddp
Black Forest, S-Germany	<1,000	Gneisses	5.0E-10	5.0E-08	8.0E-09	ddp

Analyses at outcrop scale result a clear hydraulic anisotropy with the highest hydraulic conductivity parallel to and the lowest perpendicular to the fault foliation, these differing by more than one order of magnitude (Figure 11.6, Winkler *et al.*, 2010; Koch *et al.*, 2011). This hydraulic anisotropy can also be observed where fault domains are intersected by the exploration boreholes (Figure 11.6). The results correlate well with the range of the hydraulic conductivity from the packer tests in which the sampled fault domains occur. The drill core samples include phyllites/schists and quartzites/arkoses protoliths. These lithology groups exhibits hydraulic anisotropies indicating possible additional hydraulic properties to the packer test results. The hydraulic conductivity perpendicular to the fault zones is nearly one order of magnitude lower than that parallel to the fault foliation (Figur 11.6).

The triaxial tests results at laboratory scale and the packer tests results at borehole scale exhibit similar medians of the permeability of 7,2E-16 m² and 4,7E-16 m² both data sets log-normally distributed (Figure 11.7). In addition the packer test data were sub-divided into test intervals with more or less than 15% cataclasite content. In general the trixial tests results are in the same order of magnitude as the packer tests but correlate better with the packer test intervals with more than 15% cataclasite content (Figure 11.7).

A scale dependency of the hydraulic permeability as derived from other data for crystalline rock masses (Clauser, 1992, 2001) is not seen in the Semmering data. It is noteworthy that the literature samples at laboratory scale were mostly measured

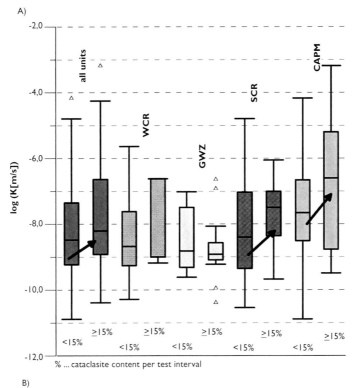

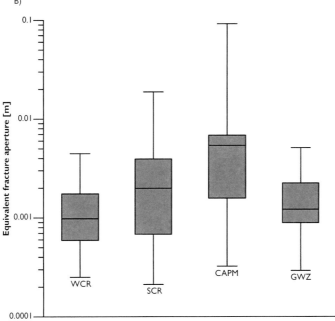

Figure 11.5 A) hydraulic conductivity variability of the individual tectonic units in dependence of the cataclasite content threshold within the test intervals; B) equivalent fracture aperture distribution of the tectonic units.

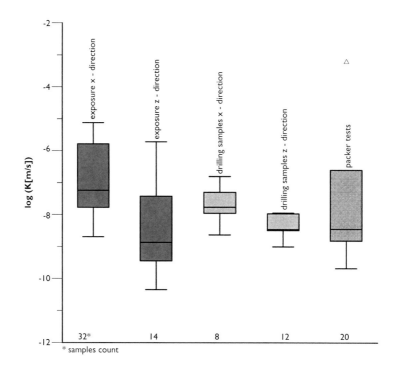

Figure 11.6 Hydraulic anisotropy parallel (*x* and *y*-direction) and perpendicular (*z*-direction) to the sampled fault domains at the exposure and the exploration drillings. The drill core samples were taken from 20 packer test intervals also presented in this diagram.

under simulating in situ pressure. The trixial cell samples of the Semmering area range at the upper section of the laboratory scale data (Figure 11.7). This might be caused by the fact that the drill core samples are taken out of cohesionless fault domains reflecting the most permeable sections of the packer tests and thus, enhance a preferential flow parallel to the fault foliation. This is supported by the observation that the packer tests show an increasing hydraulic conductivity with increasing fault rock content in the test intervals (Figure 11.5). In addition the Semmering triaxial samples were measured with a pressure of 36 kPa based on ÖNORM B 4422-1 (1992) while the other test sites were measured mostly with higher pressures (in situ pressure). Preliminary results from additional laboratory measurements with pressures up to 240 kPa indicate a non-reversible decrease of hydraulic conductivity with increase pressure.

11.5   CONCLUSION AND OUTLOOK

Complex rock masses such as faulted crystalline rocks require multiple scale investigations to answer hydrogeological questions with respect to that aquifer type. In the Semmering area combined analyses were applied to characterise these rock masses

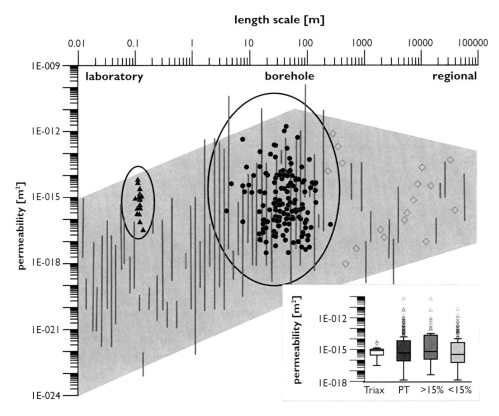

Figure 11.7 Semmering data (black dots: packer tests, black triangles: triaxial cell tests, both data sets are surrounded by black ellipses) with summarised hydraulic properties variability based on literature data of crystalline test sites (grey lines and symbols), grey background show the scale dependence of the literature data (modified after Clauser, 2001), small Figure 11.7: triangles: whisker box plot outliers, Triax: drill core samples analysed with triaxial cells, PT: all packer tests, <15%: packer test intervals with less than 15% cataclasite content, >15%: packer test intervals more than 15% cataclasite content.

hydraulically and to identify fault domains (core zones) as preferential flow paths in intensively faulted crystalline rock masses. In addition to hydraulic field scale tests as packer tests investigation methods at laboratory scale provided the quantification of hydraulic anisotropies of preferential flow zones. It is shown that the hydraulic anisotropy investigated at a well exhibited fault zone can be extended to fault domains encountered by boreholes in the same tectonic unit. The results indicate that these fault domains are preferential flow paths parallel to the fault foliation but are much less permeable perpendicular to it. It has to be considered that there is a scale lag of several orders of magnitude between the two investigation methods. However, the results provide a hydraulic characterisation of faulted rock masses at different scales. Nevertheless the bridging between the hydraulic properties at various scales is still challenging. Additional data are necessary to investigate the hydraulic heterogeneities of fault zones three dimensionally from laboratory over exposure to borehole scale (hydraulic field tests).

Regional flow models, however, require up-scaled, bulk properties of the fault zone. Thus, the up-scaling of the fault domain properties and determination of the fault bulk permeability is one main issue of the hydraulic characterisation of fault zones. Overviews of techniques for up-scaling permeabilities are provided by Wen & Gomez-Hernandez (1996); Renard & de Marsily (1997). Rawling *et al.* (2001) showed the influence of fault architecture on the fault bulk permeabilities of two different fault types, 'deformation band faults' and 'faults with fractures' using equivalent permeability calculations based on permeability data from the individual fault domains, which were assumed to be internally isotropic and homogeneous. Yet the permeability within individual fault domains can be anisotropic and heterogeneous. Numerical flow models that are able to represent the heterogeneous permeability distribution within the fault domains can be used for the up-scaling. Jourde *et al.* (2002), for instance, used a continuum flow model to infer bulk permeabilities from high-resolution fault-zone maps. Similarly, Lunn *et al.* (2008) applied a continuum flow model for detailed maps of fault zone architecture. The use of a statistical approach to the characterisation of fault architectural components, which might be useful for estimating bulk permeabilities of unexposed faults, is recommended. Thus, one important prospective research aspect for the investigation area will focus on the up-scaling of laboratory data to borehole and field scale.

ACKNOWLEDGMENT

The authors thank the Austrian Federal Railways ÖBB Infrastruktur AG for providing the hydraulic data of the investigation area and the financial support. We gratefully acknowledge Michaela Koch, Otto Leibniz and the colleagues of the Soil Mechanics Laboratory of the University of Technology Graz for their support with the laboratory tests.

REFERENCES

Antonellini M., Aydin A. (1994) Effect of faulting on fluid flow in porous sandstones; petrophysical properties. *AAPG Bulletin* 78, 355–377.

Bourdet D., Ayoub J.A., Pirard Y.M. (1984) Use of pressure derivative in well test interpretation. *SPE Paper* 12777 (4), 431–441.

Bühler Ch., Thut A. (1999) Hydraulische und felsmechanische Bohrlochversuche. In: Löw O.,Wyss O. (editors) *Vorerkundung und Prognose der Basistunnels*. Balkema, Rottterdam

Caine J.S., Evans J.P., Forster C.B. (1996) Fault zone architecture and permeability structure. *Geology (Boulder)* 24, 1025–1028.

Chakrabarty C., Enachescu C. (1997) Using the deconvolution approach for slug test analysis: Theory and application. *Ground Water* 35 (5), 797–806.

Clauser C. (1992) Permeability of crystalline rocks. *Eos Transactions AGU* 73(21), 233–237.

Clauser C. (2001) Update of the permeability of crystalline rocks, RWTH Aachen University, URL: http://www.eonerc.rwth-aachen.de/go/id/tsm/: site last visited 25/11/2012.

Crawford B.R., Myers R.D., Woronow A., Faulkner D.R., Rutter E.H. (2002), Porosity-permeability relationships in clay-bearing fault gouge. *SPE – Society of Petroleum Engineers, SPE/ISRM 78214*, 1–13.

Decker K., Meschede M., Ring U. (1993) Fault slip analysis along the northern margin of the Eastern Alps (Molasse, Helvetic nappes, North and South Penninic flysch, and the northern Calcareous Alps). *Tectonophysics* 223, 291–312.

Decker K., Peresson H. (1996) Tertiary kinematics in the Alpine-Carpathian-Pannonian system: links between thrusting, transform faulting and crustal extension. In: Wessely, G., Liebl, W. (editors) *Oil and Gas in Alpidic Thrustbelts and Basins of Central and Eastern Europe.* EAGE Special Publication, 5, 69–77.

Evans J.P., Forster C.B., Goddard J.V. (1997) Permeability of fault-related rocks, and implications for hydraulic structure of fault zones. *Journal of Structural Geology* 19, 1393–1404.

Faulkner D.R., Lewis A.C., Rutter E.H. (2003) On the internal structure and mechanics of large strike-slip fault zones, field observations of the Carboneras Fault in south-eastern Spain. *Tectonophysics* 367, 235–251.

Flodin E.A., Aydin A. (2004a) Evolution of a strike-slip fault network, Valley of Fire State Park, southern Nevada. *Geological Society of America Bulletin* 116, 42–59.

Flodin E.A., Aydin A. (2004b) Faults with asymmetric damage zones in sandstone, Valley of Fire State Park, southern Nevada. *Journal of Structural Geology* 26, 983–988.

Freeze R.A., Cherry J.A. (1979) *Groundwater.* Prentice-Hall, Englewood Cliffs, NJ.

Frisch W., Dunkl I., Kuhlemann J. (2000) Post-collisional orogen-parallel large-scale extension in the Eastern Alps. *Tectonophysics* 327, 239–265.

Frohlich R.K., Fisher J.J., Summerly E. (1996) Electric-hydraulic conductivity correlation in fractured crystalline bedrock: Central Landfill, Rhode Islands, USA. *Journal of Applied Geophysics* 35, 249–259.

Gringarten A.C., Bourdet D.P., Landel P.A., Kniazeff V J. (1979) Comparison between different skin and wellbore storage type curves for early time transient analysis. *SPE Paper 8205* (9), 11 pp.

Halihan T., Sharp Jr. J.M., Mace R.E. (1999) Interpreting flow using permeability at multiple scales. Karst Modeling: Proceedings of the symposium held Feb 24–27, 1999, Charlottesville, VA, Palmer, Palmer, and Sasowsky (editors), *Special Publication 5, Karst Waters Institute*, 82–96.

Halihan T., Mace,R.E., Sharp Jr., J.M. (2000) Flow in the San Antonio segment of the Edwards aquifer: matrix, fractures, or conduits? In: Wicks I.D., Sasowsky C.M. (editors) *Groundwater flow and contaminant transport in carbonate aquifers*, AA Balkema, 129–146.

Hesthammer J., Fossen H. (2001) Structural core analysis from the Gullfaks area, northern North Sea. *Marine Petroleum Geology* 18, 411–439.

Horne R.N. (1995) Modern Well Test Analysis – A Computer-Aided Approach. Petroway Inc., Palo Alto, CA

Jourde H., Flodin E.A., Aydin A., Durlofsky L.J., Wen X. (2002) Computing permeability of fault zones in eolian sandstone from outcrop measurements. *AAPG Bulletin* 86, 1187–1200.

Koch M., Schubnel A., Birk S., Winkler G. (2011) Architecture and hydraulic properties of a brittle fault core zone – example from the Semmering area (Austria). *Geophysical Research Abstracts* 13.

Lachassagne P., Pinault J.L., Laporte P. (2001) Radon 222 emanometry: a relevant methodology for water well sitting in hard rocks. *Water Resources Research* 37, 333–341.

Lee S.H., Lough M.F., Jensen C.L. (2001) Hierarchial modeling of flow in naturally fractured formations with multiple length scales. *Water Resources Research.* 37, 443–456.

Linzer H.G., Decker K., Peresson H., Dell'mour R., Frisch W. (2002) Balancing Lateral Orogenic Float of the Eastern Alps. *Tectonophysics* 354, 211–237.

Lunn R.J., Shipton Z.K., Bright A.M. (2008) How can we improve estimates of bulk fault zone hydraulic properties?. *Geological Society of London Special Publication* 299, 231–237.

Maclay R.W., Small T.A. (1987) Hydrostratigraphic subdivisions and fault barriers of the Edwards aquifer south central Texas, USA. *Journal of Hydrology* 61, 127–146.

Neuman S.P. (2005) Trends, prospects and challenges in quantifying flow and transport through fractured rocks. *Hydrogeology Journal* 13, 124–147.

National Research Council (1996): *Rock Fractures and Fluid Flow*. National Academy Press, Washington D.C.

Odling N.E., Harris S.D., Knipe R.J. (2004) Permeability scaling properties of fault zones in siliclastic rocks. *Journal of Structural Geology*, 26, 1727–1747.

ÖNORM B 4422-1 (1992) Erd – und Grundbau – Untersuchungen von Bodenproben – Bestimmung der Wasserdurchlässigkeit – Laborprüfungen.

Peres A.M.M., Onur M., Reynolds A.C. (1989) A new analysis procedure for determining aquifer properties from Slug test data. *Water Resources Research* 25(7), 1591–1602.

Priest S.D. (1993) *Discontinuity Analysis for Rock Engineering*, Edmundsbury Press Ltd, Burg St. Edmunds, Suffolk.

Ratschbacher L., Frisch W., Neubauer F., Schmid S.M., Neugebauer J. (1989) Extension in compressional orogenic belts: The eastern Alps. *Geology* 17, 404–407.

Ratschbacher L., Frisch W., Linzer H.-G. & Merle O. (1991) Lateral Extrusion in the Eastern Alps. Part 2: Structural Analyses. *Tectonics* 10(2), 257–271.

Rawling G.C., Goodwin L.B., Wilson J.L. (2001) Internal Architecture, Permeability Structure, and Hydrologic Significance of Contrasting Fault-Zone Types. *Geology* 29, 43–46.

Reichl P., Frieg B., Domberger G., Leis A., Winkler G. (2006) Hydraulic borehole tests in combination with hydrochemical and isotopic water sampling. *Felsbau (Rock Soil Eng.)* 24, 30–38.

Renard P., de Marsily G. (1997), Calculating equivalent permeability; a review; Upscaling in porous media; theories and computations, *Advances in Water Resources 20*, 253–278.

Riedmüller G., Brosch F.J., Klima K., Medley E. (2001) Engineering geological characterization of brittle faults and classification of fault rocks. *Felsbau (Rock Soil Eng.)* 19, 13–19.

Schmid S.M., Fügenschuh B., Kissling E., Schuster R. (2004) Tectonic map and overall architecture of the Alpine orogen. *Eclogae geologicae Helvetiae* 97, 93–117.

Schuster R., Kurz W. (2005) Eclogites in the Eastern Alps: High-pressure metamorphism in the context of Alpine orogeny. *Mitteilungen der Österreichischen Mineralogischen Gesellschaft* 150, 183–198.

Shipton Z.K., Evans J.P., Robeson K.R., Forster C.B., Snelgrove S. (2002) Structural heterogeneity and permeability in faulted eolian sandstone, implications for subsurface modeling of faults. *AAPG Bulletin* 86, 863–883.

Singhal B.B.S., Gupta R.P. (1999) *Applied Hydrogeology in Fractured Rocks*. Kluwer Academic Publishers, Dodrecht/The Netherlands.

Stober I. (1986) Strömungsverhalten in Festgesteinsaquiferen mit Hilfe von Pump – und Injektionsversuchen. Geol. Jahrb. C., Heft 42, Hannover.

Stober I., Bucher K. (2007) Hydraulic properties of crystalline basement. *Hydrogeology Journal* 15, 213–224.

Storti F., Billi A., Salvini F. (2003) Particle size distributions in natural carbonate fault rocks; insights for non-self-similar cataclasis. *Earth Planetary Science Letters* 206, 173–186.

Wang X., Neubuer F. (1998) Orogen-parallel strike-slip faults bordering metamorphic core complexes: the Salzach-Enns fault zone in the Eastern Alps, Austria. *Journal of Structural Geology*, 20, 799–818.

Wen X., Gomez-Hernandez J.J. (1996) Upscaling hydraulic conductivity in heterogeneous media; an overview; Effective parameter estimation for flow and transport in the subsurface. *Journal of Hydrology*, 183, ix–xxxii.

Winkler G., Kurz W., Hergarten S., Kiechl E. (2010) Hydraulische Charakterisierung von Störungskernzonen in kristallinen Festgesteinen am Beispiel der Talhof-Störung (Ostalpen). *Grundwasser* 15(1), 59–68.

Woodcock N.H., Dickson J.A.D., Tarasewics J.P.T. (2007) Transient permeability and reseal hardening in fault zones: Evidence from dilation breccia textures. *Geological Society Special Publication*, 43–53.

Chapter 12

Methodology to generate orthogonal fractures from a discrete, complex, and irregular fracture zone network

Stefano D. Normani, Jonathan F. Sykes & Yong Yin
Department of Civil and Environmental Engineering,
University of Waterloo, Waterloo, Ontario, Canada

ABSTRACT

Dual continuum computational models that include both porous media and discrete fracture zones are valuable tools in assessing groundwater flow and pathways in fractured rock systems. Fracture generation models can produce realisations of discrete fracture networks that honour geological structures and fracture propagation behaviours. A methodology and algorithm are presented to incorporate a discrete, complex and irregular fracture zone network, represented as a triangulated two-dimensional mesh, within an orthogonal three-dimensional finite-element mesh. Orthogonal fracture faces, between adjacent finite-element blocks, were used to represent the irregular discrete-fracture zone network. A detailed coupled density-dependent groundwater flow analysis of a hypothetical 104 km^2 portion of the Canadian Shield has been conducted using the discrete-fracture dual continuum finite-element model FRAC3DVS-OPG to investigate the importance of large-scale fracture zone networks on flow and transport. Surface water features and a Digital Elevation Model (DEM) were applied in a GIS framework to delineate the sub-watershed and to populate the finite-element mesh. The crystalline rock between the structural discontinuities was assigned properties characteristic of those reported for the Canadian Shield. Total Dissolved Solids (TDS) concentrations of 300 g/l are encountered at depth. A comparison is made to a model which represents the fracture zones as an Equivalent Porous Medium (EPM). Differences in freshwater heads, TDS, and pore water velocities in fracture zones are demonstrated. The discrete fracture zone approach yields simulated TDS values which are lower at depth than TDS values simulated using the EPM approach, primarily due to higher simulated pore water velocities within the discrete fracture zones. Solute transport is sensitive to fracture zone characterisation and dispersivity values required to achieve stable solutions. Steady-state density-independent simulations using both approaches yield similar freshwater heads.

12.1 INTRODUCTION

Following from the work of Sykes *et al.* (2003) in which a 5734 km^2 regional scale model of the Canadian Shield was developed, a detailed sub-regional scale coupled density-dependent groundwater flow analysis of a 104 km^2 portion of the Canadian Shield (Figure 12.1) has been conducted. This case study is hypothetical in the sense that no site specific field data are available, but the model was constructed using an assemblage of topography and parameters which are representative of a Canadian

Shield setting. The discrete-fracture dual continuum finite-element computational model FRAC3DVS-OPG (Therrien *et al.*, 2010) is used to investigate the importance of large-scale fracture zone networks on flow and transport in deep geologic settings (Normani *et al.*, 2007b). FRAC3DVS-OPG represents the continued development of FRAC3DVS (Therrien & Sudicky, 1996) and includes coupled density-dependent flow and transport, thermal transport, groundwater age and life expectancy, one-dimensional hydromechanical coupling and sub-gridding.

Various approaches (Selroos *et al.*, 2002; Dershowitz *et al.*, 2004; Neuman, 2005) have been developed to accommodate fracture zones within modelling domains. The Discrete Fracture Network (DFN) approach includes 2-dimensional fracture elements within a grid of 3-dimensional matrix elements. The Equivalent Porous Media (EPM) or continuum approach modifies the properties of matrix elements to approximate flow and transport behaviour due to the presence of fractures. A third approach is to treat the matrix domain and the fracture domain as two independent, overlapping and linked domains, using a dual-porosity approach (Gerke & van Genuchten, 1993). These approaches are applied depending on the scale of the fracture network relative to the matrix as represented in a numerical model. Where the scale of a grid block element in a numerical model is such that it includes a multitude of fractures, then it may be reasonable to apply an EPM or continuum approach for flow and transport simulation (Dershowitz *et al.*, 2004; Neuman, 2005). In the context of this chapter,

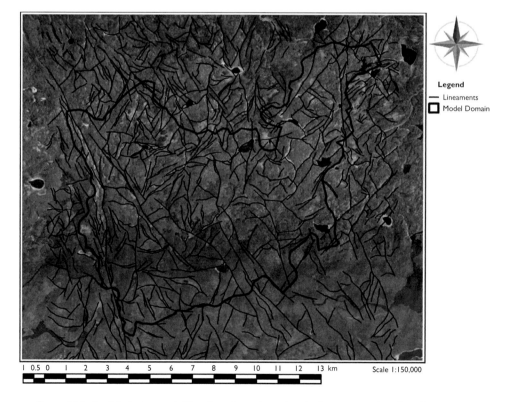

Figure 12.1 Aerial photo, modelling domain, and stochastically generated surface lineaments.

the scale of discrete features is restricted to discrete fracture zones that represent significant pathways for fluid migration on the scale of hundreds of metres to thousands of metres.

Discrete fracture networks are commonly generated stochastically within a two- or three-dimensional space. Rouleau & Gale (1987) generated line elements within a two-dimensional cross-section, while Long et al. (1985), used circular orthogonal disks in a three-dimensional space. The stochastic generation of fracture zones can also be conditioned on field data to allow site characterisation activities to continually update and enhance the discrete Fracture Zone Network Model (FZNM), whether by surface lineament data, or seismic interpretation.

A surface lineament analysis using aerial photography was conducted by Srivastava (2002) to define the major fracture features for the FZNM, which are mainly coincident with surface drainage features that exhibit linearity. Additional surface lineaments were generated to extend existing major observed lineaments and to increase the fracture density in areas where overburden cover would obscure surface lineaments. The generated surface fracture features are shown in Figure 12.1, and are based on the lineament and fracture statistics for the Lac du Bonnet region of the Whiteshell Research Area, near Pinawa, Manitoba, Canada (Sikorsky et al., 2002). The observed statistics of fracture zone orientation, fracture zone length, and area density distribution statistics are preserved. The generated fractures, conditioned on observed features, thus represent both sensible and geomechanically plausible fracture behaviour (Srivastava, 2002).

To create three-dimensional curve-planar fracture zones, the stochastically generated surface lineaments shown in Figure 12.1 are propagated to depth until one of the following conditions are met:

- The fracture zone's down-dip width reaches a prescribed length to width ratio.
- The fracture zone truncates against an existing fracture zone.
- The fracture zone reaches the edge or bottom of the modelled domain.

A network of 980 discrete curve-planar fracture zones was generated in a similar manner to Srivastava (2002) for the sub-regional domain shown in Figure 12.1. The generated FZNM is one realisation of many possible fracture zone network models that could be generated for the sub-regional domain. Fracture network density decreases with increasing depth and minor fracture features are shallower than major fracture features. The resulting FZNM contains a high degree of realism that honours many geological, statistical, and geomechanical constraints (Srivastava, 2002).

12.1.1   Methodology to generate an orthogonal Fracture Zone Network Model

The geometry of an individual curve-planar FZNM fracture is described by a mesh of triangular facets (Figure 12.2a). This approach can also be used to assign spatially variable fracture properties, although this was not done in this study. To be used in a groundwater model, the curve-planar FZNM must be represented geometrically using two-dimensional elements. One approach is to generate a discretisation which exactly honours the curve-planar FZNM using tetrahedral elements for the three-dimensional elements, and triangles for the two-dimensional elements. Alternatively, two-dimensional quadrilateral elements can be chosen which occur between adjacent hexahedral elements. This alternate approach was selected for this study as the same three-dimensional hexahedral mesh can be used for both the discrete and EPM approaches. An advantage of using the orthogonal approach is the robustness in generating the two-dimensional elements, as compared to tetrahedral mesh generator.

Software was written in Microsoft Visual Basic for Applications (VBA) to calculate the orthogonal grid block or element faces that best represents or maps onto each curve-planar FZNM fracture. The procedure used to generate the orthogonal FZNM from a triangulated FZNM (Figure 12.2a) is as follows:

- The FZNM data file is structured as one text line per triangular facet, comprised of a fracture number, and three coordinate triples (x, y, depth) representing each node of the triangular facet.
- The FZNM file is processed to produce a new FZNM file in which the triangular facets are represented by nodal coordinates and triangular element indices. Elevation is calculated by subtracting depth from the ground surface elevation at that location using

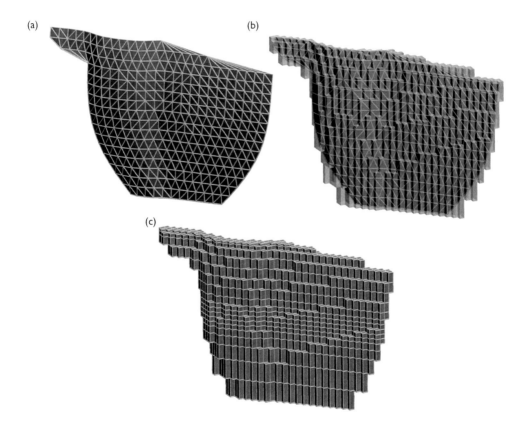

Figure 12.2 (a) View of a single FZNM fracture zone; (b) View of the single fracture zone and the model grid blocks that are intersected by the fracture zone; and (c) View of the model grid orthogonal faces that best represent the fracture zone. (See *colour plate section, Plate 35*).

a Digital Elevation Model (DEM). The ordering (clockwise or counter-clockwise) of element indices are such that they match the ordering of the first triangular element of a fracture. A fracture is comprised of numerous triangular elements.

- Process the FRAC3DVS-OPG model domain grid to determine face numbers (planar elements) associated with each hexahedral element based on an (i, j, k) ordering scheme.
- Read all triangular elements associated with a given fracture.
- Search for all grid hexahedral elements that intersect a given triangular element.
- Conduct this search for all triangular elements associated with a fracture number to produce a list of hexahedral elements which are intersected by the triangulated fracture, and output the list to a binary file. Figure 12.2b shows the selected hexahedral elements.
- Loop over chosen hexahedral elements to select all exterior quadrilateral faces.
- Select a contiguous subset of quadrilateral faces that are either on one side of the triangulated fracture or on the other side, and output to a binary file.
- Repeat the above process for all fractures in a FZNM file.
- Generate separate ASCII text files of hexahedral elements intersected by the triangulated fractures, and face numbers of quadrilateral elements representing orthogonal fracture faces for use in FRAC3DVS-OPG.

Figure 12.2c shows the orthogonal grid block faces that best represent the FZNM fracture. The stepped nature of the orthogonal discretization accommodates both the dip and orientation of the original curve-planar FZNM fracture.

Although the orthogonal FZNM contains quadrilateral fracture elements which extend to the bottom of the modelling domain, no fracture elements are included which occur within or are adjacent to the bottom layer of hexahedral elements. This was done to ensure the bottom layer of elements in the modelling domain were solely hexahedral. Horizontal quadrilateral fracture elements were also removed from the top of the modelling domain due to their high permeabilities and resulting high fracture velocities. Transport simulations did not yield stable solutions in these portions of the domain and thus these fracture elements were removed to enhance numerical stability when simulating density-dependent flow and transport.

12.2   NUMERICAL MODEL DEVELOPMENT

The governing equations for density-dependent groundwater flow, solute transport, and constitutive relationships for density and concentration are presented by Bear (1988) and Frind (1982). The implementation of these equations in the FRAC3DVS-OPG computational model using the control volume finite-element method is described by Therrien *et al.* (2010).

The sub-regional modelling domain outlined in Figure 12.1 is defined along surface water divides associated with topographic highs for the eastern, northern, and western portions of the modelling domain, and along a river to the south. The three-dimensional sub-regional domain was discretised into approximately 850 000 nodes, and 790 000 brick elements, covering an area of 104 km^2. The numerical grid is orthogonal, with each grid block having the same lateral dimensions of 50 m by 50 m, and discretised vertically into 19 layers. This numerical model has been previously used to investigate fluid migration (Normani *et al.*, 2007b), paleohydrogeologic behaviour (Normani *et al.*, 2006, 2007a; Normani, 2009; Normani & Sykes, 2010, 2011), and implementation of mean life expectancy (Normani *et al.*, 2007; Park *et al.*, 2008).

Within the model domain, surface water features, such as lakes, rivers, and wetlands, were defined using Dirichlet boundary conditions. The elevation of these water features and of the top layer of the numerical model were interpolated from a Digital Elevation Model (DEM). An average steady-state net recharge of 1 mm/a was applied to the top of the modelling domain. Higher values of net recharge yielded water tables significantly above topography. The use of net recharge reflects the fact that a large fraction of the point recharge will discharge at rivelets, swales and other small surface water features that cannot be defined within the 50 m by 50 m grid blocks used in the study. Very low recharge rates of a few millimetres per year in crystalline rock settings have been reported by Singer & Cheng (2002). A zero-flux boundary condition was used for the bottom and sides of the model domain.

A steady-state freshwater simulation is first executed without Total Dissolved Solids (TDS). The resulting freshwater heads from the steady-state simulation are adjusted to account for the presence of TDS where present. The adjusted freshwater heads and initial TDS distribution, described below, form the initial conditions for a 1 Ma transient simulation, whose final state at 1 Ma represents a pseudo steady-state between freshwater heads and the TDS distribution. Due to the lack of a source term for TDS, running the model to steady-state will linearise the TDS distribution between the Dirichlet boundary condition at the bottom of the model set to a TDS of 300 g/l and the top of the model, which is continually flushed with fresh meteoric water. The resulting linear distribution in a steady-state simulation would not represent measured TDS versus depth profiles. The 1 Ma time period is sufficient to flush the near surface system of TDS and to allow fresh meteoric waters to enter fracture zones, but the time period is not sufficient to significantly alter the TDS versus depth profile in the rock matrix.

A matrix porosity of 0.003 was chosen based on the work of Stevenson (1996). Fracture zones are assumed to be comprised of highly fractured rock; with a discrete fracture aperture spacing of 0.1 m, and a fracture aperture that varies from 40 microns near surface to 4 microns at depth. Fracture apertures are computed based on uniformly spaced discrete fractures within a fracture zone, the matrix permeability, and the fracture zone permeability (Normani *et al.*, 2007b). Fracture zone porosities are calculated based on the fracture zone permeability and the same fracture aperture spacing of 0.1 m within the fracture zone; resulting fracture zone porosities range from 0.003 to 0.0042 (Normani *et al.*, 2007b). Matrix properties for solute transport include a longitudinal dispersivity of 50 m, a transverse horizontal dispersivity of 5 m and a transverse vertical dispersivity of 0.5 m. A matrix bulk density of 2642 kg/m^3 is also used. Horizontal and vertical permeabilities for the matrix are defined with respect to depth d as (Normani *et al.*, 2007b):

$$k_H = 10^{-14.5-4.5(1-e^{-0.00247d})}$$

$$k_V = \begin{cases} 10\,k_H & \text{for } d \leq 300 \text{ m} \\ (0.09(400-d)+1)k_H & \text{for } 300 < d \leq 400 \text{ m} \\ k_H & \text{for } d > 400 \text{ m} \end{cases} \tag{1}$$

The uppermost 10 m thick layer is set to $k_H = 1.0\text{e-}13$ m^2 and the adjacent 20 m thick layer is set to $k_H = 1.0\text{e-}14$ m^2. An initial TDS distribution is required for the pseudo steady-state model. Based on Figure 2b in Frape & Fritz (1987), an upper bound for TDS as a function of depth can be developed (see Normani, 2009) whereby:

$$TDS = 10^{0.001982d} \quad \text{for } d \leq 1250 \text{ m}$$
$$TDS = 300 \text{ g/L} \quad \text{for } d > 1250 \text{ m} \tag{2}$$

Setting an upper bound for the TDS distribution as an initial condition allows recharge waters to reduce the TDS concentrations in fracture zones, and adjacent rock matrix during the pseudo steady-state simulations. The free solution diffusion coefficient for brine calculations is 1.5e-9 m^2/s representing NaCl at 1 mol/l (Weast, 1983).

Fracture zone permeabilities are defined on a depth dependent basis, ranging from 1.0e-16 m^2 at depth to 1e-13 m^2 at ground surface using the 50th percentile of a cumulative density function developed to describe fracture zone permeability versus depth as described in Normani *et al.* (2007b), Park *et al.* (2008) and Normani (2009). Fracture zone widths are arbitrarily assumed to be 1 m; the method can readily accommodate spatially varying fracture zone width in the transmissivity. Fracture zone properties for solute transport include a longitudinal dispersivity of 250 m, and a transverse dispersivity of 25 m. The longitudinal fracture zone dispersivity was increased until stable numerical solutions for density-dependent flow and transport were obtained. Higher dispersivities tend to enhance the stability of the FRAC3DVS-OPG numerical solution and to reduce the computational runtime.

In an Equivalent Porous Medium (EPM) approach, both the fracture zone porosity and fracture zone permeability contribute to a blended porosity and permeability, respectively, to represent both domains. The blended porosity n_{EPM} is calculated as the volumetrically weighted average of the porosity for the matrix block width w_M and the fracture zone width w_F. The blended permeability k_{EPM} is calculated using an arithmetic mean independent of fracture element orientation because any number of fracture zones can intersect a single EPM block (hexahedral element) at any angle. The conservative approach is to assume the permeability along each principal direction of the single EPM block is affected and each of k_x, k_y, and k_z are modified using the arithmetic mean approach. Subscripts F and M indicate fracture zone and matrix respectively.

$$n_{EPM} = \frac{w_F n_F + (w_M - w_F)n_M}{w_M} \qquad k_{EPM} = \frac{w_F k_F + (w_M - w_F)k_M}{w_M} \tag{3}$$

12.3   DISCUSSION

The discrete FZNM, and the equivalent porous media model are compared using the performance measures of freshwater head, total dissolved solids concentration, and pore-water velocity magnitude in Figures 12.3–12.5, respectively. Figure 12.6 presents freshwater heads for steady-state density-independent models. The upper images in Figures 12.3–12.6 represent the planar elements of the fracture network model as

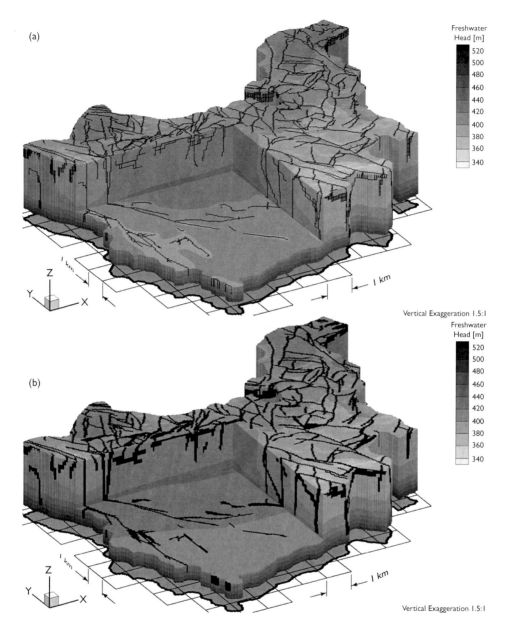

Figure 12.3 Freshwater head plots for (a) discrete fracture zone network model, and (b) equivalent porous media model.

lines, whereas the lower images represent the EPM or block-like nature of the fracture network model.

Figure 12.3 shows the freshwater heads for the two models. The block-cut view extends to a depth of approximately 1000 m. The results are generally quite similar, however, differences are apparent in the vicinity of fracture zones. In the case of

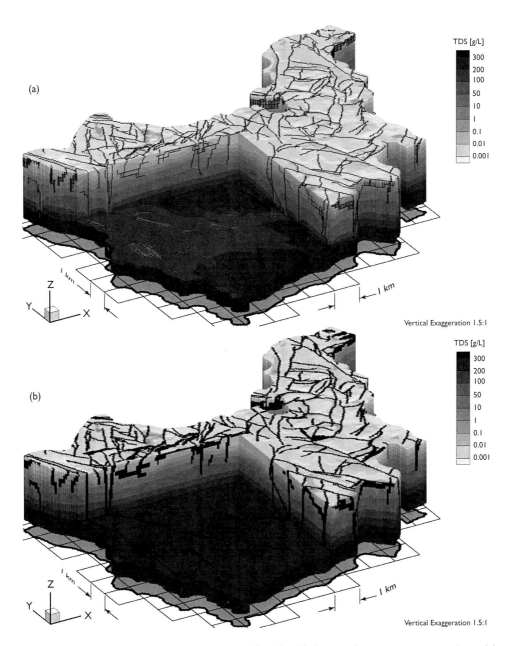

Figure 12.4 Total dissolved solids concentration plots for (a) discrete fracture zone network model, and (b) equivalent porous media model.

Figure 12.3a, representing the discrete FZNM, fracture elements occur between adjacent hexahedral elements and are shown using lines. In Figure 12.3b, the EPM blocks containing both matrix and fracture zones are indicated by darker coloured blocks. The freshwater heads are affected by the TDS distribution with higher freshwater

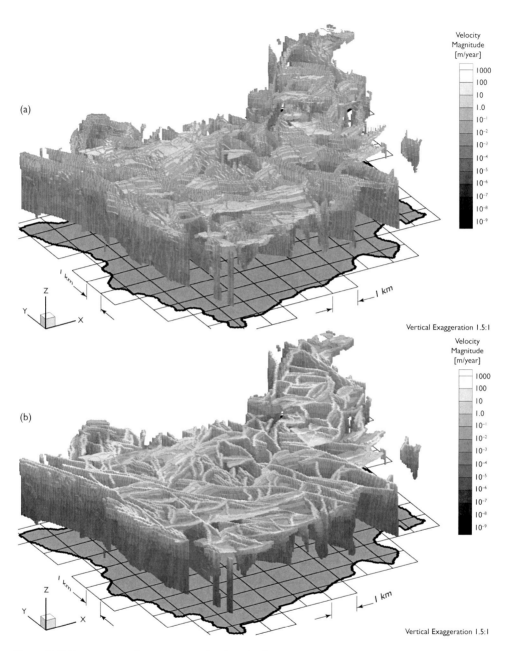

Figure 12.5 Pore-water velocity magnitude plots for fracture zones associated with (a) discrete fracture zone network model, and (b) equivalent porous media model.

heads for the EPM model as compared to the discrete FZNM. The lower freshwater heads, shown in Figure 12.3a, at a depth of approximately 1000 m are due to reduced total dissolved solids concentrations in the vicinity of the fracture zones as shown in Figure 12.4a, while the EPM approach in Figure 12.4b does not appear to show an

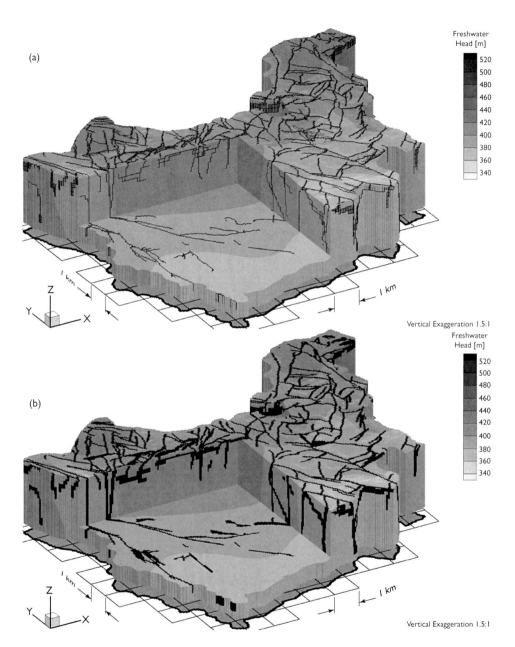

Figure 12.6 Freshwater head plots for steady-state and density-independent (a) discrete fracture zone network model, and (b) equivalent porous media model.

impact on TDS. Both up-coning and down-coning of TDS, while visible in the discrete FZNM, is absent in the EPM model. Fracture zone pore water velocities are shown in Figure 12.5a for the discrete FZNM and in Figure 12.5b for the EPM model. Higher velocities occur in the discrete FZNM as compared to the EPM model, especially in the lower half of the modelling domain; these higher velocities will tend to flush out the TDS in the fracture system more quickly than the EPM model.

Figure 12.6 shows freshwater heads for both the discrete FZNM and EPM block approaches using steady-state density-independent models. The freshwater heads are nearly identical for both approaches. In comparison to Figure 12.3, the freshwater heads in Figure 12.6 are not affected by the total dissolved solids distribution, which is different between the two approaches as shown in Figure 12.4. Although freshwater heads are very similar between both approaches in the no-TDS case, significant differences in freshwater heads occur due to density-dependent coupling of flow and solute transport, primarily due to higher simulated fracture zone pore fluid velocities in the discrete FZNM approach. These higher fracture zone velocities affect the distribution of TDS both in fracture zones and the matrix.

12.4   CONCLUSIONS

A methodology for mapping realistic fracture zone networks into discrete fracture zones using orthogonal elements, and an equivalent porous media approach is described. The importance of characterising high conductivity pathways is demonstrated using freshwater heads, total dissolved solids, and pore water velocity magnitudes. Total dissolved solids migration, governed by advection-diffusion processes, is sensitive to higher velocity pathways that can reduce both concentrations within fracture zones and the adjacent matrix (see Figure 12.4) and also reduce the transit time for an average water particle through fracture zones to the surface water system. A steady-state density-independent simulation for both approaches yields simulated freshwater heads that are nearly identical, as shown in Figure 12.6. This can give the misleading impression that a coupled density-dependent groundwater system behaves in a similar manner. Freshwater heads in a density-dependent groundwater system depend on the TDS distribution, which itself depends not solely on head gradients, but also on pore water velocities and the characterisation of fracture zone permeabilities.

Variations in freshwater heads, total dissolved solids concentrations, and pore-water velocities between the discrete approach and the EPM approach are related to discretisation and parameter values used to populate the numerical model. In the case of the discrete Fracture Zone Network Model (FZNM), large longitudinal dispersivities of 250 m for fracture elements were needed to provide a stable numerical solution due to higher groundwater flow velocities in the discrete FZNM as compared to the EPM model. The lower velocities in EPM blocks combined with lower dispersivities in the EPM model will result in more TDS remaining in the fracture zone network, and higher freshwater heads at the end of the simulation. Using the same computational grid as the discrete approach yields EPM grid blocks that are large. This tends to reduce effective hydraulic conductivities, and hence groundwater flow velocities that are important in solute transport. A calibration exercise focussed on determining EPM parameters that yield an equivalent behaviour to the discrete approach can be attempted. Large dispersivities may be reduced by considering alternate spatial discretisations and permeability depth relationships for fracture zones. Alternate numerical schemes can be attempted to minimize the need for large dispersivities to dampen oscillatory behaviour in simulating advective transport in fracture zones. As shown in this work, the use of a discrete FZNM requires consideration of the importance of advective transport and adequate parameterisation of fracture zones to yield representative and stable simulations.

ACKNOWLEDGEMENTS

Funding for this research was provided by the Nuclear Waste Management Organization (NWMO) of Canada.

REFERENCES

Bear J. (1988) *Dynamics of Fluids in Porous Media*. (Dover editor), Dover Publications Inc.

Dershowitz W.S., La Pointe P.R., Doe T.W. (2004) Advances in Discrete Fracture Network Modeling. Proceedings of the US EPA/NGWA Fractured Rock Conference, Portland, Maine, 882–894.

Frape S.K., Fritz P. (1987) Geochemical trends for groundwaters from the Canadian Shield, in Saline Water and Gases in Crystalline Rocks, edited by P. Fritz and S.K. Frape, no. 33 in Geological Association of Canada Special Paper, pp. 19–38.

Frind E.O. (1982) Simulation of long-term transient density-dependent transport in groundwater. *Advances in Water Resources* 5(2), 73–88.

Freeze R.A., Cherry J.A. (1979) *Groundwater*. Prentice-Hall, Inc., Englewood Cliffs, N.J.

Long J.C.S., Gilmour P., Witherspoon P.A. (1985) A Model for Steady Fluid Flow in Random Three-Dimensional Networks of Disc-Shaped Fractures. *Water Resources Research* 21(8), 1105–1115.

Neuman S.P. (2005) Trends, prospects and challenges in quantifying flow and transport through fractured rocks. *Hydrogeology Journal* 13, 124–147.

Normani S.D., Sykes J.F., Sudicky E.A., McLaren R.G. (2003) Modelling strategy to assess long-term sub-regional scale groundwater flow within an irregular discretely fractured Canadian Shield setting, in 4th Joint IAH-CNC and CGS Groundwater Specialty Conference, 8pp., Winnipeg, Canada, Sep. 29–Oct. 1, 2003.

Normani S.D., Sykes J.F., Sudicky E.A., Jensen M.R. (2006) Effects of paleoclimate boundary conditions on regional groundwater flow in discretely fractured crystalline rock, in Proceedings of the XVIth International Conference on Computational Methods in Water Resources, edited by P.J. Binning, P. Engesgaard, H. Dahle, G.F. Pinder, and W.G. Gray, 8 pp., Copenhagen, Denmark, Jun. 18–22, 2006, paper 223.

Normani S.D., Sykes J.F., Sudicky E.A. (2007a) Effects of glaciation on groundwater flow in discretely fractured crystalline rock, in 60th Canadian Geotechnical Conference & 8th Joint CGS/IAH-CNC Groundwater Conference, 8pp., Ottawa, Canada, Oct. 21–24, 2007, paper 148.

Normani S.D., Park Y.-J., Sykes J.F., Sudicky E.A. (2007b) Sub-regional modelling case study 2005–2006 status report, Technical Report NWMO TR-2007–07, Nuclear Waste Management Organization, Toronto, Canada, 136pp., Nov. 2007.

Normani S.D. (2009) Paleoevolution of pore fluids in glaciated geologic settings, Ph.D. thesis, University of Waterloo, Waterloo, Ontario, Canada.

Normani S.D., Sykes J.F. (2010) Paleoclimate analyses of density-dependent groundwater flow with pseudo-permafrost in discretely fractured crystalline rock settings, in XVIII International Conference on Water Resources – CMWR 2010, edited by J. Carrera, Barcelona, Spain, Jun. 21–24, 2010.

Normani S.D., Sykes J.F. (2011) Sensitivity analysis of a coupled hydro-mechanical paleo-climate model of density-dependant groundwater flow in discretely fractured crystalline rock, in Waste Management, Decommissioning and Environmental Restoration for Canada's Nuclear Activities: *Current Practices and Future Needs*, Canadian Nuclear Society, Toronto, Canada, Sep. 11–14, 2011.

Park Y.-J., Cornaton F.J., Normani S.D., Sykes J.F., Sudicky E.A. (2008) Use of groundwater lifetime expectancy for the performance assessment of a deep geologic radioactive waste repository: 2. Application to a Canadian Shield environment. *Water Resources Research* 44, W04407.

Rouleau A., Gale J.E. (1987) Stochastic discrete fracture simulation of groundwater flow into an underground excavation in granite. *International Journal of Rock Mechanics and Mining Sciences & Geomechanics* 24(2), 99–112.

Selroos J.-O., Walker D.A., Ström, B., Gylling A., Follin F. (2002) Comparison of alternative approaches for groundwater flow in fractured rock. *Journal of Hydrology* 257, 174–188.

Singer S.N., Cheng C.K. (2002) An assessment of the groundwater resources of Northern Ontario: Areas draining into Hudson Bay, James Bay and Upper Ottawa River, Hydrogeology of Ontario Series: Report 2, Ministry of the Environment: Environmental Monitoring and Reporting Branch, Toronto, Canada.

Sikorsky R.I., Serzu M., Tomsons D., Hawkins J. (2002) A GIS-based lineament interpretation method and case study at the Whiteshell Research Area, Technical Report 06819-REP-01200–10073-R00, Ontario Power Generation, Nuclear Waste Management Division, Toronto, Canada.

Srivastava R.M. (2002) The discrete fracture network model in the local scale flow system for the Third Case Study, Technical Report 06819-REP-01300–10061-R00, Ontario Power Generation, Nuclear Waste Management Division, Toronto, Canada.

Stevenson D.R., Brown A., Davison D.D., Gascoyne M., McGregor R.G., Ophori D.U., Scheier N.W., Stanchell F., Thorne, A., Tomsons D.K. (1996) A revised conceptual hydrogeologic model of a crystalline rock environment, Whiteshell Research Area, southeastern Manitoba, Canada, Technical Report AECL-11331, COG-95–271, Atomic Energy of Canada Limited, Whiteshell Laboratories, Pinawa, Manitoba, Canada.

Sykes J.F., Normani S.D., Sudicky E.A. (2003) Regional scale groundwater flow in a Canadian Shield setting, Technical Report 06819-REP-01200–10114-R00, Ontario Power Generation, Nuclear Waste Management Division, Toronto, Canada, 46pp., Sep. 2003.

Therrien R., Sudicky E.A. (1996) Three-dimensional analysis of variably-saturated flow and solute transport in discretely-fractured porous media. *Journal of Contaminant Hydrology* 23(1–2), 1–44.

Therrien R., McLaren R.G., Sudicky E.A., Panday S.M., Guvanasen V. (2010), FRAC3DVS-OPG – A Three-dimensional Numerical Model Describing Subsurface Flow and Solute Transport. Groundwater Simulations Group, University of Waterloo, Waterloo, Ontario, Canada.

Weast R.C. (Ed.). 1983. *CRC Handbook of Chemistry and Physics*. 64th edition. CRC Press, Inc., Boca Raton, Florida, USA.

Chapter 13

Remote sensing, geophysical methods and field measurements to characterise faults, fractures and other discontinuities, Barada Spring catchment, Syria

Florian Bauer & Jens Harold Draser
Department of Applied Geosciences, Berlin Institute of Technology,
Berlin, Germany

ABSTRACT

The Syrian capital Damascus obtains its drinking water from several sources: inner-city wells, the Figeh Spring and the Barada Spring with its surrounding wells. The catchment of the Barada Spring consists of limestone and dolomite outcrops at the core of two Palmyrides folds. The thickness of these formations can exceed 1000 m. The Serghaya Fault to the east of the springs is the fundamental tectonic feature in this area. It is a branch fault of the Dead Sea Transform Fault. A sinistral movement along this fault has been active since the Neogene period. Tectonic measurements and the analysis of high resolution remote sensing data show the transform fault network and block rotation in the surroundings of the main fault and the Horst structures. These can be explained by the Riedel model for simple shear. The estimated angles of block rotation rise to 53° in a counter-clockwise direction. In the vicinity of the Barada Spring the formations are intensively faulted and block rotated, whereas the south is less faulted and lacks this tectonic feature. Surrounding the Barada Spring four tracer injections, three Very Low Frequency-Electro-Magnetic (VLF-EM) soundings, two geoelectrical profiles and one geoelectrical tomography in the dried spring lake have been carried out by different authors. A VLF (Very Low Frequency) sounding carried out in 2007 and an Audio Magneto Telluric (AMT) investigation in 2004. Geophysical structures trend in south east to north west and west to east directions and possess high electrical conductivities, indicating the probability of a water-bearing structure. An intersection of both trends occurs in the Barada Spring Lake. In combination, almost all methods show a dependency of possible groundwater flow towards the main faults in the vicinity of the spring. Further investigations are to be done to delineate the borders of the catchment. AMT is the most efficient geophysical method for the deep water table in the carbonatic rocks. Ancillary geoelectrical methods yield interpretable results within shallow water tables. All geophysical and remote sensing methods in combination with in situ measurements of strike and dip give an understanding of faults, fissures and discontinuities. A proved flow direction can only be estimated, field measurements of the head of the water table could lead to a more precise interpretation.

13.1 INTRODUCTION

Water quality in fissured and karstified aquifers is a significant problem especially in water scarce regions, due to the high velocity of groundwater flow and surface infiltration through a thin soil cover (Zwahlen, 2003; Andreo *et al.*, 2006). The arid climate coupled with a population growth rate between 1.3% for Damascus city and 3.4% for rural Damascus in the last decade (SSB, 2011) makes drinking water management and the protection of the sources essential.

The Syrian capital city, Damascus, obtains its drinking water from several sources: inner-city wells, Figeh Spring and Barada Spring with its surrounding wells (JICA, 1997). The purpose of the hydrogeological investigations in the last years was to protect these sources (Hennings & Wolfer, 2008; JICA, 1997).

In fissured aquifers the direction of flow is linked to the direction and the hydraulic aperture of the fissures which depend on tectonic forces, both current and historical. Understanding tectonic forces is a useful aid and may provide some meaningful site data.

There are several investigations using geophysics and tracer tests that realized, with different results, that the hydraulic connections of the fissured rock of the catchment are complex and up to now relatively unexplored (Wolfer, 2008; JICA, 1997; Alammareen, 2010; Zwahlen, 2010). This chapter focuses on the tectonic generation of the fissures and faults and the resulting groundwater flow directions. By combining several methods an understanding some of the local groundwater flow pattern can be gained and data deficits identified.

13.2 STUDY AREA

13.2.1 Location

The Anti-Lebanon Mountain belt is the principle recharge area in this region and feeds several rivers, including the Barada River, which flows through Damascus city (Figure 13.1). The catchment area of the Barada Spring is located in the west of the Zabadani Valley which lies in the mountain range about 20 km north west of the Syrian capital Damascus. The Barada Spring itself has its spring lake in front of the Chir Mansour Mountain range near the small village of Rawda in the south of Zabadani city. It is surrounded by several pumping wells which supply drinking water to Damascus (JICA, 1997). The estimated size of the catchment area of the Barada Spring is more than 90 km² and covers, as a transboundary aquifer (Puri & Aureli, 2009), mostly Syrian but also Lebanese territories.

The highest elevation at roughly 1900 m a.s.l. is the Chir Mansour Mountain. The Barada Spring is at the lowest point at about 1100 m. The mountain ridge has very steep slopes at its western and eastern margin and runs from north to south. The southern catchment boundaries are not clearly defined. In the south it is adjacent to smaller catchments (Wagner, 2011). To the north it is superseded by younger deposits and their catchments.

13.2.2 Geologic setting

The Zabadani area is a sub-basin of the Damascus Basin which covers parts of the Anti-Lebanon Mountains in the north, the Damascus Plain and the basaltic areas in the south of Syria (LSI, 1968). The north of the Damascus Basin is dominated by sedimentary deposits. Strata is generally increasing in age towards north east.

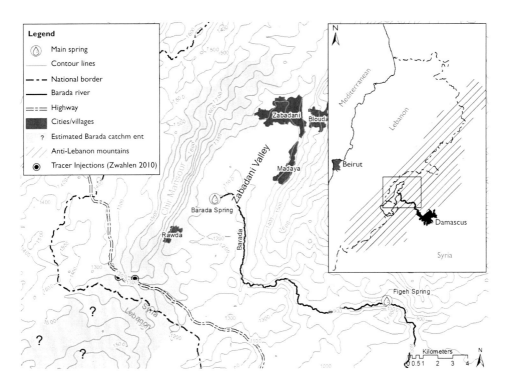

Figure 13.1 Location of the study area.

In this sequence, the Zabadani Basin is the transition between cretaceous karstified limestones of the Figeh Spring catchment and the Jurassic fissured and partly karstified limestones of the Barada Spring catchment, especially the Chir Mansour mountain ridge (Ponikarov, 1968b). The Jurassic mountains are the oldest outcropping formations in this region. Further to the north east, the age of sedimentation gets younger again (Figure 13.2).

The mountain ridge is the core of a north-north-east to south-south-west trending anticline structure. In its centre it consists of Middle Jurassic (Callovian) limestones which are Ca-rich. Dolomitisation along faults is common (Ponikarov, 1968b). The youngest Upper Jurassic deposits which outcrop at the flanks of the ridge are limestones of the Tithonian age. Between these stages all rocks are mineralogically pure limestones with minor variations in their clay content. The thickness of the Jurassic sequence is more than 1000 m, thus underlying formations were not penetrated by drilling. The overlying Cretaceous deposits consist of limestone, interbedded with thin layers of sandstones and some basaltic intrusions. The limestone have a higher clay content than the Jurassic rocks. The depression of the central Zabadani Valley is dominated by sand and gravel, derived from the surrounding Mesozoic formations.

The calcitic Jurassic rocks show a high level of karstification at the land surface which decreases rapidly in depth.

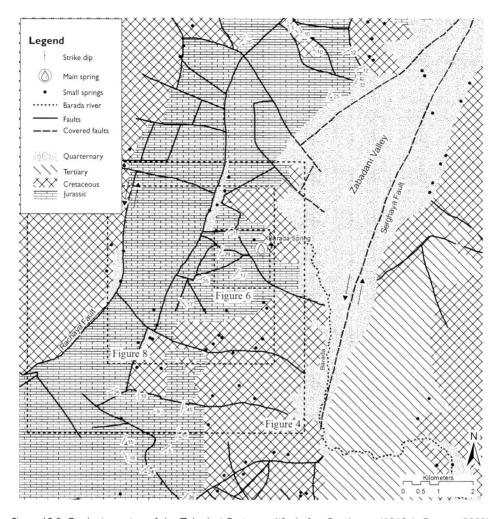

Figure 13.2 Geologic setting of the Zabadani Basin, modified after Ponikarov (1968a), Gomez (2003) and LSI (1986) and location of investigations.

13.2.3   Tectonic setting

Recent tectonic environment in the Anti-Lebanon Mountains is dominated by the Dead Sea Transform Fault zone trending from the Red Sea in the south to Turkey in the north. Its activity phase ranges from the Middle Tertiary until present (Girdler, 1990). All faults in the Zabadani area are branch faults of the Dead Sea Transform Fault, namely the Rachaya Fault in the West and the Serghaya Fault which crosses the Zabadani Valley (Figure 13.2). Still active movement along these faults causes enlargement of the basin (Gomez *et al.*, 2003). Both slip faults have left-lateral displacements, the amount of displacement along the Serghaya Fault is about 2 mm/a. The ratio between the slip and dip of the fault is 4:1 up to 5:1 (Gomez *et al.*, 2003) and the resulting structures are, therefore, almost vertically orientated in this horizontal shear zone. The anticline structure has a flat dip and causes the oldest formations of Jurassic age to be exposed in the centre of the mountain ridge. The observed dips of the layers are facing east because it is the western part of the anticline structure. The mean dip direction is approximately 131°, towards the east-south-east with an average dip of 16°.

The most important tectonic feature linked to the shear strength of the vertically moving faults is block rotation in highly tectonically penetrated regions of the investigation area. For this reason the dip of the layers is not orientated in east-south-east. Some blocks show a rotation leading to higher amount of dip (up to 45°) and irregularity in direction, especially in higher elevation regions. Because of the high strength toward deformation of the rocks, folding is subordinat and can be found only locally in the form of flexures with a small amount of dislocation. The direction of the mean stress in this region is nearly 157° to 175° (north-north-east to south-south-west) (Heidbach et al., 2008).

13.2.4 Climate and hydrogeology

The Anti-Lebanon Mountains are located in a moderate subtropical area with high variations between a dry summer season and wet winter. The strong increase of rainfall according to orographic location gives the area importance as one of the principle recharge areas of Syria. In the rainy season the precipitation is snow which gives an impulse of recharge and also discharge in the springs during melting in spring time. The average precipitation is 500–600 mm/a and rises up to 800 mm/a on top of the Chir Mansour Mountain. Infiltration in conjunction with groundwater recharge occurs rapidly over the strongly karstified structures without soil cover along tectonic structures (Ponikarov, 1968b). An analysis of the discharge graph of the Barada Spring reveals the fast reaction of the amount of discharge in the spring after rainfall. According to Maillet's relationship between the decrease of discharge in time at a spring site to the retention of the aquifer as well as to the dischargeable volume of water, the aquifer can be characterised as a fractured aquifer with a total water volume of roughly 2.1–3.3 km³ depending on the effective porosity of 2.5–4% (Bauer, 2005).

The chemical composition of the groundwater in this area depends on the mineral composition of the aquifer and the travel time of the water. The Barada catchment aquifer is composed of Jurassic limestone with a low clay content so the water is of HO_3-type with low mineralisation (electric conductivity of 260–280 µS/cm). In a few locations, an exchange with groundwater from the Cretaceous aquifer can be observed, where groundwater has electric conductivities of 500–900 µS/cm. Analysis of the Ca saturation index reflects the groundwater flow system. The higher the elevation, the more negative is the saturation index (Bauer, 2005). This correlates with the increase of precipitation with the height of the land surface. The average elevation of recharge is 1250 ± 50 m a.s.l. (Kattan, 1997).

Beside the Barada Spring there are many small intermittently discharging springs with low discharge rates (LSI, 1986). The location of the springs is mostly along faults, in some places they occur at the transition of two layers with different hydraulic conductivities such as the boundary between Jurassic and Cretaceous formations.

To define local groundwater flow paths close to Barada spring two tracer tests have been carried out. The first was very close to the Barada Spring (Alammareen, 2010). The tracers arrived from different locations in the vicinity of the spring lake with velocities up to 0.02 m/s. The second tracer test was further from the lake near the highway to Beirut (Figure 13.1). The tracers never arrived in the Barada Spring (Zwahlen *et al.*, 2010). Therefore, the distal areas of the catchment were not explored.

13.3 ANALYSIS OF THE ORIENTATION OF FISSURES

The strike and dip of all exposed tectonic features were measured at 71 locations. Three directions could be differentiated. The dip of all the fissures is steep, almost vertical. Because of the steep dip, the structures are visible in high resolution remote sensing data. This complies with the high karstification of these structures. The ridges along the karstified tectonic features are almost 1 m in width. For the interpretation of the lineaments a monochromatic remote sensing data set with a resolution of approximately 0.3 m was used.

The kinematic model used for the interpretation is the Riedel model (Eisbacher, 1996). The model is an explanation of the structures which occur during simple shear as the angle between the fissures depends on the friction angle of the material. The Serghaya Fault is the major fault in this area and strikes in an average angle of about 15° (north-north-east to south-south-west). The friction angle of limestone ranges between 35°–42° (Goodman, 1980). According to the pure shear model the angle between the sinistral fault and the direction of maximum principle stress is nearly 25°, which means it strikes roughly 170°. This value fits the direction of the regional stress field (section 13.2.3).

The simple shear mechanism produces a couple of tectonically induced joint faces (Figure 13.3). Riedel shears are acutely angled to the main fault, for a friction angle of nearly 35° to 42° it is between 17.5° to 21° anti-clockwise. The moving direction along the face is the same as the principle fault. This extends to the P-shears, which are clockwise to the principle fault and of clockwise moving behaviour (Figure 13.3, first subfigure). Both of these structures have conjugated fissures, which have the opposite moving direction to the main fault and are, therefore, antithetic. The angle between synthetic and antithetic shears for the available friction angle is nearly 60°.

Approximately 3800 lineaments were extracted from remote sensing data. The resulting strike directions of the fissures fit the Riedel model well (Figure 13.3). The intensity of the directions in the rose diagrams varies with the amount of lineaments in one direction, and is a scale for the strength of karstification. The two synthetic structures (R and P (Figure 13.3)), Riedel shears and P-shears are the most visible structures on the land surface.

The field measurements show a different distribution. In general, the structures are reliable, but the angles between the structures are smaller. Moreover the variation of the distribution is larger. This can be explained by the nearly 50-fold fewer measurements that were made which increase the statistical distribution of the measurements. Because of the strong karstification, direct measurement of the structures is coupled with sharp surfaces which mostly exist at tectonic faults which in turn appear

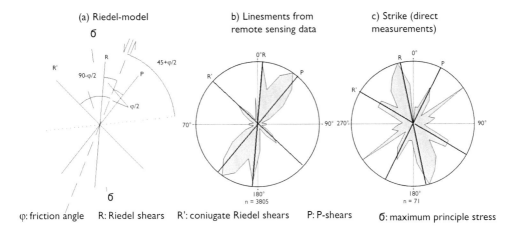

Figure 13.3 Analysis of the relationship between tectonic measurements and lineaments. Left) The basic concept of generation of the discontinuities. Center) Interpretation of the lineaments obtained by remote sensing data. Right) Interpretation of the lineaments obtained by direct measurements of the strike (compass).

in the vicinity of block rotation. This can also be observed by an overall rotation of the shear structures in a counter-clockwise direction which as a mean value of all lineaments is about 10°.

A closer inspection of the distribution at a single block was carried out (Figure 13.4) and compared with the other observations and the Riedel model. The principle evolution of structures during an induced stress caused by simple shear is displayed in Figure 13.4 (detail figure). After the generation of fissures, a pull-apart basin is created and block rotation takes place. This is replicable on the basis of the lineament directions. The southern part of the Jurassic outcrop (Figure 13.2) is almost unfaulted. The measured lineaments fit into the shear model with a direction of maximum principle stress from north-north-west to south-south-east. Near the Barada Spring Lake, the observed angles of the fissures have different directions. Rotations of the observed strike directions in relation to the stress model results in rotation angles of up to 53°. This is a hint for a displacement which takes place along some structures observed in remote sensing data (Figure 13.4). According to the model these structures should be moved in a dextral sense and the blocks rotated anti-clockwise. With this model the observed structures can be conceptually explained.

13.4   GEOPHYSICAL INVESTIGATIONS

13.4.1   Direct Current electrics (DC)

The choice of applied electrode array is a crucial influence on obtaining lateral and vertical resolutions. A Wenner array and a combined Wenner-Schlumberger array were used (Knödel *et al.*, 2005; Lowrie, 2007). A principal advantage of the Wenner-Schlumberger array compared to the classical Wenner array is a higher resolution due

Figure 13.4 Strike of the observed fissures (remote sensing lineations) and block rotation. See Figure 13.2. Subfigure: General generation of fissures in a transtension system (Eisbacher, 1996).

to an increasing number of measurements in the same time (Henning, 2008). South of the Barada Spring Lake two parallel DC soundings with a Syscal Junior Switch 72 device were measured in the spring of 2011.

Inversion results show a relative uniform upper layer with low resistivities of $\rho_S \leq 50$ Ωm followed by a sharp rise on resistivities of up to 2200 Ωm in profile 1 and the hard underlying bedrock. Both profiles show a similar anomaly with ρS in a range of 150 up to 200 Ωm from 150 m to 210 m in profile length and vertical extent (Figure 13.5).

Low RMS errors of 8.88%, respectively 5.46% combined with a good compliance of inverted ρ_S to the lithology of a nearby water well, enable a lithologic interpretation. According to published values of ρ_S by Davis *et al.* (1989) and Knödel *et al.* (2005) the upper layer is an excellent fit to loam or marl and the deeper areas

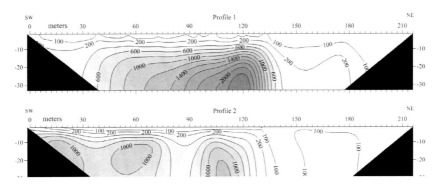

Figure 13.5 Inversion result from the two DC soundings south of Barada Spring (isovalues in Ωm, locations in Figure 13.6).

with high ρ_S a fit to dry and hard rock. Between the upper loamy layer and the hard rock an area with 100 Ωm ≤ ρ_S ≤ 300 Ωm can be interpreted as partly saturated and weathered limestone.

In the summer of 2010 three parallel resistivity profiles were carried out within the dry Barada Spring lake (Figure 13.6). The Wenner array with an electrode spacing of 40 m was utilised. The penetration depth of a Wenner array of 40 m spacing is about 13 m (Knödel *et al.*, 2005). Conspicuous within the Barada Spring Lake is that maximum resistivities range up to 200 Ωm (Figure 13.6). This indicates a lack of dry and non-fractured hard rock in this depth. Lithologically the southern part with ρ_S from 10 to 70 Ωm might correspond to clayey, loamy or marly facies.

Additionally, in 2004 another two resistivity profiles, 300 and 350 m west of Barada spring, were carried out by Alammareen (2010). In each profile it was possible to detect an anomaly and each equates the lateral trend of ρ_S. Alammareen (2010) interpreted both anomalies as a likely fault trace and delineated the connection line as a fault direction from east to west (Figure 13.6).

13.4.2 Very-Low-Frequency (VLF) and Audio-Magneto-Telluric (AMT)

Electromagnetic methods use either natural or anthropogenic sources as their primary field. This primary field induces a secondary one in a conductive body, which in turn, can be recorded by a detector (Telford *et al.*, 1990).

VLF methods use low-frequency transmitter signals from highly energy military radio stations, which are used for submarine communication. These transmitters consist of high antennas and emit signals in a frequency range of 15–25 kHz (Milsom, 2002). Variations in amplitude or phase of the detector signal reveal a conductive structure in the bedrock, visible through the existence of a secondary magnetic field (Stiefelhagen, 1998). The resulting magnetic field is elliptically polarised and consists of the primary magnetic field and one or more tilted secondary magnetic fields (Telford *et al.*, 1990). VLF dip-angle anomalies occur when there is an electrical conductivity contrast between local geological units. Applicable fields are mapping fault and small scale fracture zones, contaminant or groundwater mapping and mineral exploration. As VLF measurements are sensitive to linear conductive objects, degrading topographic effects and ionospheric conditions, the interpretation is in general qualitative. The main benefits of VLF measurements are low costs and speed (Stiefelhagen, 1998).

In the autumn of 2011, three parallel south-west to north-east striking VLF-EM profiles were measured south of Barada Spring. The Russian transmitter UMS near Moscow, with a frequency of 17.1 kHz at 0.2 MW, was used. Measurements were made with T-VLF from Iris Instruments with a maximum investigation depth of 50 m. Measuring results are In-Phase values of Dip-angles.

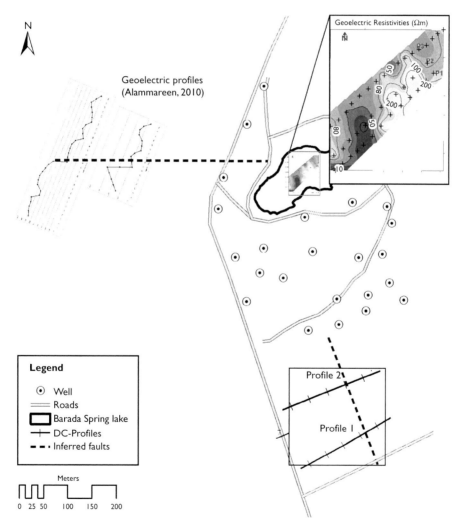

Figure 13.6 Map of geoelectrical measurements in the surrounding area of Barada Spring and location of the VLF measurements (Figure 13.7).

The first step is to Fraser-filter the observed Dip-angles before plane interpolation. Radial basis function interpolation of measured Dip-angles reveals several anomalies. In general, a north-south contrast, with high values in the south and very low ones in the north, is visible. Possible delineated anomalies trend in north-west to south-east directions (Figure 13.7), but they are not clearly present.

In 2004 several Audio-Magneto-Telluric soundings (Carvalho, 2009) were carried out in the adjacent area of Barada Spring (Figure 13.8) (Alammareen, 2010). The result

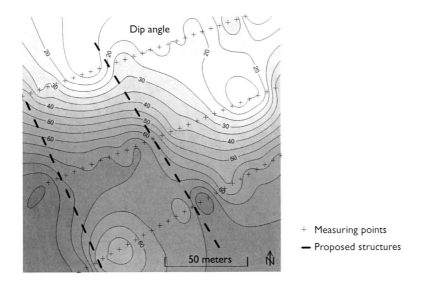

Figure 13.7 Map of interpolated Dip-angles from VLF measurements south of Barada Spring and possible location of structures.

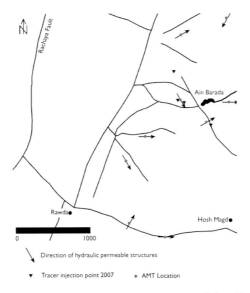

Figure 13.8 AMT measurement and tracer injection points (after Alammareen, 2010).

of the directional measurements in 30° steps from 0° to 150° and a frequency domain of 35–12 700 Hz (with 10 steps) is the direction of the highest hydraulic conductivity within the Jurassic and Cretaceous karstified and fractured aquifers (Alammareen, 2010). The investigation depth of electromagnetic waves is inversely proportional to the observed frequency. Surveying points were selected on the basis of the geological map of Ponikarov (1968a). As fractured or karstified aquifers feature high anisotropy factors, measuring of apparent resistivity and phase difference was carried out in six directions from 0° to 150°. According to the results of the AMT readings by Dussel (2005) the frequency with the highest factor of anisotropy was chosen as the reference horizon. The tendency of hydraulic permeable structures is represented by the direction of electrodes with the highest observed apparent resistivity (Alammareen, 2010). According to the dependency of the high resistivity and hydraulic permeability some of the faults seems to be permeable, and in some cases the permeability is perpendicular to the orientation of it (Figure 13.8). There is no clear dependency of direction and hydraulic permeability.

13.5   CONCLUSION AND DISCUSSION

The distribution of fissures in combination with geophysical investigations provides an image of the tectonic regime and the local distribution of permeabilities. A high grade of faulting in the vicinity of the Barada Spring, intensified by block rotation, increases the transmissivity and drains the catchment along the faults. Tracer test results show high velocities in groundwater flow which cannot be associated with one preferred tectonic direction as Riedel shears or P-shears. The crucial permeabilities detected by AMT measurements do not follow the principal faults in every location. The direction of karstification differs from the direction of the key faults.

For investigation of anomalies close to the Barada lake geoelectrical methods provide useful results. For more distant regions where the depth of water table is large, the AMT method is preferred due to its high skin depths. The use of VLF in the vicinity of Lake Barada is limited by the amount of electric cables which supply the station as well as the pumps, which induce an interfering electromagnetic field for the sensitive method.

In the southern part of the spring lake, geoelectrical profiles identify main anomalies consistent with the major fault direction. This is not consistent with the AMT measurements. Direct tracer injection tests in the groundwater system provide a different picture. To the west of the lake the flow direction has been investigated with direct tracer tests which is consistent with resistivity profiles, and the measured strike dip of the fissures and the karstification indicate a west to east trending flow path. In the lake itself, an intersection of two faults occurs which may allow a high amount of discharge, due to assumed greater transmissivities. The structural anomalies in the less faulted southern area of the catchment are not clear.

More quantitative field data are required to define precisely the water flow to the spring. To detect and quantify open fractures with a high hydraulic conductivity, hydraulic testing and in-situ flow logging can be conducted. Tracer tests are also necessary to define flow pathways and measure effective flow velocities.

The investigations show that there is a need for field testing hydraulic methods to quantify flow paths and directions.

REFERENCES

Alammareen A.M. (2010) Groundwater Exploration in Karst Examples for Shallow Aquifers Using Microgravity Technique in Paderborn-Germany and the Reconnaissance of Deep Aquifers in the Catchment of Barada Spring-Syria. PhD thesis at Berlin Institute of Technology.

Andreo B., Goldscheider N., Vadillo I., Vias J.M., Neukum C., Sinreich M., Jimenez P., Brechenmacher J., Carrasco F., Hotzl H., Perles M.J., Zwahlen F. (2006) Karst groundwater protection: First application of a Pan-European Approach to vulnerability, hazard and risk mapping in the Sierra de Libar (Southern Spain). *Science of the Total Environment* 357, 54–73.

Bauer F. (2005) Hydrochemical exploration and correlation in the Zabadani basin – Syria. In German language. Diploma thesis at the Berlin Institute of Technology.

Dussel M. (2005) Hydrotectonic and groundwater-dynamics in the recent stress field with the example of karstified rocks in zentral Algarve (Portugal). In German language. PhD thesis at the Berlin Institute of Technology, Germany. 162p.

Eisbacher (1996) *Introduction to tectonics.* In German language. 2nd Edition. Stuttgart, Ferdinand Enke Verlag.

Gomez F., Meghraoui M., Darkal A.N., Hijazi F., Mouty M., Suleiman Y., Sbeinati R., Darawcheh R., Al-Ghazzi R., Barazangi M. (2003) Holocene faulting and earthquake recurrence along the Serghaya branch of the Dead Sea Fault system in Syria and Lebanon. *Geophysical Journal International* 153, 658–674.

Goodman R.E. (1980) *Introduction to rock mechanics.* New York, John Wiley & Sons.

Heidbach O., Tingay M., Barth A., Reinecker J., Kurfeß D., Müller B. (2008) The World Stress Map database release 2008 doi:10.1594/GFZ.WSM.Rel2008.

Hennings V., Wolfer J. (2008) Management, protection and sustainable use of groundwater and soil resources in the Arab Region – report on the compilation of a Groundwater Vulnerability Map Zabadani Basin, Syria. ACSAD/BGR, Damascus/Hannover.

JICA (1997) The study on water resources development in the north western and central basin in the Syrian Arab Republic. Final report, Damascus, Syria.

Kattan Z. (1997) Environmental isotopes study of the major karst springs in Damascus limestone aquifer systems: Case of the Figeh and Barada springs. *Journal of Hydrology* 193,161–182.

Knödel K., Krummel H., Lange G. (2005) *Geophysics. Handbook for exploration of bedrock from landfills and contaminated land.* In German language. Berlin: Springer, 2nd Edition. 1102p.

LSI (1986) Water Resources Use in Barada and Awaj Basins for Irrigation of Crops, Syrian Arab Republic, Feasibility Study, Leningrad State Institute for Design of Water Resources Development Projects, Moscow/Damascus.

Lowrie W. (2007) *Fundamentals of Geophysics.* Second Edition. Cambridge, Cambridge University Press, 375p.

Milsom J. (2002) *Field Geophysics.* Third Edition. West Sussex, England: John Wiley and Sons Ltd., 229p.

Ponikarov V.P. (1968a) The Geological Map of Syria; Sheet I-37-VIII-3a,c (Zabadani); Scale 1:50000. Vsesojuznoje Exportno-Importnoje Objedinenije "Technoexport"; Ministry of Geology, Moscow.

Ponikarov V.P. (1968b) The Geological Map of Syria; Sheet I-37-VIII-3a,c (Zabadani); Explanation notes. Vsesojuznoje Exportno-Importnoje Objedinenije "Technoexport"; Ministry of Geology, Moscow.

Puri S., Aureli A. (2009) *Atlas of transboundary aquifers. ISARM Programme.* United Nations Educational, Scientific and Cultural Organization, Paris. 295–297.

Schoen J.H. (2011) *Physical Properties of Rocks. A workbook.* Handbook of Petroleum Exploration and Production, 8. s.l. Elsevier, Amsterdam. 463p.

Seaton W.J., Burbey T.J. (2002) Evaluation of two-dimensional resistivity methods in a fractured crystalline-rock terrane. *Journal of Applied Geophysics* 51, 21–41.

Stiefelhagen W. (1998) Radio Frequency Electromagnetics (RF-EM): Continuous measuring broadband-VLF, expanded for hydrogeological problems. In German language. PhD Thesis, University of Neuchâtel, Switzerland. 225p.

SSB (2011) Statistical abstract, chapter 2: Population and demographic indicators. http://www.cbssyr.org/yearbook/2011/chapter2-EN.htm. – Syrian Statistics Bureau Syrian Arab Republic, Office of prime minister, Damascus. Cited 16 January 2013.

Telford W.M., Geldart L.P., Sheriff R.E. (1990) *Applied Geophysics*. Second Edition. Cambridge: Cambridge University Press. 751p.

Wagner W. (2011) *Groundwater in the Arab Middle East*. Springer, Heidelberg, pp. 63–137.

Zwahlen F. (2003) Vulnerability and Risk Mapping for the Protection of Carbonate (Karst) Aquifers, COST Action 620, Final Report.

Zwahlen F., Schnegg P., Babic D. (2010) First results of the March 2010 Barada Spring tracing test. Report at University Neuchâtel, Switzerland.

Chapter 14

Using heat flow and radiocarbon ages to estimate the extent of recharge area of thermal springs in granitoid rock: Example from Southern Idaho Batholith, USA

Alan L. Mayo[1] *& Jiri Bruthans*[1,2]
[1]*Department of Geosciences, Brigham Young University, UT, USA*
[2]*Faculty of Science, Charles University in Prague, Czech Republic*

ABSTRACT

The extent of subsurface flow and recharge areas of thermal groundwater systems are difficult to estimate. Two simple methods for making such estimates are described. The methods are based on heat flow and on the ^{14}C age of thermal groundwater discharge. The two methods were applied to thermal groundwater systems in the southern Idaho Batholith and the results suggest that although thermal groundwater only accounts for about 1.5% of the drainage basin base flow, about one-third of the drainage basin surface area is involved in thermal groundwater system recharge. The results also suggest that hard rock hydrothermal heating does not simply occur in discrete fault and fracture systems, but rather in the vast volume of granitic mass. In the southern Idaho Batholith the total thermal system flow area is ~2100 km^3 of granite and the fault and fracture systems are the final collectors of thermal groundwater prior to surface discharge. The methods presented rely on readily obtainable data for most of hydrothermal systems.

14.1 INTRODUCTION

In granitoid rock the myriad network of surface and near surface fractures are the recharge avenues for meteoric-hydrothermal groundwater systems. Recharge water then flows laterally and several to tens of metres below ground surface. Most recharge water discharges nearby a cold system springs or is lost to evapotranspiration. A small portion of the recharge water migrates to depth along a progressively decreasing network of fissures and ultimately discharges at a limited number of thermal springs. Thermal groundwater flow system geometries are poorly understood (Ferguson & Grasby, 2011). Because thermal water is commonly accessible only at the discharge point (i.e., thermal spring) it is difficult to evaluate the thermal system recharge area.

The general lack of deep hard rock boreholes means that thermal groundwater flow systems are typically evaluated using indirect indicators such as chemical and isotopic analysis of spring and shallow borehole water, geophysical techniques, and by numerical modelling. Groundwater circulation depths are often inferred from chemical geothermometry (Allen *et al.*, 2006; Ferguson *et al.*, 2009). Early attempts to numerically model hydrothermal systems include work by Forester & Smith (1986, 1988, 1989), and Lowell (1991). Flow and mixing in vertical and dipping fault zones have been modelled by López & Smith (1995, 1996), and Ferguson *et al.* (2009), and flow in volcanic edifices has been modelled by Hurwitz *et al.* (2003). Coupled groundwater flow-heat flow models are often used to evaluate geothermal systems (Gvirtzman *et al.*, 1997; Serban *et al.*, 2001; Ferguson *et al.*, 2009; Heilweil *et al.*, 2012). However, most of the information needed for the construction of coupled models must be estimated (Gvirtzman *et al.*, 1997). Many coupled numerical models of heat and groundwater flow are, therefore, designed to simulate conceptual thermal flow rather than site specific thermal groundwater systems. In recent years the need for a better understanding of geothermal systems has been driven by the increased use of geothermal resources (e.g. Galván *et al.*, 2011; Zaher *et al.*, 2012).

In this chapter two approaches for calculating a first order estimate of the flow and recharge areas of fracture flow for site specific thermal groundwater systems in crystalline bedrock are presented. One approach is based on convective heat flow transfer and the other method is based on the relationship between thermal discharge volume and groundwater age. These approaches are tested on thermal groundwater systems in the southern Idaho Batholith, USA and the results are compared.

14.2 APPROACHES

14.2.1 Heat flow transfer

It is well established that groundwater flow (free and forced convection) redistributes heat (Gvirtzman *et al.*, 1997; Meert *et al.*, 1991) if the hydraulic conductivity exceeds ~2.10^{-9} m/s (Smith & Chapman, 1983). Perturbations in the thermal gradient may be a powerful indicator of groundwater flow (Mongelli & Pagliarulo, 1997). Where deep circulation of groundwater occurs, groundwater recharge and discharge can respectively cause negative and positive terrestrial heat flow anomalies (Jones & Marorowicz, 1987). Because thermal springs convectively transfer terrestrial heat from depth to the surface, it is possible to calculate the convective heat transferred using the following equation (Jetel, 1972):

$$H = Q \, c \, \rho \, (T_{spring} - T_{surface}) \tag{1}$$

where H = heat output that is convectively transferred from the bedrock to the spring discharge (W); Q = spring discharge (m³/s); c = water specific heat capacity (J kg⁻¹ K⁻¹); ρ = water density (kg/m³); T_{spring} = mean annual spring discharge temperature (K); and $T_{surface}$ = mean annual surface temperature (K). Mean annual air temperature can be used as a surrogate.

If a terrestrial heat flow is the only source for groundwater heating and the heat flow is steady state the minimum extent of the deep flow area associated with a spring can be calculated using the following equation (Ehlers & Chapman, 1999; Ferguson & Grasby, 2011):

$$S_{min} = H/q \tag{2}$$

where S_{min} = minimum flow area (m²); and q = basal terrestrial heat flow (W/m²).

Equation 2 is applicable for natural groundwater discharges, such as undisturbed thermal springs. Removal of thermal water from deep wells at a discharge rate or temperature greater than the natural drainage violates the steady-state assumption in equations 1 and 2 and makes this concept inapplicable for any setting where more heat is removed than is produced by terrestrial heat flow. Equations 1 and 2 are also based on the assumption that all heat is produced by terrestrial heat flow. This condition is met for most thermal springs as sulfide oxidation and other processes are mostly negligible heat sources in the several km thick uppermost zone of Earth's crust where the water flux is small relative to basal terrestrial heat flow (Jokinen & Kukkonen, 1999; Derry *et al.*, 2009).

Equation 2 assumes that all terrestrial heat flow will be fully utilised by groundwater heating and that this heat ($T_{spring} - T_{surface}$) will be transferred to the surface via spring discharge. In reality, only part of the terrestrial heat flow is transferred to groundwater and most of the heat flow continues to the surface via bedrock conduction in the form of residual heat flow (Figure 14.1). Therefore, the actual groundwater heat flow area is several times greater than S_{min}. When residual heat flow can be measured or estimated it is possible to estimate the actual groundwater flow area:

$$S_H = H/(q - q_{residual}) \qquad (3)$$

where S_H = estimation of the actual flow area (m^2); and $q_{residual}$ = residual terrestrial heat flow (W/m^2).

Because equation 3 is based on heat balance, it is not necessary to know the thermal groundwater flow pattern, flow system geometry, and groundwater circulation depth.

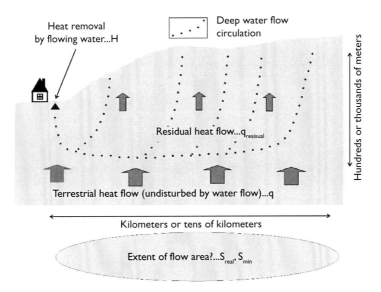

Figure 14.1 Interaction between deep groundwater flow and decrease of terrestrial heat flow. The minimum flow area of deeply circulating groundwater can be estimated if basal heat flow is compared with the amount of heat carried out by spring water.

Neither S_{min} nor S_H are direct calculations of the recharge surface area; however the recharge area occupies part or all of the S_H. Assuming nearly vertical downward groundwater movement from the recharge area to depth, the calculated S_H can roughly be equated to recharge area in many granitoid rock thermal systems because: 1) the bottom of the valley where many thermal springs are situated (point discharge, commonly associated with faults) has a negligible area compared to the valley flanks and ridges where most recharge occurs, and 2) confining units do not commonly occur in granitic rocks, thus recharge can occur throughout much of the upland surface (Foster & Smith, 1983).

14.2.2   Discharge – radiocarbon age

The volume of thermal groundwater in storage associated with a single thermal spring can be calculated by multiplying the radiocarbon age of water by the mean spring discharge rate (Ciezkowski *et al.*, 1992):

$$Vs = MRT\ Q \tag{4}$$

where Vs = volume of groundwater in storage (m^3); and MRT = mean groundwater residence time at the spring discharge(s). For example, if the mean ^{14}C age is 9700 years and the average thermal discharge is 0.002 m^3/s the volume of water in storage may be calculated as:

$$Vs = 9700\ y \times 0.002\ m^3/s = 6.1\ 10^8\ m^3$$

Because granitic rocks have very low effective porosity (close to 1% of total rock volume) the calculated volume of saturated bedrock (Va) in the spring catchment is approximatley100 times greater than volume of water in storage or ~610 km^3 for the above example. If the effective porosity is 0.5% the calculated Va is ~1220 km^3 and if the effective porosity is 2% the calculated Va is 305 km^3.

Assuming mostly vertical downward movement of thermal groundwater, the flow area can be estimated if the depth of groundwater circulation is known or can be calculated according to:

$$S_R = Va/d \tag{5}$$

where S_R = estimated flow area (km^2); Va = saturated bedrock volume in spring catchment (km^3); and d = average circulation depth (km).

The thermal groundwater circulation depth may be calculated by dividing the adjusted maximum aquifer temperature by the geothermal gradient. The adjusted maximum aquifer temperature is the maximum aquifer temperature minus the mean annual air temperature ($T_{spring} - T_{surface}$). The maximum aquifer temperature may be estimated by means of geothermometer calculations or using the discharge temperature if there is reason to believe that this temperature is a good approximation. For the example above, if the geothermometer temperature is 76°C, the mean annual air temperature is 7°C, and the geothermal gradient is 35°C/km the calculated circulation depth is 1.97 km. Using equation 5 the calculated flow area is ~358 km^2 (i.e., 705 km^3/1.97 km).

As with the heat flow method the discharge-age method is not a direct calculation of the surface recharge area. Assuming nearly vertical downward groundwater movement from recharge areas to depth, the calculated results can be considered as a first order approximation of recharge area.

14.3 APPLICATION TO THE SOUTHERN IDAHO BATHOLITH

Both the heat flow and the discharge-age methods have been applied to the thermal systems in the Payette and Boise River drainages in the southern Idaho Batholith (Figure 14.2). The southern portion of the Idaho batholith, known as the

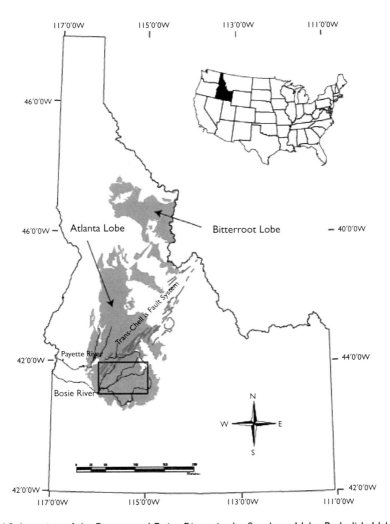

Figure 14.2 Location of the Payette and Boise Rivers in the Southern Idaho Batholith, Idaho, USA.

Atlanta Lobe, is Late Cretaceous in age, was emplaced approximately 75–100 Ma (Armstrong *et al.*, 1977; Criss & Taylor, 1983) and represents orogenic magmatism produced by the subduction of the Farallon Plate beneath western North America (Hyndman & Talbot, 1979). The topography is typified by steep, rugged, heavily forested terrain. Ridgelines are commonly 600 to 1500 m above canyon bottoms and canyon wall gradients are typically 0.1 to 0.4. Annual precipitation (300 and 1300 mm) and mean annual temperature (2 to 10°C) are elevation dependent (Lewis & Young, 1980, 1982). Heavy snowfall in the mountains is the major source of surface water and groundwater recharge. Total surface area for Payette and Boise Rivers is 16 265 km^2.

More than 60 thermal springs with discharge rates ranging from <1 to ~50 l/s and discharge temperatures ranging from 20.5 to 87°C have been identified in the Payette and Boise River basins (Table 14.1). Most thermal springs discharge along river bottoms and are associated with major regional fault structures in Cretaceous age micaceous granite, leucogranite, and granodiorite, and in Tertiary age plutons, ranging in composition from granite to diorite (Clayton *et al.*, 1979). Thermal spring discharges account for only about 1.5% of the baseflow of the Payette and Boise Rivers (Young, 1985; Himes, 2012). Thermal spring waters are of sodium bicarbonate type, have low TDS (150–570 mg/l), and notably high pH (8.0–9.8).

Geothermometry estimates using SiO_2, Na-K, Na-K-Ca, and mineral stability suggest maximum aquifer temperatures range between 85 and 160°C (Druchsel & Rosenberg, 2001; Lewis & Young, 1982; Mariner *et al.*, 2006). Maximum aquifer temperatures <150°C are consistent with the stable isotopic compositions (Schlegel *et al.*, 2009; Truesdell & Hulston, 1980). Based on borehole data, International Heat Flow Commission (2012) found that the geothermal gradient of the Atlanta lobe ranges between 21 and 59°C/km. Assuming a mean annual air temperature of 7.5°C, a maximum aquifer temperature of 150°C, and a geothermal gradient of 21 to 59 °C/km, the maximum circulation depth is 2.4 to 6.8 km. Similar estimated circulation depths of 2.4 to 6.7 km below the topographic surface were estimated by Druchsel & Rosenberg (2001) and Schlegel *et al.* (2009). An average thermal water circulation depth of 4 km is considered representative and was used for calculations in this investigation. Mean residence times for the hydrothermal systems have been calculated as 1500 to 34 850 radiocarbon years (Holdaway, 1994; Mariner *et al.*, 2006; Schlegel *et al.*, 2009; Young, 1985). Average heat flow measured in boreholes in the area is ~0.09 W/m^2 (International Heat Flow Commission, 2012).

The calculated H and S$_{min}$ for 62 springs in the Boise and Payette drainages (Figure 14.3) are listed in Table 14.1. There are insufficient data to calculate S$_H$. Only 15% of the springs (n = 9) are required to account for approximately 56% of the total thermal discharge and 60% of the calculated S$_{min}$. The total calculated S$_{min}$ is 1149 km^2 or about 7% of the total catchment area of the two basins. In other words only 7% of terrestrial heat flow is necessary for heating all thermal springs. S$_{min}$ is an unrealistic estimate of the actual flow area, because it assumes that all conductive terrestrial heat flow in the calculated area is consumed by convective water flow. If this were the case the geothermal gradient would be zero in the areas affected by convective flow. Most of the terrestrial heat flow in the Cretaceous and Tertiary plutons is residual heat and is transferred to the surface by bedrock conductive flow.

Table 14.1 Calclated S_{min} and estimated flow areas for 62 thermal springs in the Boise and Payette Drainages, Idaho.

Spring ID (township and range)	Spring name	UTM Location Zone 11		Discharge (°C)	Q (L s⁻¹)	H (MW)	Calculated flow area (km²)			
		Northing	Easting				S_{min} (100% heat transfer)	20% heat transfer	10% heat transfer	5% heat transfer
3n 14e 28 cad	Worswick	4825500	678100	87	29.4	9.99	111	555	1,110	2,220
10n 4e 33 cbd	Deadwood 1	4889200	580500	75	31.5	9.13	101	505	1,010	2,020
5n 7e 24bdd1S				76	26.7	7.85	87	435	870	1,740
10n 10e 31 bcc	Bonneville	4890500	634900	84	22.7	7.43	83	415	830	1,660
4 N 6E 24 bcd	Ninemeyer			74	22.0	6.25	69	345	690	1,380
15n 3e 13bbc1s				34	50.3	5.91	66	330	660	1,320
9n 8e 32 cba	Kirkham	4880900	617000	64	22.1	5.37	60	300	600	1,200
4n 6e 24 bcb	Twin Springs	4835700	605200	67	18.9	4.84	54	270	540	1,080
5n 9e 7 bab	Roaring River	4849600	626000	65	18.9	4.68	52	260	520	1,040
12n 5e 22 bbc	Deadwood 3	4912600	591200	86	8.4	2.84	32	160	320	640
12n 5e 36 dba	Swim pool	4909000	595600	38	21.0	2.83	31	155	310	620
3n 10e 33 acd	Paradise Creek	4823500	639500	53	12.6	2.49	28	140	280	560
8n 6e 1 adb	Pineflat	4879300	605200	60	11.0	2.48	28	140	280	560
4n 14e 29 dcd	Skillern	4834600	676100	65	9.5	2.32	26	130	260	520
6n 10e 30 cda	Weatherby	4854000	634500	64	9.5	2.30	26	130	260	520
6n 5e 33 adc	Idaho City 2	4852000	591500	42	12.6	1.91	21	105	210	420
5n 7e 34 ccb	Troutdale	4841700	611600	60	6.3	1.43	16	80	160	320
11n 7e 16 aab	Unnamed	4905100	610600	65	5.7	1.41	16	80	160	320
5n 9e 5 aad	Granite Creek	4851000	628500	56	6.3	1.32	15	75	150	300
6n 11e 35 dbb	Powerplant 2	4852500	651000	53	6.3	1.24	14	70	140	280
10n 11e 31 add	Grand Jean	4891000	645700	67	4.7	1.21	13	65	130	260
13n 4e 31 cab	North Payette	4918500	577200	71	4.4	1.20	13	65	130	260
8n 5e 10 bdd	Unnamed	4877500	593500	55	5.7	1.15	13	65	130	260
2n 10e 5aca1S				59	4.7	1.05	12	60	120	240
	Atlanta			62	4	0.95	11	55	110	220

(Continued)

Table 14.1 Continued.

Spring ID (township and range)	Spring name	UTM Location Zone 11		Discharge (°C)	Q (L s⁻¹)	H (MW)	Calculated flow area (km²)			
		Northing	Easting				S_{min} (100% heat transfer)	20% heat transfer	10% heat transfer	5% heat transfer
8n 5e 6 dcc	Unnamed	4878000	587600	51	4.7	0.89	10	50	100	200
6n 11e 35 dad	Powerplant 1	4852300	652000	60	3.8	0.86	10	50	100	200
6n 11e 35ddb1S				54	4.1	0.83	9	45	90	180
7n 1e 8dda1s				64	3.2	0.77	9	45	90	180
4n 11e 34 dbb	Williow Creek	4834600	650700	53	3.8	0.75	8	40	80	160
2n 10e 5 ada	Dixie Hot	4821700	638100	60	3.2	0.71	8	40	80	160
6n 11e 30 adb	Queens	4854300	645200	51	3.8	0.71	8	40	80	160
3n 13e 7 dcd	Lightfoot	4829600	665500	62	2.8	0.67	7	35	70	140
9n 3e 25 bac	Banks	4882500	575900	80	2.0	0.63	7	35	70	140
13n 6e 29 dab	Unnamed	4920100	598500	53	3.2	0.62	7	35	70	140
5n 8e 10 dca	Browns Creek	4848500	622000	51	3.2	0.60	7	35	70	140
	Deer Creek			78	2.0	0.60	7	35	70	140
11n 5e 29 cdb	Deadwood 2	4900200	588400	49	3.2	0.57	6	30	60	120
8n 5e 11 baa	Unnamed	4878000	593000	60	2.2	0.50	6	30	60	120
	Dutch Frank	4849720	625805	65	1.9	0.48	6	30	60	120

12n 5e 10ddd1s				67	1.4	0.36	4	20	40	80
9n 8e 31 aca	Deadwood 5	4881300	616100	64	1.3	0.30	3	15	30	60
5n 9e 5dcb1S				54	1.3	0.25	3	15	30	60
15n 4e 21 dcc	Nude Spring	4940600	581300	36	2.0	0.25	3	15	30	60
13n 3e 13 ada	Unnamed	4923600	576100	49	1.3	0.23	3	15	30	60
7n 1e 9cdc1s				45	1.3	0.21	2	10	20	40
	Goller			51	1.0	0.20	2	10	20	40
5n 7e 34 dba	Loftus Creek	4842100	612500	55	0.9	0.19	2	10	20	40
9n 9e 22dca1s				54	0.8	0.17	1.8	9	18	36
3n 14e 30aa1S				61	0.6	0.15	1.6	8	16	32
8n 5e 6 dcb	Unnamed	4878300	587600	42	0.9	0.13	1.5	7.5	15	30
12n 5e 2 ccc1s				47	0.8	0.13	1.4	7	14	28
8n 5e 10 add	Grimes	4877500	593000	54	0.6	0.13	1.4	7	14	28
12n 5e 2 dac	Deadwood 4	4916700	593800	50	0.6	0.12	1.3	6.5	13	26
3n 12e 7 dcd	Baumgarter	4829200	655800	50	0.6	0.12	1.3	6.5	13	26
4n 7e 8 cbb	East Twin Springs	4839900	608100	61	0.5	0.12	1.3	6.5	13	26
6n 10e 30 cca1S				50	0.4	0.07	0.8	4	8	16
8n 5e 11bac1s				57	0.3	0.07	0.7	3.5	7	14
6n 5e 33 abc	Idaho City 1	4852500	591500	41	0.4	0.05	0.6	3	6	12
9n 7e 35aaa1s				37	0.2	0.02	0.3	1.5	3	6
3n 14e 19 ddb	Little Smokey	4826900	675200	41	0.1	0.02	0.2	1	2	4
8n 5e 1bcc1s				40	0.1	0.02	0.2	1	2	4
Total					439.7	103.1	1,149	5,747	11,494	22,988

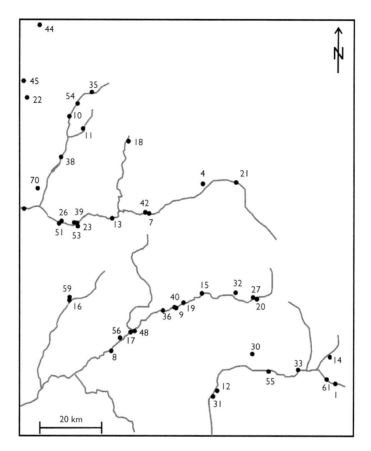

Figure 14.3 Position of the thermal springs in the Boise and Payette River basin in the southern part of Idaho Batholith. Spring numbers correspond to numbers in Table 1.

Discharge volume/radiocarbon age calculations have been performed on 20 springs with ^{14}C calculated ages ranging from 1500 to 34850 years (Table 14.2). The table includes data from several sites collected as part of this investigation. The springs have measured discharge rates of 1 to 32 l/s. Based on equations (4) and (5) and a thermal water circulation depth of 4 km the calculated S_R for individual springs range from 6 to 528 km^2.

14.4 DISCUSSION

The minimum flow area S_{min} associated with individual springs using the heat flow approach is only about 20% as calculated using the discharge-age approach (Figure 14.4). The heat flow approach provides insight into the minimum extent of the flow area, but the results are unrealistically low as it assumes all of the terrestrial heat is transferred to convective heat. Assuming S_{min} represents the sum of the total convective heat flow and the convective heat flow captures 20, 10 and 5% of this heat,

Table 14.2 Calculated minimum thermal water flow area based on ^{14}C ages.

Spring name	[14]C age (years)	Discharge (°C)	Q (L s[−1])	Water volume (km³)	Minimum extent of flow area (km²)	Reference for radiocarbon age
Worswick	22,800	86	29	21.1	528	Mariner et al. (2006)
Bonneville	20,400	85	23	14.6	365	Mariner et al. (2006)
Kirkham	16,100	64	22	11.2	280	Mariner et al. (2006)
Ninemeyer	9,700	74	22	6.7	168	Unpublished (sampled 2009)
Boiling	18,850	86	8	5	125	Mariner et al. (2006)
Idaho City 2	10,500	42	13	4.2	104	Holdaway (1994)
Deadwood 1	4,100	75	32	4.1	102	Holdaway (1994)
Twin Sprigs	6,000	67	19	3.6	89	Unpublished (sampled 2005)
Deadwood 3	12,000	86	8	3.2	79	Holdaway (1994)
Roaring river HS	5,250	65	19	3.1	78	Unpublished (sampled 2005)
Atlanta	18,700	62	4	2.2	56	Mariner et al. (2006)
Deer Creek	34,850	78	2	2.2	55	Mariner et al. (2006)
9N-3E-25BAC1S	28,800	80	2	1.8	45	Young (1985)
Brown Creek	4,500	50	12.6	1.8	45	Unpublished (sampled 2009)
8N-5E-6DCC1S	9,000	51	5	1.3	33	Young (1985)
Queens River	7,800	50	4.7	1.2	30	Unpublished (sampled 2009)
Powerplant 1	7,000	60	4	0.8	21	Holdaway (1994)
Dutch Franks	6,300	65	1.9	0.4	10	Unpublished (sampled 2009)
Goller	10,800	51	1	0.3	8	Mariner et al. (2006)
Grand Jean	1,500	67	5	0.2	6	Holdaway (1994)
Total			237.2	89	2,227	

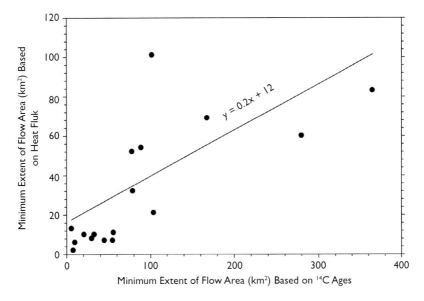

Figure 14.4 Comparison of the calculated extent of flow areas for the heat flux and ^{14}C methods. Best fit line has an r² of 0.64.

the total convective flow areas would be ~5750, 11 500 and 23 000 km², respectively (Table 14.1). The calculated 5% heat transfer is too low because the calculated area is greater than the total surface area. The flow area calculated assuming 20% heat transfer is about twice as large as the flow area calculated using the ^{14}C method (Table 14.2). Because the total heat flow area calculated by the ^{14}C method only includes about 60% of the total thermal discharge and the heat contained in the convective flow (H), 5000 km² is a reasonable first order approximation for the subsurface heat flow area and of the surface thermal spring recharge catchment area. A thermal catchment area of 5000 km² means that about one-third of the drainage area contributes to thermal groundwater recharge.

Both Young (1985) and Druchsel & Rosenberg (2001) postulated that deeply circulating hot-water systems feeding thermal springs in the Idaho Batholith developed in fault-controlled conduits. Druchsel & Rosenberg (2001) presented a conceptual model of thermal water flow in the southern Idaho Batholith where flow occurs along fracture zones in 'small, separate convection cells located within fracture networks peripheral to the river'. This model is based on similar isotopic composition of O and H in nearby springs, although there are differences between individual groups of springs. According to Druchsel & Rosemberg (2001) these convection cells should encompass one or two thermal springs, with little or no communication between them (e.g., Pine Flats–Kirkham hot springs and Bonneville–Sacajawea hot springs).

Based on flow area calculations in this study, it can be postulate that isolated, channelised fluid flow in deep granitic systems cannot explain the amount of water stored in the catchments of individual springs (up to 21 km³; Table 14.2) and that large volumes of granite mass are involved in the total flow zone of thermal water (~2100 km³). The idea of channelised flow is also in inconsistent with the extensive area (S_H) needed to convert conductive heat flow into heat transported convectively by water. Most of the water needs to be stored in the effective porosity of granite occupying volumes of tens to few thousands of km³ of rock mass. This idea of a large rock mass is supported by oxygen isotope studies of ancient fluid flow in the Idaho Batholith granite (Criss & Fleck, 1990; Taylor, 1977) where they found that the oxygen isotopic composition of the entire granite mass has been altered by large scale ancient thermal fluid flow in the whole granite body and not by fluid flow along discrete fault and fracture systems.

Similar large volume hydrothermal systems can be expected in other areas, where thermal geothermal power is derived from terrestrial heat flow transported by thermal springs exceed 5 MW and/or where multiple of mean residence time and spring discharge exceeds 2 km³. Springs with heat output exceeding these limits are reported from many areas (e.g. Ehlers & Chapman, 1999; Ingebritzen & Mariner, 2010; Ferguson & Grasby, 2011).

The methods used here use data which are available or which are possible to obtain for most of hydrothermal systems (spring discharge, spring discharge temperature, annual air temperature, terrestrial heat flow, depth of thermal water circulation, mean groundwater residence time, granite porosity). The calculation of the extent of the flow area together with stable isotope studies provide a basis for reasonable estimates geothermal recharge areas in granitic rocks and may be helpful for source protection for geothermal resources.

14.5 CONCLUSION

Methods for estimating the extent of thermal groundwater flow in crystalline bedrock based on heat flow and groundwater age-discharge calculations are presented. Results of these methods, applied to the thermal springs in the Boise and Payette River basins, of the southern Idaho Batholith, were compared. In the southern Idaho Batholith the minimum extent of subsurface flow calculated by the heat flow method for all thermal springs was 1149 km², which accounts for about 7% of the total conductive heat flow. The extent of subsurface flow calculated by the ^{14}C method was about 2225 km². The heat flow method results are only about 20% of the area of ^{14}C method results. This difference is reasonable as the heat flow method assumes complete transfer of conductive heat flow to convective heat flow. The total convective heat flow area, which is a reasonable first order approximation of the recharge area, is estimated at about 5000 km². It is estimated that about one-third of the entire 16 264 km² drainage area contributes recharge to the thermal groundwater systems. Estimates of the area involved in thermal groundwater flow suggest that hydrothermal flow does not largely occur in discrete fault and fracture systems, but rather occupies vast volumes of granite mass to a depth of ~4 km. Fault and fracture systems are the final collector of thermal groundwater prior to discharge to the surface. The methods use parameters which are available or which are possible to obtain for most of hydrothermal systems. The methods provide a basis for reasonable estimates of recharge areas and may be helpful for the protection of geothermal resources.

REFERENCES

Allen D.M., Grasby S.E., Voormeij D.A. (2006) Determining the circulation depth of thermal springs in the southern Rocky Mountain Trench, south-eastern British Columbia, Canada using geothermometry and borehole temperature logs. *Hydrogeology Journal* 14, 159–172.

Armstrong R.L., Taubeneck W.H., Hales P.O. (1977) Rb-Sr and K-Ar geochronometery of Mesozoic granitic rocks and their Sr isotopic composition, Oregon, Washington, and Idaho, *Geological Society of America Bulletin* 88, 397–411.

Ciezkowski W., Gronig M., Lesniak P.M., Weise S.M., Zuber A. (1992) Origin and age of thermal waters in Cieplece Spa, Sudeten, Poland, inferred from isotope, chemical and noble gas data. *Journal of Hydrology* 140, 89–117.

Clayton J.L., Megahan W.F., Hampton D. (1979) Soil and bedrock properties: weathering and alteration products and processes in the Idaho batholith, *USDA Forest Service,* Research Paper INT-237, 35p.

Criss R.E., Fleck R.J. (1990) Oxygen isotope map of the giant metamorphic-hydrothermal system around the northern part of the Idaho batholith, USA. *Applied Geochemistry* 5, 641–655.

Criss R.E., Taylor H.P. Jr. (1983) An $^{18}O/^{16}O$ and D/H study of Tertiary hydrothermal systems in the southern half of the Idaho batholith. *Geological Society of America Bulletin* 94(5), 640–663.

Derry L.A., Evans M.J., Darling R., France-Lanord C. (2009): Hydrothermal heat flow near Main Central Thrust, central Nepal Himalaya. *Earth and Planetary Science Letters* 286, 101–109.

Druchsel G.K., Rosenberg P.E. (2001) Non-magmatic fracture-controlled hydrothermal systems in the Idaho Batholith: South Fork Payette geothermal system. *Chemical Geology* 173, 271–291.

Ehlers T.A., Chapman D.S. (1999) Normal fault thermal regimes: conductive and hydrothermal heat transfer surrounding the Wasatch fault, Utah. *Tectonophysics* 312, 217–234.

Ferguson G., Grasby S.E., Hindle R. (2009) What do aquaeous geothermometers really tell us? *Geofluids* 9, 39–48.

Ferguson G., Grasby S.E. (2011): Thermal springs and heat flow in North America. *Geofluids* 11, 294–301.

Foster C., Smith L. (1986) The influence of groundwater flow on thermal regimes in mountainous terrain. *Proceedings of 11th workshop on geothermal reservoir engineering*. Standford University, Standford 135–139.

Forster C., Smith L. (1988) Groundwater flow systems in mountainous terrain: 1. Numerical modeling technique. *Water Resources Research* 24(7), 999–1010.

Forster C., Smith L. (1989) The influence of groundwater flow on thermal regimes in mountainous terrain: a model study. *Journal of Geophysical Research* 94(B7), 9439–9451.

Galván C.A., Prol-Ledesma R.M., Flores-Márquez E.L., Canet C., Estrada R.E.V. (2011) Shallow submarine and subaerial, low enthalpy hydrothermal manifestations in Punta Banda, Baja California, Mexico: Geophysical and geochemical characterization. *Geothermics* 40, 102–111.

Gvirtzman H., Garven G., Gvirtzman G. (1997) Thermal anomalies associated with forced and free ground-water convection in the Dead Sea rift valley. *Geological Society of America Bulletin* 109 (9), 1167–1176.

Heilweil V.M., Healy R.W., Harris R.N. (2012) Noble gases and coupled heat/fluid flow modeling for evaluating hydrogeologic conditions of volcanic island aquifers. *Journal of Hydrology* 464–465, 309–327.

Himes, S.A. (2012) Self-organizing fluid flow patterns in crystalline rock: Theoretical approach to the hydrothermal systems in the Middle Fork of the Boise River, unpublished Master's Thesis, Brigham Young University, Provo.

Holdaway, B.K. (1994) The geochemical evolution of cold and thermal ground waters in the southern part of the Idaho batholith, unpublished Master's Thesis, Brigham Young University, Provo.

Hurwitz S., Kipp K.L., Ingebritzen S.E., Reid M.E. (2003) Groundwater flow, heat transport and water table position within volcanic edifices: Implications for volcanic processes in the Cascade Range. *Journal of geophysical research-solid Earth* 108(B12), 2557.

Hyndman D.W., Talbot J.L. (1979) *The Idaho batholith and related subduction complex: Field Guide* no. 4, Geological Society of America 72nd Annual Meeting, Cordilleran Section, 16p.

International Heat Flow Commission (2012) The Global heat flow data base: www.heatlfow.und.edu, accessed March 12, 2012.

Ingebritzen S.E & Mariner R.H. (2010) Hydrothermal heat discharge in the Cascade Range, northwestern United States. *Journal of Volcanology and Geothermal Research* 196, 208–218.

Jetel J. (1972) Hydrogeology of the Sokolov Basin (function of rocks, hydrogeochemistry, mineral waters). *Sbor. geol. věd, Ř. HIG*, 9, 7–146.

Jokinen J., Kukkonen I.T. (1999) Random modelling of the lithospheric thermal regime: forward simulations applied in uncertainity analysis. *Tectonophysics* 306, 277–292.

Jones F., Marorowicz J. (1987) Some aspects of the thermal regime and hydrodynamics of Western Canadian sedimentary basin. In: Coffand J., Williams J. (eds.): *Fluid flow in sedimentary basins and aquifers*. Geological Society special publication 34, 79–85.

Lewis R.E., Young H.W. (1980) *Thermal springs in the Payette River basin, West-Central Idaho*. USGS Water-Resources Investigations 80–1020. 23p.

Lewis R.E., Young H.W. (1982) *Thermal springs in the Boise River basin, South-Central Idaho.-* U.S. Geological Survey. Water-Resources Investigations 82–4006. 22p.

López D., Smith L. (1995) Fluid flow in fault zones: Analysis of the interplay of convective circulation and topografically driven groundwater flow. *Water Resources Research* 31, 1489–1503.

López D., Smith L. (1996) Fluid flow in fault zones: influence of hydraulic anisotrophy and heterogeneity on the fluid flow and heat transfer regime. *Water Resources Research* 32, 3227–3235.

Lowell R.P. (1991) Modeling continental and submarine hydrothermal systems. *Reviews of geophysics* 29(3), 457–476.

Mariner R.H., Evans C.E., Young H.W. (2006) Comparison of circulation times of thermal waters discharging from the Idaho batholith based on geothermometer temperatures, helium concentrations, and ^{14}C measurements. *Geothermics* 35, 3–25.

Meert J.G., Smith D.L., Fishkin L. (1991): Heat-flow in the Ozark Plateau, Arkansas and Missouri – relationship to groundwater flow. *Journal of volcanology and geothermal research* 47, 337–347.

Mongelli F., Pagliarulo P. (1997) Influence of water recharge on heat transfer in a semi-infinite aquifer. *Geothermics* 26(3), 365–378.

Schlegel M.E., Mayo A.L., Nelson S., Henderson R., Eggett D. (2009) Paleo-climate of the Boise area, Idaho from the last glacial maximum to the present based on groundwater δ^2H and δ^{18}O compositions. *Quaternary Research* 71, 172–180.

Serban D.Z., Nielsen S.B., Demetrescu C. (2001) Transylvanian heat flow in the presence of topography, paleoclimate and groundwater flow. *Tectonophysics* 335, 331–344.

Smith L., Chapman D.S. (1983) On the thermal efects of groundwater flow: 1. Regional scale systems. *Journal of Geophysical Research* 88, 593–608.

Taylor H.P. Jr. (1977) Water/rock interactions and the origin of H20 in granitic batholith. *Journal of the Geological Society of London* 133, 509–558.

Truesdell A.H., Hulston J.R. (1980) Isotopic evidence on environments of geothermal systems, in *Handbook of Environmental Isotope Geochemistry, v. 1:*, In: P. Fritz P., Fontes J. (eds.) *The terrestrial environment*. Elsevier Press, Amsterdam, p. 179–226.

Young H.W. (1985) Geochemistry and hydrology of thermal springs in the Idaho batholith and adjacent areas, Central Idaho.-USGS Water resources Investigation Report 85–4172. 44p.

Zaher M.A., Saibi H., Nishijima J., Fujimitsu Y., Mesbah H., Ehara S. (2012) Exploration and assesment of the geothermal resources in the Hamman faraun hot spring, Sinai Peninsula, Egypt. *Journal of Asian Earth Sciences* 45, 256–267.

Chapter 15

Tunnel inflow in granite – fitting the field observations with hybrid model of discrete fractures and continuum

Milan Hokr[1], *Aleš Balvín*[2], *Ilona Škarydová*[2] & *Petr Rálek*[1]

[1]*Centre for Nanomaterials, Advanced Technologies and Innovations, Technical University of Liberec, Liberec, Czech Republic*
[2]*Faculty of Mechatronics, Informatics and Interdisciplinary Studies, Technical University of Liberec, Liberec, Czech Republic*

ABSTRACT

The water inflow into a tunnel has been studied in a granite massif in Jizera Mountains, Czech Republic. The dominant inflow occurs in the shallow parts of the system and in few fault zones in the deeper part. A combination of 2D discrete fractures (planar representation of vertical conductive faults) and 3D equivalent continuum are used to model the flow around the tunnel in a kilometre-scale domain with Flow123D. The geometry is based on the digital terrain model and the positions of faults determined by geophysical methods. The meshing is difficult due to the small scale of the tunnel and intersections with the fault planes. The inflow is measured as a flow rate in a collecting canal either by either weirs or by the tracer dilution method in five segments of the tunnel. The hydraulic conductivities of upper weathered zone, lower compact massif, and the vertical faults were calibrated by fitting the inflow from these subdomains, using the inverse solver UCODE. The model is verified against a 2D vertical plane solution with a finer mesh. The fitted hydraulic parameters are consistent with data for other sites in the Bohemian massif.

15.1 INTRODUCTION

One of goals in groundwater modelling studies is to determine the physical parameters of the system and to build models that fit the field observations and, especially in the case of an inverse problem, to estimate parameters from a model calibrated to measurements. The ultimate aims are to understand the behaviour of a particular groundwater system (a choice of the model) and to obtain parameter values for a study of the system in changed conditions. One of possible observations used for rock hydraulic properties estimation is the tunnel inflow, which is a mathematical analogy of borehole pressure tests, except for the horizontal direction and larger scale. There are several studies interpreting properties of the rock and its local physical behaviour from the tunnel inflow observation (e.g., Witherspoon, 2000; Mas Ivars, 2006; Marechal, 2012; Thapa *et al.*, 2003). In contrast, studies of a tunnel effect on the unsaturated zone flow regime and surface conditions are more typical for the shallow tunnels in known rock conditions (e.g., Valentová & Valenta, 2004). Most of the deep tunnel inflow studies are based on simple models such as the analytical solution of radial flow or other analytical or numerical concepts based on the 2D representation in the plane perpendicular to the tunnel.

A crucial issue and challenge in the fractured rock study is inhomogeneity, in particular the discontinuities, from all points of view – from conceptual models, numerical solutions, and field data acquisition. Standard approaches in hydrogeological textbooks and simulation software for fractured rocks are the discrete fracture network and the equivalent continuum models (Bear et al., 1993). New approaches can combine and extend the concepts and provide better representation. Such a case can be a representation of important fractures as discrete objects and the use of an equivalent continuum for a rock matrix with a network of less important fractures. The terminology used in the literature for this concept is a hybrid or multidimensional model. The simulation codes FEFLOW and FRAC3DVS (HydroGeoSphere) are examples that implement this concept with some limitations. The use of discrete fractures in numerical models is also a challenge with regard to discretisation due to their complex geometry. The inclusion of an excavation (e.g. a tunnel) to such a model brings even more difficulties.

In this chapter, an example of a field study and the modelling of tunnel inflow in 3D is described, with an emphasis on the rock inhomogeneity – represented as a combination of the discrete fractures and the equivalent continuum using in house simulation software Flow123D (TUL, 2010) based on the mixed-hybrid finite element method that offers a more general representation and solution of multidimensional model problems (Maryška et al., 2008). Previous work (Hokr et al., 2013b) concentrated on the numerical scheme and a demonstration of an application of the multidimensional concept in the context of a numerical solution for a real-world tunnel inflow example rather than an actual interpretation of field data.

Complex geological studies have been undertaken at the Bedrichov site since 2003 that consider a 30 year old water supply tunnel in a granite massif as an industrial analogue of the conditions in the geological disposal of the spent nuclear fuel (Klomínský & Woller, 2010; Žák et al., 2009). Until 2008, the research was managed by the Czech Geological Survey with support from the Radioactive Waste Repository Authority of the Czech Republic (SURAO). Later, the measurement automation (Špánek et al., 2011) and modelling project was led by the Technical University of Liberec. The site and measured data have been used as the 'test case' in the DECOVALEX project for comparison of models and numerical codes. New studies of natural tracer sampling and modelling for groundwater dating (Šanda et al., 2011) have started in addition to the work presented in this chapter.

The objectives of the work are to develop a numerical model which can represent appropriate details of the groundwater flow around the tunnel and to calibrate the hydraulic parameters of the rock by fitting the model results to the tunnel inflow field measurements, by a full inverse problem solution (optimization algorithm). A necessary part of the modelling is preprocessing the measured data. Assumptions on a distinction of the inflow from faults and from compact rock are made (the discrete features and the equivalent continuum respectively) with respect to the differences in the shallow and deep rock zones.

15.2   SITE LOCATION AND ITS PROPERTIES

The research area is located near the city of Liberec in the Jizera Mountains in the north of the Czech Republic. A water supply tunnel was built during the period 1980 to 1982 through compact granite, 2600 m long, 3.6 m diameter, and up to 150 m deep. The azimuth is about 70°, with 1.5% slope rising from west south west to east north east. A water pipe about 0.8 m wide is installed inside that connects the Josefův Důl drinking water reservoir in one valley (east north east) with the water treatment plant near the Bedřichov village in other one (west south west – main entrance) (Figure 15.1).

Two excavation techniques were used: the first 885 m west south west of the tunnel boring machine (TBM) and the remaining part by the drill-and-blast (DB) method. The tunnel walls are mainly open bare rock, but in several less stable zones there is a shotcrete layer. The inflow water from the rock is collected into a collection canal in the concrete bottom. There are shafts with grating covers approximately every 50 m in the collection canal (Figure 15.2).

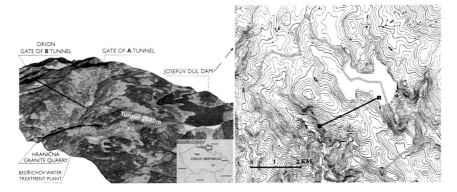

Figure 15.1 The site location and tunnel position (after Klomínský and Woller, 2010) (left), the digital terrain model (DTM) with the problem domain and the tunnel position (right). *(See colour plate section, Plate 36).*

Figure 15.2 (Left) a part of the tunnel excavated by boring machine with wet and dry strips visible, (Middle) a part by drill-and-blast method, shotcrete covered, with more visible inflow, (Right) an uncovered shaft to the collection canal (conductivity measurement during the dilution experiment). *(See colour plate section, Plate 37).*

15.3 METHOD

15.3.1 Water inflow measurement

The first water inflow measurement made after monitoring during the tunnel construction in 1981 was carried out in spring 2004 and repeated in autumn 2004; the flow rate in the collecting canal by means of weirs in the shafts and flow rate of individual inflow (springs). There were four profiles selected in 2004 that divide the tunnel into five sections (denoted as A–E, see Table 15.1 and Figure 15.3) according to the typical inflow and the rock state/properties. Together with the flow rate measurement in the entrance shaft of the tunnel (the total tunnel inflow), the inflows for each section could be obtained as differences between the neighbouring profiles. The section A was determined by the near-surface weathered zone. The middle part in the hard rock with the smaller inflow from several fault zones was divided by excavation method (section B of TBM and section C blasted). The east north east part in the weathered zone with the larger inflow was split into sections D and E, the latter already under the reservoir (Figure 15.3).

The measurement in 2004 was considered less reliable due to a leaky weir installation (Bělohradský & Burda, 2005) and imprecise water level measurement. In the middle part of the tunnel, the differences of the flow rate were generally less accurate due to subtraction of very similar values (Table 15.1, Figure 15.4). A new measurement of the flow rate at the same profiles was carried out in February 2012 by means of the tracer dilution method (Hokr *et al.*, 2012, Table 15.1). A solution was injected with a known concentration and by a controlled flow rate into the collection canal in the upstream end (east north east part). The conductivity was measured in the canal shafts downstream until steady state was derived from the mixing. Here, a part of the measurement uncertainty and error is common for all profiles, so the section inflow error resulting from subtraction of similar flow rates at the profiles is less (Hokr *et al.*, 2012).

Table 15.1 The inflow rate measured and the flux per one meter length derived; a compilation of two measurements in 2004 by the weir water level and one in 2012 by the tracer dilution method. The Apr 2004 is in high water stage while Oct 2004 and Feb 2012 in low (section A difference).

Tunnel section		Measurement at weirs				Tracer dilution method	
Code	Position (chainage) [m]	Flux April 2004 [l/s]	Flux per length Apr 2004 [m³/s/m]	Flux October 2004 [l/s]	Flux per length Oct 2004 [m³/s/m]	Flux Feb 2012 [l/s]	Flux per length Feb 2012 [m³/s/m]
A	0–150	1.26	8.4×10^{-6}	0.61	4.1×10^{-6}	0.56	3.7×10^{-6}
B	150–885	0.12	1.6×10^{-7}	0.05	6.8×10^{-8}	0.06	8.1×10^{-8}
C	885–1995	0.13	1.2×10^{-7}	0.09	8.1×10^{-8}	0.09	8.1×10^{-8}
D	1995–2424	0.17	4.0×10^{-7}	0.06	1.4×10^{-7}	0.64	1.4×10^{-6}
E	2424–2600	1.65	9.4×10^{-6}	1.84	1.5×10^{-5}	1.35	7.7×10^{-6}
	Sum/aver.	3.33	1.2×10^{-6}	2.65	1.5×10^{-5}	2.69	1×10^{-6}

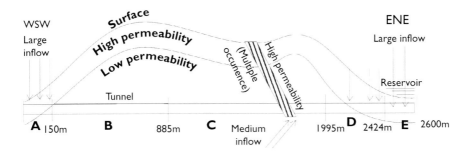

Figure 15.3 The conceptual model of hydraulic conductivity zones with the division of the tunnel to the five measured sections, the subvertical high permeability zones are considered in both B and C sections shown in Figure 15.5.

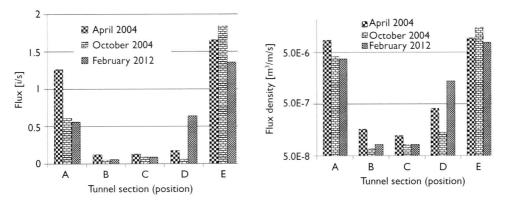

Figure 15.4 The total inflow to the tunnel sections defined in Figure 15.3 and the inflow per one meter length (density) at various times and methods.

Besides the total inflow, the individual groundwater discharges from the tunnel wall were measured at set of locations since 2006 till present, originally manually at 14 day intervals. Recently the data have been measured by automatic loggers. The discharges were classified according to their intensity as dripping discharges and the discharges with continual fluxes. The flow rates vary in time within a low range, except in the A section. Other quantities (e.g. temperature, conductivity, pH) were measured together with these discharge volumes. Some of the measurements were taken as a reference for the estimation of the inflow from the faults within the B and C sections, but commonly only part of the inflow from a particular fault can be collected directly.

Both the total inflow methods (Figure 15.4) gave similar inflow values for the B and C sections, the difference for A is due to different water stage (snow melt in April 2004). The sum of D and E is also consistent. The crucial difference is for the D interval, where the dilution method results are more accurate, which is confirmed by a visual observation of a strong inflow in contrast with the B and C sections.

15.3.2 Theory of the multidimensional model and software

The basic model is a combination of discrete fractures and equivalent continuum, both of which are individual standard methods for a fractured rock modelling. The model uses software developed in house over the last decade (Maryska *et al.*, 2008; TUL, 2012), for both the fractured hard rock and the fractured porous rock. The Darcy-type flow is considered in the combination of a 3D domain, set of 2D domains (polygons) and set of 1D domains (lines). The 3D domain is either the rock matrix (with the Darcy flow) or the equivalent continuum of a subset of smaller fractures. The 2D domain can mean either a fracture (i.e. a void between two parallel rock surfaces) or a planar representation of fault-like structures (i.e. a layer of a conductive porous rock or equivalent continuum of a small thickness with respect to the model scale). In the former case, the flow is governed by the Hagen-Poiseuille (cubic) law, which is equivalent to the Darcy law if the two-dimensional hydraulic conductivity K is set appropriately. In the latter case, the structure is characterized by the transmissivity that (in terms of the model concept) is determined as a product of the hydraulic conductivity (local) of a more conductive material and the real thickness of the layer. The 1D domain (with assigned physical cross-section area) represents a channel structure in 3D or a fracture in a 2D model (with two possible interpretations). There are more options for the mathematical formulation of the 3D-2D-1D interactions, e.g. concerning pressure/flux continuity over the interface. Depending on the numerical discretisation geometry, the interaction is introduced either as a source/sink or a boundary condition (Maryška *et al.*, 2008).

The software Flow123D (TUL, 2012) is an open-source under active development. The code is written in C++ and uses several established numerical libraries such as PETSc. It is running in the command line regime. The implemented phenomena include steady/unsteady flow in saturated media, multicomponent advective solute transport, non/equilibrium mobile-immobile exchange, sorption, single-component reaction, and an interface to geochemical codes. The numerical method for the flow is the mixed-hybrid finite element method (MHFEM) (Maryška *et al.*, 2008).

15.3.3 Inverse problem solution procedure

In the model, three values of hydraulic conductivity are calibrated by fitting the model to observed values of inflow. In such a combination, the inverse problem should have unique solution with a possibly zero residuum. The UCODE program available from the US Geological Survey was used (Poeter *et al.*, 2005). It allows running any code for the direct problem solution with input/output by means of text files. Normally the requirement for a model calibration procedure is that the number of observations is larger than the number of fitted model parameters (an over-determined problem). Such software is not intended for the type of problems commonly encountered in fractured rocks, but the limitation is only technical. The optimisation algorithm can only converge to the correct solution for a unique problem (a simpler one).

The software can be conveniently used in the following manner. Start with one fixed parameter and calibrate two free parameters against three inflow observations. Then one of the previously calibrated parameters is fixed and the remaining two are calibrated. The third (or more) such iteration is by analogy (i.e. exchanging the fixed and calibrated parameters). Three iterations are enough to drop the residuum to a negligible value.

15.4   CONCEPTUAL MODEL AND REFERENCE DATA PROCESSING

To define and parameterise a model appropriate to the coarse resolution of the observed data, the two major features in the tunnel inflow distribution are considered. First the larger inflow from the shallow part of the tunnel (sections A + D + E) through the weathered rock layer and second the inflow from the deep part (B + C), mostly in compact rock, concentrated to several short (meters) intervals where typically the tunnel crosses fault zones (with almost no inflow elsewhere). Therefore, the model consists of three subdomains (Figure 15.3): '3D shallow' (a weathered shallow layer – 3D equivalent continuum), '3D deep' (deep blocks of hard rock – 3D equivalent continuum), and '2D faults' (a set 2D planes, 'discrete fractures' in modelling sense, representing the vertical conductive faults within the massif). Each subdomain has homogeneous hydraulic conductivity (transmissivity).

Time changes of inflow related to seasonal or single-event water infiltration changes are not considered (i.e. the models are steady-state, which is accurate enough to compare measured data in a particular time). Further, the model is simplified regarding the unsaturated zone. The water table is taken to be at ground surface for convenient boundary condition definition. The experience at the site is that the water table elevation is several meters under the surface but there were no direct level measurement except observation of several permanent swamps along the profile. The depth of the tunnel is much larger then the thickness of the unsaturated zone under the surface, so the inaccuracy in the water level has little impact on the model inflow results (this is confirmed by stable inflow rate), except in the shallowest part of the tunnel next to the west south west entrance, section A. Here the model assumption corresponds to high water stage conditions. In the D and E sections, the water level is controlled by the reservoir which is represented in the model. These assumptions are no more restrictive than those applied in the reported deep tunnel inflow studies and still allow benefit from the 3D model concept (e.g. in a more realistic effect of the topography and mutual hydraulic/flow distribution influence between neighbouring zones of different conductivity along the tunnel).

The thickness of the weathered layer derived from the observed interface in the tunnel is 25–30 m from the west south west tunnel entrance, which also corresponds to data from boreholes in the region. For the model, the boundary is defined at a depth of 40 m depth that better fits the sections of measured inflow, A–E, and it is easier to implement to discretisation mesh.

Two sources of data are used to define the position of vertical permeable faults in the model. The profile of electrical resistivity between the tunnel and the surface (Figure 15.5a) is interpreted from stray current, spontaneous polarisation, and vertical electric sounding methods by Bárta *et al.* (2010). The lineament analysis (Figure 15.5b) from the terrain morphology and the aerial radiometric data from (Klomínský *et al.*, 2008) and estimates of tectonic structures in the region after Žák *et al.* (2009). These data are used for the model geometry up to small shifts for easier meshing. The positions of lineaments along the tunnel profile are only partly coupled with the geophysics and with the observations inside the tunnel. The four intersections in the model are illustrated in Figure 15.5 with their counterparts from the resistivity field, which are sufficient for the model purposes.

The measured inflow data presented in Section 15.3.1 are adjusted to a form suitable for the model calibration. The first issue is to split the measured inflow between the '3D shallow' and '3D deep' domains of the model consistently. The inflow in the section D, which drains both the weathered layer and hard rock massif, is attached to the '3D shallow' model domain as the former has the larger contribution.

The second issue is a distribution of the measured inflow in the B and C sections between the faults and the hard rock blocks (counterparts of the '3D deep' and '2D faults' model domains). This splitting is not explicitly caught in the measured data and the possible canal flow rate difference between the shafts upstream and downstream the fault intersection could not be accurately derived from the tracer dilution data. The range is between 0.05 l/s and 0.15 l/s. The individual discharges that can be collected in a measuring vessel at two obvious tunnel/fault intersections are about 0.05 l/s in total. The real inflow at those places can be estimated as twice as much due to non-measurable flow behind the shotcrete cover and to the tunnel floor. To cover this wide range of uncertainty, four model distributions between the fault inflow and the hard rock blocks (continuum) inflow, as the ratios 20:1, 3:1, 1:1, 1:3 are used (Table 15.2) and the calibration of the model towards all of these 'measured' values is interpreted as the uncertainty of the hydraulic parameter determination. In particular, the estimate above 0.1 l/s inflow for the '2D fault' domain corresponds to the f > r ratio for the February 2012 measurement (0.15 l/s for B + C) and in between the f = r and f < r for the April 2004 measurement (0.25 l/s for B + C).

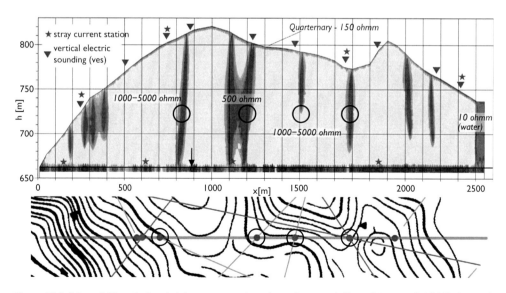

Figure 15.5 (Upper) Electrical resistivity cross-section along the tunnel (from Bárta et al., 2010), (Lower) Tectonic lineaments in the map (Klomínský et al., 2008). Circles denote the corresponding locations from both sources, represented in the model. (See colour plate section, Plate 38).

Table 15.2 Variants of the ratio between the flow rate from the compact rock and from the fault zones.

| Symbol | Ratio of flow rate | | Description |
	Fault planes	Massif	
f >> r	20	1	Major inflow is from the faults.
f > r	3	1	More water flows from the faults.
f = r	1	1	Inflow is equal.
f < r	1	3	More water flows from the hard rock.

15.5   NUMERICAL MODEL SETTING

15.5.1   Geometry and meshing

The model is established by means of the topographic boundaries such as ridges and valleys. The scheme in the plan view is given in Figure 15.1. The whole domain covers as much as possible from the two watersheds of about 20 km² size and fits the area 5000 × 6000 m. The top surface is determined by the Digital Terrain Model (DTM) of 100 m resolution (Figure 15.6a), which still contains the main features but it is kept coarse enough to avoid uselessly fine meshing. The DTM triangulation is deformed and extended to follow the important terrain structure lines (valleys, the dam, and the shore of the reservoir). Also the elevation at the tunnel ends are adjusted to avoid overlapping of the terrain surface with the tunnel. The bottom is flat at the 400 m altitude, the maximum extent of the terrain ranges from 550 to 880 m.

The tunnel is represented in a simplified and coarsened form. In contrast with the previously constructed 3D model without the faults (Hokr, 2013b, Table 15.3), where the circular shape of the tunnel was represented with a regular octagon of the actual size, the calculation of the intersections between the tunnel and the fault planes required coarsening the tunnel to avoid fine-scale shapes causing numerical problems. The adopted tunnel shape is a square with 7.1 m sides (10 m diagonal). The tunnel is represented as a hollow volume of a square-base prism. The effect of the coarsening and the size difference has been justified and corrected by the 2D model comparison (Section 15.5.3). The vertical conductive fault zones are represented by planes (2D) according to the multidimensional concept. There are 30 fault planes in the model and four of them cross the tunnel (Figure 15.6b). For simplicity a 1 m width is used for all, so the numerical value is the same for the hydraulic conductivity in m/s and for the transmissivity in m²/s.

The mesh is generated with 100 m resolution in the peripheral parts of the domain and it is automatically refined close to the fault plane intersections and along the tunnel, where the mesh of 7.1 m corresponds to the model tunnel size but is finer at the intersections between the tunnel and the vertical planes (Figure 15.7).

15.5.2   Boundary conditions

The boundary conditions (Figure 15.8) are set according to the conceptual model and the assumptions stated above. The zero flow is prescribed on the bottom (impermeable) and on the lateral vertical boundaries, expressing the symmetry

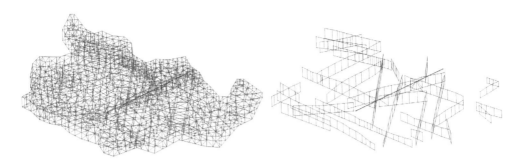

Figure 15.6 (a) Model geometry based on the digital terrain model, (b) Positions of fault zones and of the tunnel in the model (the same scale as (a)).

Table 15.3 Overview of related tunnel inflow models applied for the Bedřichov site.

Notation	Dimension/ tunnel shape	Features	Reference/objective
2D vertical	500 × 300 m 24-sided polygon	Homogeneous (single tunnel section)	Hokr et al. 2012 (Feb 2012 inflow – basic interpretation)
	500 × 300 m both 24-sided and square	Homogeneous (single tunnel section)	Hokr et al. 2013a (evaluation of meshing influence)
3D homogeneous	5000 × 6000 m octagon and square profile	Homogeneous 3D + faults 2D	Hokr et al. 2013b (demonstration of numerical technique)
3D full-feature	5000 × 6000 m square profile	Two 3D subdomains + faults 2D	This chapter – model calibration to real data

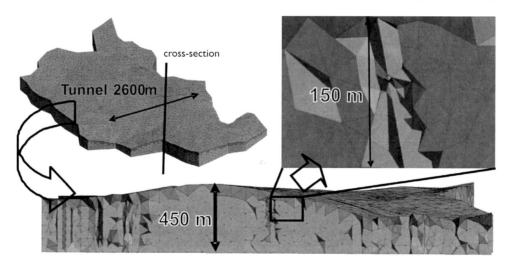

Figure 15.7 Discretisation of the model with tetrahedra in 3D – (Upper left) the full view, (Lower) a vertical cross-section, and (Upper right) a detail around the tunnel. The colour corresponds to the subdomains of different hydraulic conductivity. *(See colour plate section, Plate 39).*

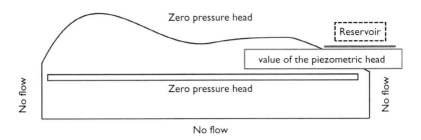

Figure 15.8 Boundary conditions of the three-dimensional model of the tunnel inflow.

of flow from both sides of ridges and valleys (for simplicity this condition is also extended to small part of the vertical boundary in the south east where some outflow due to low topography should exist, but with no tunnel effects. The zero pressure is prescribed on the top surface, then the infiltration or drainage is calculated as a model result. On the part of the surface on the reservoir bed, the constant piezometric head of the mean reservoir water level is prescribed. The zero pressure is prescribed on the tunnel wall (i.e. a free discharge at atmospheric pressure). The conditions are applied consistently to both the 3D and the 2D subdomains, in particular the same values are applied for the line boundaries of the vertical fault planes.

15.5.3 Auxiliary 2D models

The 3D model is related to other models of the site used either in earlier work or supporting the current model verification. Only the basic features and their role in the analysis are presented here. The list of the models is given in Table 15.3. Solution of the tunnel inflow problem by means of a 2D vertical cross-section model is a simpler reference case for the 3D model check and validation. The rectangular vertical cross-section domain, 500 m × 300 m (sufficiently large with respect to the tunnel hydraulic reach) with appropriate analogy of the conceptual assumptions is used: homogeneous hydraulic conductivity, the impermeable bottom boundary, the zero pressure top boundary (the water table at the ground surface), and the corresponding hydrostatic lateral boundaries. Five such models are defined, each with a different tunnel depth, representing the average depth of the respective tunnel intervals A-E with the evaluated inflow.

This set of models was applied in (Hokr et al., 2012) for basic evaluation of the tracer dilution measurement (February 2012), calibrating the hydraulic conductivity for each tunnel section. The model has been used to study effects of a different tunnel size, a simplified tunnel shape, and different mesh density. One of the results of Hokr *et al.* (2013a) was a correction factor between the tunnel inflow rate evaluated from the most accurate reference model (real shape and maximum mesh density) and from selected coarse models, in particular for the geometry corresponding to the 3D model presented in this paper (the value is 1.12). The data are already corrected with this factor.

15.6 EVALUATION AND DISCUSSION OF RESULTS

A single value of the hydraulic conductivity was calibrated for each of the model sub-domains. The tunnel inflow within the model (i.e. boundary flux) was calculated in the analogous division of the tunnel intersections with the respective subdomains and these three inflow values were fitted to their measured counterparts by the optimisation algorithm in Section 15.3.3.

The results of the hydraulic conductivity estimation are given in Table 15.4 (for February 2012 inflow measurement only) and visualised in Figure 15.9 (for April 2004 and February 2012 inflow). The calibrated hydraulic conductivity of the shallow weathered layer is almost the same for both methods, consistent with only small difference of the total measured inflow in the A, D, and E sections (Table 15.1). As

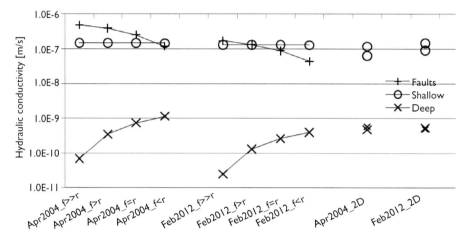

Figure 15.9 Comparison of the calibrated hydraulic conductivities for the model subdomains. The measurement is determined by the date and the assumption or model type by abbreviation on the horizontal axis. The values for 2D are each from a different model (Section 15.5.3) and the deep variants include the effect of faults.

Table 15.4 Inflow rates for the variants in Table 15.2, based on the February 2012 measurement, and the respective calibrated hydraulic conductivities for the three model subdomains.

Inflow [l/s]	Most from faults [l/s]	More from faults [l/s]	Inflow is equal [l/s]	More from rock [l/s]
Ratio fault/cont.	20:1	3:1	1:1	1:3
Shallow inflow	2.55	2.55	2.55	2.55
Deep inflow	0.0071	0.0375	0.075	0.1125
Fault inflow	0.1428	0.1125	0.075	0.0375
Sum	2.7	2.7	2.7	2.7
Calibrated conductivity	*[m/s]*	*[m/s]*	*[m/s]*	*[m/s]*
$K_{2Dfaults}$	1.72×10^{-7}	1.35×10^{-7}	8.95×10^{-8}	4.43×10^{-8}
$K_{3Dshallow}$	1.32×10^{-7}	1.31×10^{-7}	1.31×10^{-7}	1.3×10^{-7}
K_{3Ddeep}	2.47×10^{-11}	1.31×10^{-10}	2.63×10^{-10}	3.98×10^{-10}

expected, the differences are related to the applied assumptions on the distribution of the inflow between the 2D faults and the deep 3D continuum. As the measured inflow from the deep part (B + C) of the tunnel in total is smaller in February 2012 as opposed to April 2004, the estimated hydraulic parameters of both the faults and continuum are smaller, but only slightly less than half an order of magnitude. The calibrated difference between the fault conductivity and that of the deep rock equivalent continuum is the same for both measurement methods.

Taking into account the estimation of the fault inflow based on single discharge (e.g. 0.1 l/s), model variants were selected with calibrated fault conductivities similar to each other, but with the calibrated 3D deep rock conductivities in the margins of the range – see April 2004 f > r and February 2012 f > r plots (Figure 15.9). The range of the equivalent conductivity of the rock massif with excluded known conductive faults is almost two orders of magnitude based on these measurements. The 3D model results were compared with the set of 2D vertical cross-section (Figure 15.9). The models for A and E sections are assigned to the shallow subdomain of the 3D model and the models for B and C sections to the deep subdomain. The results fit well for the respective subdomains.

The average recharge in the model was evaluated by dividing the total negative (inwards) flux through the terrain boundary by half of the model area (simplest estimate of recharge/discharge partition). The value 100 mm per year is almost the same for all model variants and is 3 to 4 times smaller then a possible real value. It implies that the hydraulic conductivity in the shallow domain can be underestimated, one of the reasons can be that it is calibrated from inflow at higher model water level than the real one.

The conductivity for the 3D deep domain in the model is at the lower limit of the range of hydraulic conductivities in Bohemian massif (Rukavičková, 2009) or corresponds to bigger depths. A continuum without the main conductive structures can partly explain this, but it could also be accounted for the scale difference between this study and that by Rukavičková (op cit.), which is based on tests conducted in boreholes.

Figures 10 and 11 show the piezometric head and the velocity fields separately for the 3D continuum and the 2D fault subdomains. The combined views were constructed with each subdomain selected to offer a view 'inside' the model to either the tunnel or the fault planes. The transition between the converging flow field to the tunnel and the flow field controlled by the topography is interesting as the flow is either downward or upward along the faults or the faults act as barriers between the two flow-field patterns.

15.7   CONCLUSIONS

The multidimensional fractured rock model for interpretation of the tunnel inflow observations has been successfully applied. In the given scale, the 2D planar features represent the vertical conductive faults and the 3D domain represents the equivalent continuum of the remaining rock volume, which is split between the shallow weathered section and the deep hard rock section. The geometric complexity of the data forced a compromise for the model resolution, nevertheless, reasonable results were

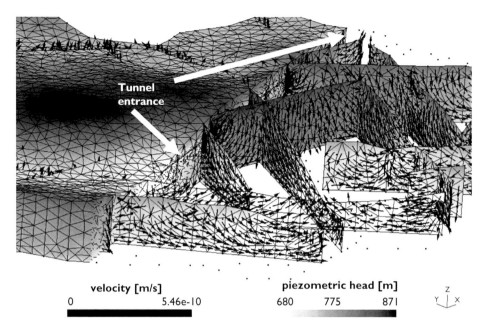

Figure 15.10 Piezometric head and velocity field visualisation – the section along the tunnel separates a part with 3D domain visible and a part with only 2D fault structures visible. *(See colour plate section, Plate 40).*

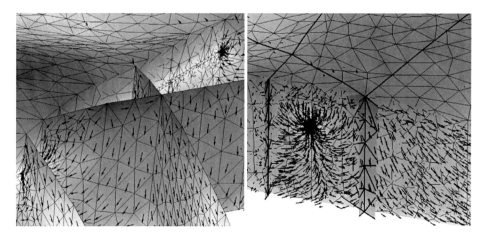

Figure 15.11 Details of the flow field: in the left the flow along the fault structures, in the right the flow in the 3D continuum separated by the fault structures between the tunnel-controlled and the topography-controlled parts. The scale is the same as in Figure 15.10. *(See colour plate section, Plate 41).*

obtained, as confirmed by the comparison with the 2D model of finer resolution, in terms of relation between the hydraulic conductivity and the tunnel inflow.

The calibrated hydraulic conductivity values are in agreement with the data for the Bohemian massif (Rukavičková, 2009) taking into account the scale difference.

In general, the range corresponding to the variants of the inflow measurements and the distribution assumptions define the uncertainty of the hydraulic properties. The model resolution and the accuracy of fit could be improved in the future with continuous flow rate measurements in the tunnel sections, including more comprehensive measurement of individual discharge points. The overall understanding of the water regime on site could be extended with the water dating techniques.

ACKNOWLEDGEMENTS

This work has been supported by the Ministry of Industry and Trade within the research project FR-TI1/362. The research reported in this paper was supported in part by the Project OP VaVpI Centre for Nanomaterials, Advanced Technologies and Innovation CZ.1.05/2.1.00/01.0005.

The work described in this chapter was conducted within the context of the international DECOVALEX Project (DEmonstration of COupled models and their VALidation against EXperiments). The authors are grateful to the Funding Organisation who supported the work: the Radioactive Waste Repository Authority contract SO 2011–017. The views expressed in the chapter are, however, those of the authors and are not necessarily those of the Funding Organisations. The authors also thank to SČVK a.s., the water treatment company, for providing the access to the tunnel.

REFERENCES

Bárta, J., Kněz, J., Budínský V., Jirků, J. (2010), Seven years of experience in experimental testing of granite rocks in the gallery serving as water conduit from the Josefův Důl hydraulic structure to water treatment plant in Bedřichov (Northern Bohemia). *Exploration Geophysics, Remote Sensing and Environment (EGRSE)*, 2010/2, pp. 1–13, Czech Association of Geophysicists.

Bear, J., Tsang C.F., de Marsily, G. (1993) *Modelling Flow and Contaminant Transport in Fractured Rocks*, USA Academic Press, San Diego, CA.

Bělohradský V., Burda J. (2005) Hydrogeologie vodárenských tunelů a jejich okolí povrchu (Hydrogeology of the water supply tunnels and their surroundings), In (Klomínský J., ed.) Geologická a strukturní charakteristika granitoidů z vodárenských tunelů v Jizerských horách Etapa 2004–2005, SURAO (Radioactive Waste Repository Authority), in Czech.

Hokr M., Rálek P., Balvín A. (2012) Channel flow dilution measurement used for tunnel inflow evaluation, In: *Latest trends in environmental and manufacturing engineering (Proceedings of WSEAS EG '12)* (Ponis S *et al*, eds.), WSEAS Press, pp. 171–176.

Hokr, M., Balvín A., Frydrych D, Škarydová, I. (2013a) Meshing issues in the numerical solution of the tunnel inflow problem, In: *Mathematical Models in Engineering and Computer Science* (Marascu-Klein, ed.), NAUN, 2013, pp. 162–168.

Hokr M., Škarydová I., Frydrych D. (2013b) Modelling of tunnel inflow with combination of discrete fractures and continuum, *Computing and Visualization in Science* 15(1), 21–28.

Klomínský J. (ed.) (2008) Studium dynamiky puklinové sítě granitoidů ve vodárenském tunelu Bedřichov v Jizerských horách – Etapa 2006–2008 (Study of the fracture network dynamics in granitoids of the Bedrichov water supply tunnel in Jizera mountains), SURAO (Radioactive Waste Repository Authority), in Czech, 172p.

Klomínský J., Woller F. (eds.) (2010): Geological studies in the Bedrichov water supply tunnel. RAWRA Technical Report 02/2010. 103p., Czech Geological Survey, Prague. ISBN 978-80-7075-760-4.

Maréchal J.-C. (2012) Les tunnels alpins: observatoires de l'hydrogéologie des grands massifs montagneux (The Mont-Blanc road tunnel: example of groundwater observatory in the Alps), La Houille Blanche, n°1, 2012, pp. 44–50, in French.

Maryška J., Severýn O., Tauchman M., Tondr D. (2008) Modelling of processes in fractured rock using FEM/FVM on multidimensional domains. *Journal of Computational and Applied Mathematics*, 215/2, pp. 495–502.

Mas Ivars D. (2006). Water inflow into excavations in fractured rock – a three-dimensional hydro-mechanical numerical study. *International Journal of Rock Mechanics and Mining Sciences* 43, 705–725.

Poeter E.P., Hill M.C., Banta E.R., Mehl S., Christensen (2005) UCODE_2005 and Six Other Computer Codes for Universal Sensitivity Analysis, Calibration, and Uncertainty Evaluation: U.S. Geological Survey Techniques and Methods 6-A11, 283p.

Rukavičková L. (2009): Hydraulická vodivost granitů Českého masivu – srovnání dat detailního a regionálního měřítka (Hydraulic conductivity of granites in the Bohemian Massif – comparison of a detailed and a regional scale), *Geoscience Research Report for 2008*, pp. 252–254, Czech Geol. Survey. (in Czech)

Šanda M., Vitvar T., Hokr M., Balvín A. (2011): Tritium-Helium-3 Dating Technique in Granitic Structures of Catchments, In the Northern Czech Republic, presented at ERB workshop 2011 *Geochemical, isotope and innovative tracers: Challenges and perspectives for small catchment research*.

Špánek R., Hernych M., Hokr M., Svoboda P., Tyl P., Řimnáč M., Štuller J. (2011) Bedřichov Tunnel – Continual Automated Measurement of Physical Quantities, *Exploration Geophysics, Remote Sensing and Environment*, 2011/2, pp.73–82, Czech Association of Geophysicists (ČAAG), ISSN 1803–1447.

Thapa B.B., Nolting R.M., Teske M.J., McRae M.T. (2003) Predicted and Observed Groundwater Inflows into Two Rock Tunnels. *Rapid Excavation and Tunneling Conference Proceedings* 2003, SME.

TUL (2012) FLOW123D version 1.7.0, Documentation of file formats and brief user manual, NTI TUL, Online: https://dev.nti.tul.cz/trac/flow123d.

Valentová J, Valenta P. (2004) Modelling of groundwater flow in the vicinity of tunnel structures. *Journal of Hydrology and Hydromechanics* 52, 91–101.

Witherspoon P.A. (2000) The Stripa project. *International Journal of Rock Mechanics and Mining Sciences* 37, 385–396.

Žák J., Verner K., Klomínský J., Chlupáčová M. (2009) "Granite tectonics" revisited: insights from comparison of K-feldspar shape-fabric, anisotropy of magnetic susceptibility (AMS), and brittle fractures in the Jizera granite, Bohemian Massif. *Int. Journal of Earth Science (Geol. Rndsch.)* 98(5), 949–967.

Chapter 16

Uranium distribution in groundwater from fractured crystalline aquifers in Norway

Bjørn S. Frengstad[1] & David Banks[2]
[1]*Geological Survey of Norway, Trondheim, Norway*
[2]*Holymoor Consultancy Ltd., Chesterfield, Derbyshire, UK*

ABSTRACT

Uranium is a heavy metal which is omnipresent in nature as a trace element. Solubility is high over a wide pH range in oxidising groundwater systems. In Norway, uranium in groundwater is particularly linked to granitic and gneissic aquifers. The specific radioactivity of naturally occurring uranium is rather low. Epidemiological studies indicate that long-term intake of drinking water with elevated uranium content affects the kidneys due to the chemo-toxicity of the element. A survey of 476 private bedrock boreholes in South Norway showed a median uranium concentration of 2.5 µg/l with a maximum of 750 µg/l. 12% exceeded the guideline value for drinking water of 30 µg/l set by WHO. No drinking water limit is so far defined by the European Union. A further survey of public waterworks and wells used for food production and other industries in Norway showed a median concentration, 75th percentile and maximum value for fractured crystalline aquifers of 2.04 µg, 6.8 µg and 246 µg/l, respectively (N = 346). Of these, 7.5% exceeded 30 µg/l.

16.1 INTRODUCTION

Natural uranium consists of three isotopes, which are radioactive and emit α-particles. The isotope ^{238}U with a half-life of 4.47 billion years is absolutely dominant with a share of 99.2745%. ^{235}U with a half-life of 704 million years constitutes 0.72%, while the remaining 0.0055% comprises ^{234}U with a half-life of 245 000 years. The specific radioactivity of naturally occurring uranium (mainly the ^{238}U isotope) reflects these long half-lives and is thus rather low. Enriched uranium has undergone an isotopic separation to increase the percent composition of ^{235}U and is used in nuclear weapons and fission reactors for energy production. The residue from this enrichment is referred to as 'depleted uranium' (depleted in ^{235}U) and has a very low specific radioactivity. Depleted uranium is used as ballast and in armour-piercing ammunition, due to its very high density (19 kg/dm^3). The potential health effects of depleted uranium have received concern. Incidents of leukaemia among Italian war veterans returning from the Balkan conflicts were suspected to be caused by exposure to depleted uranium and the international press published alarming reports in 2001 (Agence France-Presse, 2001; Norton-Taylor, 2001; CNN, 2001), which were followed up by more considered studies by, among others, the United Nations Environment Programme (2002). Swedish soldiers were also checked for uranium concentrations in their urine after they had served in United Nations forces in Kosovo. Surprisingly, the investigations revealed that the control group of soldiers who had not yet left Sweden had significantly higher uranium levels in their urine (Lagercrantz, 2003). Further investigations pointed towards the drinking water supply at the military training camp in Sweden as the main source for ingested uranium. A later study in Italy showed that the number of cancer incidents (leukaemia included) among the Balkan veterans was lower than among the general Italian population (Lagercrantz, 2003).

The health personnel at the Swedish Armed Forces Headquarters realised what had been known among Scandinavian geochemists and hydrogeologists for some years, namely that several types of Fennoscandian crystalline bedrock aquifer may yield groundwater with elevated uranium concentrations (Asikainen & Kahlos, 1979; Salonen, 1994; Banks *et al.*, 1995; Reimann *et al.*, 1996; Midtgård *et al.*, 1998; Frengstad *et al.*, 2000).

16.1.1   The chemistry and geology of uranium

Uranium is a heavy metal which does not occur in its elemental state in nature. Two oxidation states, +IV and +VI are common; the third, +V, state is usually unstable and is of little importance in the groundwater environment (Åström *et al.*, 2009). Typical uranium minerals are uraninite ($U^{IV}O_2$) and its partially oxidised form (U_3O_8 – sometimes, though not unambiguously, referred to as pitchblende), brannerite ((U^{IV},Ca)(Ti,Fe^{III})$_2O_6$) and carnotite ($K_2(U^{VI}O_2)_2(VO_4)_2 \cdot 3H_2O$), but uranium is more widespread as an accessory element in zircon, apatite and monazite (Altschuler *et al.*, 1958; Reimann *et al.*, 1998). The average concentration in the upper continental crust is estimated to be 2.5–2.8 mg/kg (Wedepohl, 1995) or ≈ 0.0003 weight percent which makes it a rather common element. Sea water contains around 3 µg/l U (Åström *et al.*, 2009).

The mobility of uranium in groundwater is strongly redox-sensitive and also pH-sensitive. Solubility is high over a wide pH range in oxidising groundwater systems, where uranium occurs in its hexavalent (U^{+VI}) state. In oxidising conditions, at low pH, uranium may occur as its uranyl cation, UO_2^{2+}, while at higher pH, hydroxyuranyl (e.g. UO_2OH^+) or carbonate (e.g. $UO_2(CO_3)_2^{2-}$) species predominate (Figure 16.1). Complexation with fluoride, sulphate, phosphate and dissolved humic substances is documented and can significantly affect solubility and mobility (Åström *et al.*, 2009). U^{+VI} is strongly sorbed to certain metal oxides, and its immobilisation by ferric oxyhydroxide and amorphous aluminium oxide is documented at pH > 5 (Skeppström & Olofsson, 2007; Åström *et al.*, 2009).

Under reducing conditions uranium occurs in its tetravalent (U^{+IV}) state which typically has a very low solubility in water, co-precipitating with other metal oxides on mineral grain coatings (Casas *et al.*, 1998; Skeppström & Olofsson, 2007). Thus, uranium, mobilised in groundwater, can re-precipitate on fracture surfaces under reducing conditions, providing a source of radium and, ultimately, radon, to the groundwater within the fractures.

16.1.2   Uranium and Health

Naturally occurring uranium is not particularly radioactive and the concern for uranium in drinking water is linked to its chemotoxicity rather than to its radiotoxicity (Milvy & Cothern, 1990). Indeed, the German Umweltbundesamt (2011) suggests that

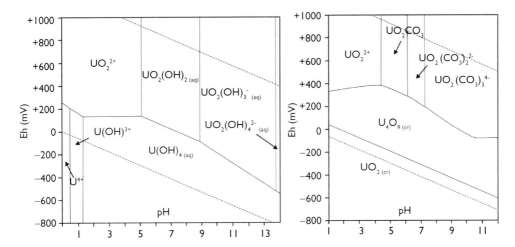

Figure 16.1 Eh-pH (Pourbaix) diagrams for aqueous uranium solutions: (Left) the U-O-H system at low uranium concentrations at 298.15 K and 10^5 Pa, with $\Sigma U = 10^{-10}$ M = 0.0238 µg/l (modified after Takeno 2005); (Right) the U-O-H-carbonate system at a high uranium and carbonate concentration at 298.15 K, with $\Sigma U = 10^{-5}$ M = 2.38 mg/l and $\Sigma C_{carbonate} = 10^{-2}$ M = 10 mmol/l (modified after public domain diagram from Wikimedia Commons http://commons.wikimedia.org/).

radiotoxicity may not become a significant concern until concentrations of 60–90 µg/l are reached, whereas there is a general consensus that chemotoxicity becomes a concern in the range 10–30 µg/l. The Canadian Federal-Provincial-Territorial Committee on Drinking Water (2006) estimate that a drinking water concentration of 20 µg/l would correspond to 0.25 Bq/l of [238]U, 0.01 Bq/l of [235]U, and 0.25 Bq/l of [234]U, assuming equilibrium concentrations of the isotopes. They note, however, that, in groundwater, [238]U and [234]U may be in disequilibrium by factors of around 2. Skeppström & Olofsson (2007) and Ek *et al.* (2008) suggest that a drinking water concentration of 100 µg/l uranium may result in a dose of 0.1 mSv/year.

Some epidemiological studies (Zamora *et al.*, 1998) suggest that prolonged consumption of uranium-rich water may lead to progressive or irreversible renal injury but not necessarily to kidney failure. Wagner *et al.* (2011) used GIS techniques to combine data on groundwater uranium concentrations in South Carolina, USA (N = 4 600) with census data on residential water use and with cancer incidence rates (N = 134 685) from the Central Cancer Registry. This comprehensive study found elevated incidence of colorectal, breast, kidney and total cancer in census tracts with elevated uranium concentrations in the groundwater and with frequent use of groundwater as a drinking water source. Sensitivity analysis of the data showed that the association between uranium concentrations in groundwater and cancer incidents became stronger when groundwater use per capita increased. However, the study was not able to distinguish between uranium's radiological and chemical effects and the potential effect of possible radioactive decay products of uranium, e.g. [226]Ra, [228]Ra, [222]Rn (and presumably, other elements or hydrochemical species that may co-vary with uranium).

16.1.3   Drinking water legislation

With the introduction of the inductively coupled plasma-mass spectrometry (ICP-MS) method, analysis of trace elements in water, including uranium, became both rapid and convenient (Houk & Thompson, 1988). Nevertheless, relatively few waterworks have ever checked their water for uranium. Despite many years of health concerns over uranium in drinking water (Health Canada, 1996; World Health Organisation, 1998), very few European national health authorities have yet established maximum admissible concentrations of uranium in their drinking water legislation. The National Food Agency in Sweden became aware of the possible implications of uranium in drinking water through their investigations of Balkan veterans (Lagercrantz, 2003) and promotes an action limit of 15 µg/l (Ek *et al.*, 2008). German health authorities have introduced differentiated drinking water limits, taking into account the larger water intake of small children, relative to body weight. Water used for preparation of baby nutrition should thus have a uranium content below 2 µg/l. For the remainder of the population, a limit of 10 µg/l was specified in 2011 by Umweltbundesamt (2011). In Finland, there is no legislated maximum permissible uranium concentration in drinking water, although the Radiation and Nuclear Safety Authority recommends a maximum ^{238}U concentration of 100 µg/l (Vesterbacka, 2007).

The World Health Organisation initially suggested a guideline value for drinking water of 2 µg/l (World Health Organisation, 1998) although, following further consideration, this has now been set at 30 µg/l (World Health Organisation, 2011). The US Environmental Protection Agency has also set a maximum admissible concentration of 30 µg/l on the basis of a cost-benefit analysis (United States Environmental Protection agency, 2000). Table 16.1 shows an overview of the current drinking water standards for uranium in different countries.

It should be noted that private wells and waterworks serving less than 50 people are not regulated in Norway, although a very rough estimate of circa 60 000 people who may be affected by significant uranium concentration in potable groundwater is far from negligible from a national health perspective, at least in Norway.

16.1.4   Geological, hydrological and demographic characteristics of Norway

More than half of mainland Norway is underlain by Precambrian rocks, mainly granitic gneisses, amphibolites, granites and migmatites. In addition, another 15% is made up of late Precambrian sedimentary rocks, mainly meta-sandstones. Metasedimentary and metavolcanic rocks from the Caledonian orogeny cover roughly 30% of the land area and strike through most of the country from the Stavanger area in southwest to North Cape in northeast. Finally, igneous and volcanic rocks related to the Permian Oslo graben account for 2% of the land area (Oftedahl, 1981). All rocks, with very few exceptions, display negligible primary porosity and are regarded as fractured aquifers.

A combination of tectonic uplift during the Tertiary and repeated glacial cycles in the Quaternary resulted in a pronounced topography with deep valleys, fjords and countless lakes. Unconsolidated Quaternary sediments include glacial till over much of the country with glaciofluvial outwash deposits in specific locations and in many valleys. Marine clays and silts, exposed during postglacial isostatic uplift, occur in

Table 16.1 Some drinking water standards for uranium.

Country/Institution	Limit	Value (µg/l)	Introduced	Comment
World Health Organization (2011)	Provisional guideline	30	2011	2 µg/l 1998 9 µg/l 2003 15 µg/l 2004
USA, USEPA (2000)	Max. contaminant level	30	2000	Raised from 20 µg/l
Russian Federation Kirjukhin *et al.* (1993)		1700	1993	May be outdated?
Canada, Health Canada (2012)	Max. Acceptable Concentration	20	1999	
Australia, National Health and Medical Research Council (2011)	Drinking water guideline	17	2011	Lowered from 20 µg/l
Germany, Umweltbundesamt (2011)	Max. value	10	2011	2 µg/l for preparation of baby nutrition
Sweden, National Food Agency. Livsmedelsverket (2011)	Recommended action level	15	2005	
European Union, The Council of the European Union (1998)		No standards		
Norway, Helse og omsorgsdepartementet (2001)		No standards		

coastal areas and below the marine limit in valleys. These deposits are overlain in valleys by more recent alluvial sediments.

The Norwegian Institute of Public Health (Myrstad *et al.*, 2010) publishes statistics concerning the water sources utilised by public waterworks serving more than 50 persons. From these statistics, we estimate that 73% of the Norwegian population are served by lake water, while 8% derive their potable water from rivers. Larger groundwater well-fields, mainly based on river bank infiltration in Quaternary aquifers, but including some productive drilled well-fields in fractured crystalline aquifers, serve 9% of the population. The remaining 10% of the population ($\approx$500 000 people) receive their water supply from small private solutions: traditionally, these included dug wells, springs or streams, but since 1950 the majority of these have been replaced by boreholes drilled in fractured crystalline aquifers. These drilled boreholes typically have low yields, with a median of 600–700 l/hr being typical of Norway, Sweden and Finland (Banks *et al.*, 2010).

16.2   URANIUM DISTRIBUTION IN GROUNDWATER IN NORWAY

There have been four main surveys of crystalline bedrock groundwater in Norway, which have included uranium as an analytical parameter.

16.2.1   Survey 1

The first survey was carried out in 1992–93 on a limited number (N = 28) of bore-holes in the Oslofjord and Trøndelag regions of Norway (Banks *et al.*, 1995). Median and maximum uranium concentrations of 7.6 µg/l and 170 µg/l were recorded. The lowest values were generally from the Caledonian rocks of the Trøndelag area and the highest values from the Precambrian Iddefjord Granite of outer Oslofjord (N = 11, median = 15 µg/l).

16.2.2   Survey 2

This 1994 survey encompassed boreholes in crystalline bedrock in the Vestfold (Oslofjord, N = 89) and Hordaland (N = 56) regions of Norway. The 145 samples were analysed for uranium by ICP-MS techniques and median concentrations were found of 3.38 µg/l for the Vestfold subset and 3.72 µg/l for the Hordaland subset. In both regions, the highest values typically came from granitic aquifers: in the Vestfold area, the median concentration in granitic groundwaters (dominated by the Permian Drammen Granite) was around 100 µg/l (Reimann *et al.*, 1996, Morland *et al.*, 1997). The maximum uranium concentration recorded was 2.0 mg/l from a source in the Drammen Granite (although a duplicate analysis of the same sample only returned 0.9 mg/l).

16.2.3   Survey 3

The most extensive (nationwide) survey referred to in this article was performed in the period 1996–1997 and covered 1604 sources (mainly private drilled boreholes) in fractured crystalline aquifers throughout Norway (Banks *et al.*, 1998a). All of the samples were analysed by ICP-AES, ion chromatography and liquid scintillation techniques (for ^{222}Rn). Of these 1604 samples, a subset of 476 samples was carefully selected so as to be as representative of the range of Norwegian lithologies and of geographical distribution as possible. These 476 samples were analysed by ICP-MS techniques for a range of trace elements (including uranium) at sub-µg/l concentrations (Frengstad *et al.*, 2000). The detection limit for uranium in this survey was 0.001 µg/l. The set of 476 samples exhibited a median uranium concentration of 2.5 µg/l with a maximum of 750 µg/l (Figure 16.2). 12% of the samples exceeded the current WHO guideline of 30 µg/l.

16.2.4   Survey 4

The most recent survey was carried out in the period 2004 to 2008 of public water-works based on groundwater and of wells used for food production and other industries in Norway (Seither *et al.*, 2012). 346 of these samples were from wells in fractured crystalline aquifers, and these samples exhibited median, 75th percentile and maximum concentrations of 2.04 µg/l, 6.8 µg/l and 246 µg/l uranium, respectively. For comparison, the 314 samples from the same survey, derived from unconsolidated Quaternary aquifers exhibited median, 75th percentile and maximum concentrations of 0.12 µg/l, 0.34 µg/l and 77.6 µg/l, respectively (Figure 16.3).

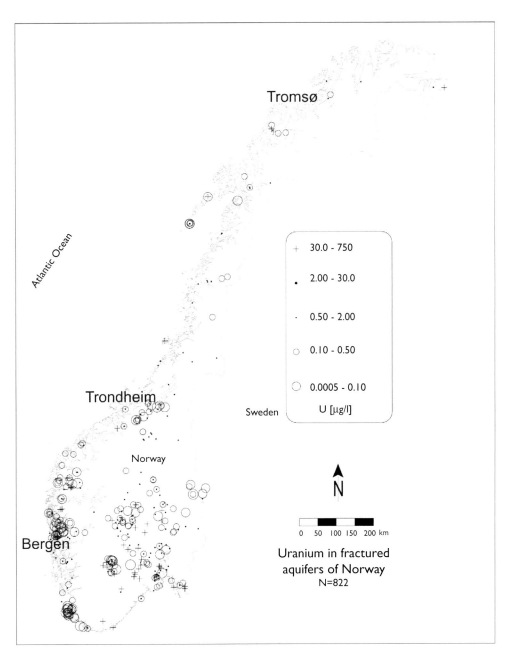

Figure 16.2 Map of Norway showing the distribution of uranium concentrations in groundwater from fractured aquifers. The map shows the uranium data sets for Surveys 3 and 4 (see text), combined (two data points not plotted). The median concentration for the combined dataset was 2.18 µg/l with 25- and 75-percentiles of 0.53 µg/l and 8.93 µg/l respectively.

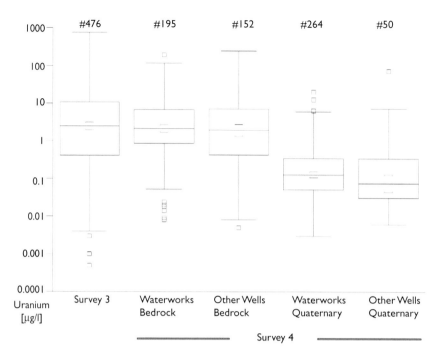

Figure 16.3 Boxplot comparing uranium data from Survey 3 (crystalline fractured aquifers only) with waterworks and other wells in fractured crystalline aquifers and unconsolidated Quaternary aquifers, respectively (Survey 4).

16.2.5   Survey synopsis

The surveys revealed that the uranium content in Norwegian groundwater is, to a large extent, dependent on the lithology of the aquifers, with elevated uranium concentrations broadly coinciding with acidic igneous or metamorphic rocks enriched in uranium (granites can contain around 5 ppm uranium, or more, compared with a crustal average of around 2.7 ppm (Skeppström & Olofsson, 2007)). A degree of co-variation with radon was found in the Norwegian data: this is to be expected, as both solutes are ultimately derived from uranium in rock matrices or fracture coatings. The correlation is modest, however (Figure 16.4), for the reasons that: (i) radon's immediate parent is ^{226}Ra and the mobility of radium is very different to that of uranium: the former is soluble in reducing, sulphate-depleted groundwaters, but immobilised in alkaline, sulphate-rich oxidising environments; (ii) uranium and radium can remain in immobile mineral phases in a rock matrix or fracture coating, while still releasing radon, via diffusion, to solution in fracture groundwater.

Figure 16.5 compares the distribution of uranium in all fractured crystalline aquifers from Survey 3, with the distribution of groundwater uranium in the lithological subsets representing Precambrian granites (uranium strongly enriched) and Precambrian anorthosites (uranium at very low concentrations in groundwater). Within each lithological group, however, there is a wide variation in uranium concentrations (in the entire data set, concentrations vary over at least six orders of magnitude). It is also

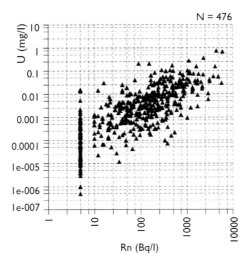

Figure 16.4 x-y plot of radon versus uranium concentrations in Norwegian crystalline bedrock groundwater from Survey 3 (from Frengstad *et al.*, 2000). Note the log scales.

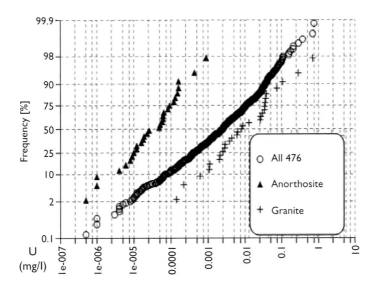

Figure 16.5 Cumulative frequency distribution diagram for uranium showing the distribution in groundwater samples from Survey 3 (N = 476). The diagram also shows cumulative distributions in the subsets representing lithological groups 'Precambrian anorthosite' (N = 30) and 'Precambrian granite' (N = 29) (from Frengstad *et al.*, 2000).

of interest that the overall distribution is approximately log-normal (straight line on the log-probability plot of Figure 16.5). As Misund *et al.* (1999) and Skeppström & Olofsson (2007) have pointed out, the variety of factors governing uranium mobilisation (e.g. bulk rock matrix U content, mineralisation type, fracture mineralisation, the presence of acidic U-rich pegmatites or dykes in otherwise U-poor rocks, redox conditions, lack of solubility controls for U^{+VI} under oxidizing conditions, complexation with other ions and kinetic factors) all conspire to render the deterministic prediction of uranium concentrations in groundwater impossible. The large concentration span implies that sampling of only a small selection of wells in an area may provide misleading results. Thus, all potable groundwater supplies derived from crystalline silicate aquifers should be analysed for uranium.

Figure 16.6 illustrates the dependence of uranium concentration on pH in the data set in Survey 3. As there was a significant clustering of pH values around 8–8.3, the data set was ranked according to pH and sub-divided into five equal cohorts of data. Within each pH cohort, there is substantial variation in U concentration, demonstrating the rather weak correlation between pH and U (see Figure 16.7). It

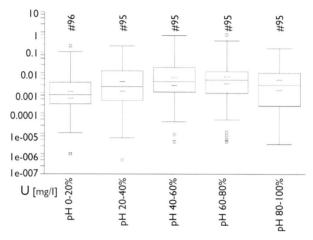

Figure 16.6 Boxplots showing the distribution of uranium in Norwegian crystalline bedrock groundwater (Survey 3, N = 476) dependent on the pH of the samples. The dataset is divided into five percentile groups according to pH: 0–20% (pH 6.17–7.73); 20–40% (pH 7.74–8.01); 40–60% (pH 8.01–8.13); 60–80% (pH 8.14–8.22); and 80–100% (pH 8.22–9.58). (From Frengstad *et al.*, 2001).

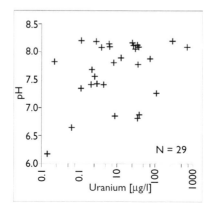

Figure 16.7 x-y plot of uranium concentrations versus pH in groundwater from Precambrian granites, a subset of the data from Survey 3 (see text). Note the log scale of uranium concentrations.

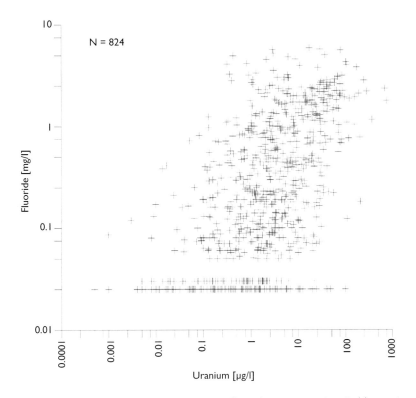

Figure 16.8 x-y plot of uranium concentrations versus fluoride concentrations in Norwegian crystal-line bedrock groundwater from the combined data sets of Surveys 3 and 4. Concentra-tions less than detection limit are plotted at a value of ½ × detection limit. Note the log scales and slightly different lower detection limits for the datasets.

appears, however, that the lowest pH cohort also contains the lowest uranium con-centrations. This observation may reflect kinetic factors (low pH in groundwaters with a low-degree of water rock interaction and low residence time) or complexation with other ions at higher pH.

Very few other significant correlations with uranium were noted in the Norwegian groundwater data sets – a rather weak correlation with fluoride being the only one of note (Figure 16.8). This reflects the fact that fluorite fracture mineralisations are rather common in the same Norwegian granites (e.g. the Precambrian granites of Oslofjord) and also the co-occurrence of uranium and fluoride in minerals such as apatite.

Figure 16.9 shows the uranium concentrations recorded in Survey 3, sorted accord-ing to borehole depth. While we must remember that the depth of the borehole or well may not reflect the depth of the fracture(s) yielding the bulk of the groundwater, there appears to be relatively little depth-dependence in groundwater uranium concentrations. Only in the very shallowest and the very deepest depth intervals, do uranium concentra-tions seem to be systematically lower. Bearing in mind that the populations of these two subsets are rather small, raising questions over statistical significance, one may speculate that, in the shallowest depth interval, uranium may have been depleted via weathering or that there is simply little opportunity to accumulate U-concentrations due to limited

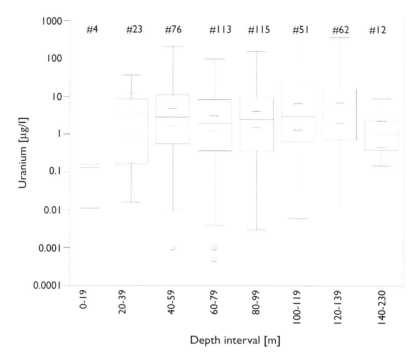

Figure 16.9 Boxplot showing the distribution of uranium concentrations in Norwegian crystalline bedrock groundwater (Survey 3, see text) in subgroups sorted according to increasing borehole depth.

water-rock interaction (kinetic factors, short residence time). In the deepest interval, one might speculate that reducing conditions suppress U-concentrations. For comparison with depth dependence of other elements, see Frengstad & Banks (2003).

16.2.6   Comparison with other data-sets

Large surveys have also been carried out in Sweden and Finland of uranium in crystalline bedrock groundwaters.

A survey of crystalline bedrock groundwater from 683 drilled boreholes throughout Sweden (Ek *et al.*, 2008) found that around 20% of the samples exceeded the Swedish guideline value of 15 µg/l, and 4% exceeded 100 µg/l. The median concentration was 3.3 µg/l. The highest values typically came from east Central Sweden, parts of Gävleborg County, Dalarna, the Bohus (Iddefjord) Granite and, especially, the Siljan ring complex, believed to represent a 360 million year old meteorite impact. The highest recorded concentration of 1.3 mg/l was from this complex. Wells drilled into alum shales did not, typically, yield especially high uranium concentrations. Åström *et al.* (2009) found that, in groundwaters and surface waters in black shale areas, there was a clear positive correlation of uranium concentrations with redox potential. Thus, provided a reducing environment is maintained within the carbonaceous shale, uranium remains immobile. Upon oxidation of the shales, however, uranium can be released in high concentrations (over 100 µg/l) in run off.

In Finland, a survey of 1388 boreholes into crystalline bedrock aquifers, Juntunen (1991) and Karro & Lahermo (1999) report median and arithmetic mean concentrations of 5 and 73 µg/l, respectively. The highest values were recorded in granitic rocks of southern Finland, especially the Wiborg rapakivi granite area. Salonen (1994) reports a nationwide survey of 1 162 drilled boreholes in fractured crystalline aquifers. She reports median activities of 0.17 Bq/l for ^{238}U and 0.27 Bq/l for ^{234}U, with maxima of 150 and 288 Bq/l, respectively. Salonen reckons that this total activity of c. 440 Bq/l in the most U-rich sample represents a uranium concentration of 12.4 mg/l, assuming a $^{234}U/^{238}U$ activity ratio of 1.9. Even higher uranium concentrations are reported from Finnish literature, however. Asikainen & Kahlos (1979) report a concentration of over 14 mg/l from granites near Helsinki, while values in excess of 20 mg/l are claimed by other studies in Finland (Juntunen, 1991; Salonen, 1994; Karro & Lahermo, 1999). A later study of around 460 drilled boreholes in Finland found that 5% and 18% of the samples exceed activity-equivalent concentrations of 100 µg/l and 15 µg/l ^{238}U, respectively (Vesterbacka, 2007).

Given the relatively high uranium concentrations found in crystalline bedrock in Fennoscandia, it is reasonable to question whether such concentrations are also found in similar aquifers further south in Europe. Smedley et al. (2006) considered a dataset of 1556 groundwater samples in Great Britain (not just from crystalline bedrock) and found uranium concentrations in the range <0.01 to 67 µg/l, with a median of 0.29 µg/l. The highest concentrations were associated with the Devonian Old Red Sandstone and the Permo-Triassic sandstone aquifers, with a single elevated concentration of 6.6 µg/l from the Precambrian Torridonian Sandstone of Scotland. They note that all of these are red-bed sandstones, where it is believed that sorbed uranium is released by desorption from ferric oxyhydroxide grain coatings and cements, facilitated by mobilisation due to carbonate complexation in solution. Elevated concentrations were associated with unconfined, oxidising portions of these aquifers: in reducing, confined aquifers, uranium concentrations tended to be low (<1µg/l).

Banks et al. (1998b) compared the hydrochemistry of the granitic (Carboniferous age) Scilly Islands of South West England with the granitic Hvaler Islands of SE Norway. The latter are part of a province (the Precambrian Iddefjord/Bohus granite) known for elevated uranium, radon and fluoride concentrations. The 11 Hvaler samples exhibited high pH and alkalinity and uranium concentrations ranging up to 170 µg/l, with a median of 15 µg/l. The Scilly samples exhibited very low pH and alkalinity and all uranium concentrations were <4 µg/l, with a median of <2 µg/l. The authors suggest that the 'immature' hydrochemistry of the Scilly waters and the low uranium concentrations reflect a long history of subaerial weathering, during which the majority of basic minerals and those species mobilised under oxidising conditions (e.g. U) have been removed from the zone of groundwater circulation. In contrast, the Hvaler islands have been glacially scoured (leaving fresh bedrock exposed) and have only emerged from the sea (due to post-glacial isostatic uplift) during the past several thousand years, leaving a mineral assemblage with basic and oxidisable mineral phases intact. This conclusion is supported by Smedley et al. (2006) with findings of typically <10 µg/l uranium (and a maximum reported 11.6 µg/l) in the granites of South West England.

It is outside the scope of this paper to carry out a complete literature review of uranium in groundwater from fractured crystalline aquifers. Table 16.2 provides core statistics from a selection of surveys of uranium concentrations in drinking water

Table 16.2 Examples of uranium concentration distributions in groundwater from fractured crystalline bedrock aquifers.

Area and Rock Type	Min [µg/l]	Max [µg/l]	Arithmetic mean * or median# [µg/l]	Reference
Sweden Granites, gneisses and migmatites. Some sedimentary rocks. (N = 683)	–	1328	3.3#	Ek et al., 2008
Finland Granites, gneisses and migmatites (N = 1162) Bedrock (N = 263)	– <0.005	12400 643	32* 0.64#	Salonen, 1994 Åström et al., 2009
Valais canton, Switzerland Granites, gneisses, schists and amphiboles (N = 1473)	0.025	92.02	1.94#	Stalder et al., 2011
Kolar district, India Granites, gneisses and schists (N = 52)	0.3	1443		Babu et al., 2008
Korea Granitoids, gneisses, volcanics, some sed. rocks (N = 498) Granites (N = 145)	– –	402 215	3.72* 1.72* 0.19#	Kim et al., 2004 Cho et al., 2002
Fujian Province, China Granite (N = 45)	–	13.4		Zhuo et al., 2001
Rift Valley, Ethiopia Volcanics, some sediments, springs and rivers (N = 138)	0.005	48	1.84#	Reimann et al., 2003
Canada, Nova Scotia Various bedrock (N = 840)	<0.1	200	0.7* 4.3#	Kennedy and Finlayson-Bourque, 2011
USA East-Central Massachusetts Various bedrock (N = 344) New England, New Jersey, NY Crystalline bedrock (N = 556)	<0.02 <1	817 3640	– 2.3#	Colman, 2011 Flanagan et al., 2011

mainly derived from crystalline bedrock aquifers – it could not be entirely avoided that some data from other lithologies also have been included. There is a growing awareness of uranium being a potential health problem in groundwater (Nriagu *et al.*, 2012 and references herein), and many geographical areas are still not investigated. In some cases, surveys have been performed by chemists, health authorities or radiophysicists (e.g. Nriagu *et al.*, 2012; Hakam *et al.*, 2001) and the lithology of the aquifers has unfortunately not been evaluated. Granitic aquifers are not the only possible source of uranium-containing drinking water, but seem to have a prominent position on the list of main 'suspects'.

16.3   SUMMARY AND CONCLUSIONS

- Uranium is naturally present in the hydrogeological environment.
- Health concerns are linked to the chemical toxicity of uranium rather than to its radioactive properties. Chronic intake of drinking water with elevated uranium concentrations may lead to kidney injury.
- EU and Norwegian drinking water legislation does not yet include uranium, although Germany has recently introduced a limit of 10 μg/l. USEPA operates a Maximum Contaminant Level of 30 μg/l, which is also the updated guidance value of WHO.
- Surveys of the chemical quality of drinking water from crystalline bedrock boreholes in Norway reveal that 12% of the investigated private boreholes (Survey 3) yield water with uranium concentrations above 30 μg/l. 7.5% of the larger public waterworks, based on fractured crystalline aquifers, breach this limit (Survey 4).
- Elevated uranium concentrations are often found in groundwater in fractured crystalline aquifers, especially those composed of granites and felsic gneisses or intercepted by pegmatites.
- The solubility and mobility of uranium are highly dependent on redox conditions and are also affected by pH, complexation and sorption.
- A brief literature study shows that groundwater from fractured crystalline aquifers in Sweden, Finland, Canada and USA also display relatively high levels of uranium. Comparison of Nordic granitic environments with those in southern England suggests that uranium may be depleted in the zone of groundwater circulation following a long history of sub-aerial erosion, without glacial scouring, resulting in modest groundwater uranium concentrations. In emergent, glacially scoured granitic terrain (e.g. much of Fennoscandia) generally elevated groundwater uranium concentrations are observed.
- Distributions of uranium concentrations span up to four log cycles within a single lithology and the concentration in a given well cannot be deterministically predicted. Uranium cannot be detected in drinking water by taste, smell or appearance. Owners of wells in fractured crystalline aquifers are recommended to get their drinking water analysed, as effective water treatment methods are available (e.g. Water Systems Council, 1997; Nova Scotia Environment, 2008).

ACKNOWLEDGEMENTS

The authors would like to thank: Ulrich Siewers and his colleagues at BGR laboratory in Hannover, Germany and Tomm Berg and his colleagues at NGU laboratory in Trondheim, Norway for analysing the water samples; Aase Kjersti Skrede and Anna Seither at NGU for managing data; Terje Strand and Bjørn Lind at the Norwegian Radiation Protection Authority and Morten Nicholls at the Norwegian Food Safety Authority for project cooperation; representatives of the local food safety authorities, waterworks managers and well owners for water sampling; Clemens Reimann at NGU for fruitful discussions; and Sylvi Gaut at Sweco for a thorough review of the manuscript.

REFERENCES

Agence France-Presse (2001) Another Italian soldier dies of 'Balkans syndrome'. Agence France-Presse (AFP), Rome.

Altschuler Z.S., Clarke R.S., Young E.J. (1958) Geochemistry of uranium in apatite and phosphorite. U S Geol Surv Prof Paper 314-D, Washington.

Asikainen M., Kahlos H. (1979) Anomalously high concentrations of uranium, radium and radon in water from drilled wells in the Helsinki region. *Geochim Cosmochim Acta* 43, 1681–1686.

Åström M.E., Peltola P., Rönnback P., Lavergren U., Bergbäck B., Tarvainen T., Backman B. (2009) Uranium in surface and groundwaters in Boreal Europe. *Geochemical Expl. Env. Analysis* 9, 51–62.

Babu M.N.S., Somashekar R.K., Kumar S.A., Shivanna K., Krishnamurthy V., Eappen K.P. (2008) Concentration of uranium levels in groundwater. *International Journal of Environment Science Technology* 5, 263–266.

Banks D., Røyset O., Strand T., Skarphagen H. (1995) Radioelement (U, Th, Rn) concentrations in Norwegian bedrock groundwaters. *Environmental Geology* 25, 165–180.

Banks, D., Frengstad, B., Midtgård, Aa.K., Krog, J.R., Strand, T. (1998a) The chemistry of Norwegian groundwaters I. The distribution of radon, major and minor elements in 1604 crystalline bedrock groundwaters. *Science of the Total Environment* 222, 71–91.

Banks D., Reimann C., Skarphagen H. (1998b) The comparative hydrochemistry of two granitic island aquifers: the Isles of Scilly, U.K. and the Hvaler Islands, Norway. *Science of the Total Environment* 209, 169–183.

Banks D., Gundersen P., Gustafson G., Mäkelä J., Morland G. (2010) Regional similarities in the distributions of well yield from crystalline rocks in Fennoscandia. Norges geologiske undersøkelse *Bulletin* 450, 33–47.

Casas I., de Pablo J., Giménez J., Torrero E.M., Bruno, J., Cera E., Finch R.J., Ewing R.C. (1998) The role of pe, pH, and carbonate on the solubility of UO_2 and uraninite under nominally reducing conditions. *Geochim Cosmochim Acta* 62, 2223–2231.

Cho B.W., Lee I.H., Park S.K., Lee, B.D., Sung, I.H. (2002) Uranium and radon concentrations in the groundwater of South Korea. 32nd IAH Congress: Groundwater and human development, 1010–1015.

CNN (2001) Call to test Balkans soldiers. CNN news article dated 16 January 2001.

Colman, J.A. (2011) Arsenic and uranium in water from private wells completed in bedrock of east-central Massachusetts – Concentrations, correlations with bedrock units, and estimated probability maps: United States Geological Survey Scientific Investigation Report 2011–5013, 113pp.

Ek B.M., Thunholm B., Östergren I., Falk R., Mjönes L. (2008) Naturligt radioaktiva ämnen, arsenik och andra metaller i dricksvatten från enskilda brunnar [Naturally radioactive elements, arsenic and other metals in drinking water from private wells – in Swedish]. Statens strålskyddsinstitut rapport 2008, 15pp.

Federal-Provincial-Territorial Committee on Drinking Water (2006) Radiological characteristics of drinking water. Draft document for public comment. Federal-Provincial-Territorial Committee on Drinking Water, Canada, July 2006.

Flanagan S.M., Ayotte J.D., Robinson G.R. Jr. (2012) Quality of water from crystalline rock aquifers in New England, New Jersey, and New York, 1995–2007. United States Geological Survey Scientific Investigation Report 2011–5220, 104pp.

Frengstad B., Banks D. (2003) Groundwater chemistry related to depth of shallow crystalline bedrock boreholes in Norway. In: Krásný J, Hrkal Z, Bruthans J (Eds) Proceedings International Conference on Groundwater in Fractured Rocks. 15–19 Sept 2003, Prague, Czech Rep IHP-VI, Series on groundwater 7, 203–204.

Frengstad B., Midtgård A.K., Banks D., Krog J.R., Siewers U. (2000) The Chemistry of Norwegian Groundwaters III. The Distribution of Trace Elements in 476 Crystalline Bedrock Groundwaters, as Analysed by ICP-MS Techniques. *Science of the Total Environment* 246, 21–40.

Frengstad B., Banks D., Siewers U. (2001) The Chemistry of Norwegian Groundwaters: IV. The pH-Dependence of Element Concentrations in Crystalline Bedrock Groundwaters. *Science of the Total Environment* 277, 101–117.

Hakam O.K., Choukri A., Moutia Z., Chouak A., Cherkaoui R., Reyss J.L., Lferde M. (2001) Uranium and radium in groundwater and surface water samples in Morocco. *Radiation Physics and Chemistry* 61, 653–654.

Helse- og omsorgsdepartementet (2001) *Forskrift om vannforsyning og drikkevann* [Regulation of water supply and drinking water – in Norwegian] FOR 2001–12–04 1372.

Health Canada (1996) *Guidelines for Canadian Drinking Water Quality.* Sixth Edition. Minister of Health, Ottawa.

Health Canada (2012) *Guidelines for Canadian Drinking Water Quality – Summary Table.* Water, Air and Climate Change Bureau, Healthy Environments and Consumer Safety Branch, Health Canada, Ottawa, Ontario. Catalogue H129–6/2011E.

Houk R.S., Thompson J.J. (1988) Inductively coupled plasma mass spectrometry. Mass Spectrometry Review 7, 425–461.

Juntunen R. (1991) Etelä-Suomen kallioporakaivojen uraani-ja radontutkimukset [Uranium and radon in wells drilled into bedrock in southern Finland – in Finnish]. Report of Investigation 98. Geological Survey, Finland, Espoo, 22pp.

Karro E., Lahermo P. (1999) *Occurrence and chemical characteristics of groundwater in Precambrian bedrock in Finland.* In: Autio S (ed) Geological Survey of Finland, Current Research 1997–1998. Geol Surv Finland Spec Paper 27, 85–96.

Kennedy G.W., Finlayson-Bourque D. (2011) Uranium in groundwater from bedrock aquifers in Nova Scotia. Nova Scotia Department of Natural Resources, Mineral Resources Branch, Open File Map ME 2011–031, scale 1:500 000.

Kim Y., Park H., Kim J., Park S., Cho B., Sung I., Shin D. (2004) Health risk assessment for uranium in Korean groundwater. *Journal of Environmental Radioactivity* 77, 77–85.

Kirjukhin, V.A., Korotkov A.N., Shvartsev S.L. (1993) Гидрогеохимия [Hydrogeochemistry – in Russian]. Nedra, Moscow, 383pp.

Lagercrantz B. (2003) Utarmat uran en cancerrisk som försvann. Larmrapporterna om leukemi hos Balkansoldater kom av sig [Depleted uranium a cancer risk that disappeared. Leukaemia alarm regarding Balkan veterans came to nothing – in Swedish]. *Läkartidningen* 100, 219–221.

Livsmedelsverket 2011 Uran i dricksvatten [Uranium in drinking water – in Swedish] http://www.slv.se/sv/grupp1/Dricksvatten/Dricksvattenkvalitet/Uran/

Midtgård Aa.K., Frengstad B., Banks D., Krog J.R., Strand T., Siewers U., Lind B. (1998) Drinking water from crystalline bedrock aquifers – not just H_2O. *Mineralogical Bulletin December* 1998, 9–16.

Milvy P., Cothern C.R. (1990) *Scientific background for the development of regulations for radionuclides in drinking water.* In: Cothern CR, Rebers P (eds) Radon, Radium and Uranium in Drinking Water. Chelsea, Michigan. Lewis Publishers 1–16.

Misund A., Frengstad B., Siewers U., Reimann C. (1999) Variation of 66 elements in European bottled mineral waters. *Science of the Total Environment* 243, 21–41.

Morland G., Reimann C., Strand T., Skarphagen H., Banks D., Bjorvatn K., Hall G.E.M., Siewers U. (1997) The hydrogeochemistry of Norwegian bedrock groundwater – selected parameters (pH, F, Rn, U, Th, B, Na, Ca) in samples from Vestfold and Hordaland, Norway. *Norges geologiske undersøkelse Bulletin* 432, 103–117.

Myrstad L., Nordheim, C.F., Einan B. (2010) Rapport fra Vannverksregister et – Drikkevannsstatus (data 2005 og 2006) [Report from the waterworks register – Drinking water status (data 2005 and 2006) – in Norwegian]. Norwegian Institute of Public Health Water Report 114, 113pp.

National Health and Medical Research Council (2011) *Australian Drinking Water Guidelines Paper 6 National Water Quality Management Strategy*. National Health and Medical Research Council, National Resource Management Ministerial Council, Commonwealth of Australia, Canberra.

Norton-Taylor R. (2001) Italy alarmed by 'Balkan syndrome'. *The Guardian*, 4 January 2001.

Nova Scotia Environment (2008) The drop on water – uranium. Fact sheet March 2008. http://www.gov.ns.ca/nse/water/docs/droponwaterFAQ_Uranium.pdf.

Nriagu J., Nam D-H., Ayanwola T.A., Dinh H., Erdenechimeg E., Ochir C., Bolormaa T-A. (2012) High levels of uranium in groundwater of Ulaanbaatar, Mongolia. *Science of the Total Environment* 414, 722–726.

Oftedahl C. (1981) Norges geologi [Geology of Norway – in Norwegian]. Tapir, Trondheim, Norway.

Reimann C., Hall G.E., Siewers U., Bjorvatn K., Morland G., Skarphagen H., Strand T. (1996) Radon, fluoride and 62 elements as determined by ICP-MS in 145 Norwegian hard rock groundwater samples. *Science of the Total Environmemnt* 192, 1–19.

Reimann C., Äyräs M., Chekushin V., Bogatyrev I., Boyd R., Caritat P. de, Dutter R., Finne T.E., Halleraker J.H., Jæger Ø., Kashulina G., Lehto O., Niskavaara H., Pavlov V., Räisänen M.L., Strand T., Volden T. (1998) Environmental geochemical atlas of the Central Barents region. Norges geologiske undersøkelse, 745pp.

Reimann C., Bjorvatn K., Frengstad B., Melaku Z., Tekle-Haimanot R., Siewers U. (2003) Drinking water quality in the Ethiopian section of the East African Rift Valley I – data and health aspects. *Science of the Total Environment* 311, 65–80.

Salonen L. (1994) ^{238}U series radionuclides as a source of increased radioactivity in groundwater originating from Finnish bedrock. Future Groundwater Resources at Risk (Proc Helsinki Conf June 1994). *IAHS Publication* 222, 1–83.

Seither A., Eide P.E., Berg T., Frengstad B. (2012) The inorganic drinking water quality of some groundwaterworks and regulated wells in Norway. *Norges geologiske undersøkelse Report* 2012.073, 166pp.

Skeppström K., Olofsson B. (2007) Uranium and radon in groundwater. An overview of the problem. *European Water* 17/18, 51–62.

Smedley P.L., Smith B., Abesser C., Lapworth D. (2006) Uranium occurrence and behaviour in British groundwater. British Geological Survey Commissioned Report CR/06/050 N, 60pp.

Stalder E., Blanc A., Haldimann M., Dudler V. (2012) Occurrence of uranium in Swiss drinking water. *Chemosph.* 86, 672–679.

Takeno N. (2005) Atlas of Eh-pH diagrams. Intercomparison of thermodynamic databases. *Geol Surv Japan Open File Rep* 419, 285pp.

The council of the European Union (1998) Council Directive 98/83/EC of 3 November 1998 on the quality of water intended for human consumption.

Umweltbundesamt (2011) Hintergrund: Uran (U) im Trinkwasser: Kurzbegründung des gesundheitlichen Grenzwertes der Trinkwasserverordnung (10 µg/l U) und des Grenzwertes für „säuglingsgeeignete" abgepackte Wässer (2 µg/l U). Umweltbundesamt, Berlin, December 2011.

United Nations Environment Program (2002) Depleted uranium in Serbia/Montenegro post-conflict environmental assessment. United Nations Environment Program.

United States Environmental Protection Agency (2000) Technical Fact Sheet: Final Rule for (Non-Radon) Radionuclides in Drinking Water. Fact Sheet EPA 815-F-00–013 Nov 2000.

Vesterbacka P. (2007) Natural radioactivity in drinking water in Finland. *Boreal Environment Research* 12, 11–16.

Wagner S.E., Burch J.B., Bottai M., Puett R., Porter D., Bolick-Aldrich S., Temples T., Wilkerson R.C., Vena J.E., Hébert J.R. (2011) Groundwater uranium and cancer incidence in South Carolina. *Cancer causes contr* 22, 41–50.

Water Systems Council (1997) Wellcare® information for you about uranium & groundwater. Water Systems Council factsheet, updated May 1997.

Wedepohl K.H. (1995) The composition of the continental crust. *Geochim Cosmochim Acta* 59, 1217–1232.

World Health Organization (1998) *Guidelines for Drinking-water Quality*, 2nd edition, Addendum to Vol 1 Recommendations. Geneva, Switzerland.

World Health Organization (2011) *Guidelines for Drinking-water Quality*, 4th edition, 541 pp.

Zamora M.L., Tracy B.L., Zielinski J.M., Meyerhof D.P., Moss M.A. (1998) Chronic Ingestion of Uranium in Drinking Water: A Study of Kidney Bioeffects in Humans. *Toxicological Science* 43, 68–77.

Zhuo W., Iida T., Yang X. (2001) Occurrence of ^{222}Rn, ^{226}Ra, ^{228}Ra and U in groundwater in Fujian Province, China. *Journal of Environmental Radioactivity* 53, 111–116.

Chapter 17

Technical quality of Norwegian wells in crystalline bedrock related to groundwater vulnerability

Sylvi Gaut[1,2] & *Gaute Storrø*[1]
[1]*Groundwater and Urban Geology Section, Geological Survey of Norway, Trondheim, Norway*
[2]*Sweco, Trondheim, Norway*

ABSTRACT

Well construction, and especially leakage at the base of the casing, has been investigated by video camera in 213 Norwegian wells in crystalline bedrock. In addition, fractures, type of well head protection and land use has been recorded. Both wells supplying private households and waterworks were filmed. Leakage, via the base of the casing, was observed in 60 out of 150 wells. Cavities and fractures in the borehole wall from directly below the casing and down to about 10 m below surface, were observed in 126 out of 213 wells. Biofilm and mineral precipitation were common in the wells. In Norway both well drillers and well owners have been aware of groundwater vulnerability and the importance of proper well construction due to increased focus on the topic since 1995. Photos and videos from the wells illustrating the problems, have contributed to this.

17.1 BACKGROUND

Norway's groundwater resources are located in either (i) crystalline, hard, fractured rocks, mostly belonging to the Fennoscandian shield and Caledonian orogenic terrains (e.g. schists, gneisses, granites and metasediments), or (ii) the relatively thin overlying granular Quaternary sedimentary deposits (alluvial or glaciofluvial sands and gravels and moraines). The former category is referred to as bedrock in this chapter.

About 15% of the Norwegian population are supplied by groundwater either from a waterworks or a private well or spring. Most Norwegian waterworks are owned and operated by the municipality, but some are owned and operated by the end users (subscribers). These are private waterworks. Whereas a private well may supply only one household, farm or cottage, the waterworks (private and municipal) supply several households, and include wells supplying hotels, cafeterias, and diaries. Groundwater derived from crystalline bedrock is widely used in rural areas, supplying both small and medium sized (<1000 people) waterworks and single households, farms and holiday cottages. Most bedrock wells are 60–100 m deep.

At the end of the 1990s the Programme for Improved Water Supply (PROVA) focused on groundwater quality. Water samples from several of the waterworks based on groundwater from bedrock showed poor microbiological water quality not satisfying the Norwegian drinking water regulations. Although several regional investigations of the hydrogeochemistry of groundwater in Norway had been carried out (e.g., Englund & Myrstad, 1980; Bjorvatn *et al.*, 1994; Hongve *et al.*, 1994; Banks *et al.*, 1995a; Banks *et al.*, 1995b; Reimann *et al.*, 1996; Banks *et al.*, 1998; Frengstad, 2002), examination of the vulnerability of bedrock wells to microbiological contamination had not been studied. As a result the Geological Survey of Norway (NGU) initiated a PhD-study to evaluate the vulnerability of groundwater wells in bedrock to microbiological contamination (Gaut, 2005). This study showed that the contaminated wells were often located close to potential sources of contamination and/or had a poor well head protection. It was concluded that the latter could be one of the main causes of microbiological contamination. Well owners often commented that the water quality was poorer after periods of intensive rain and/or snowmelt. Wells were observed to have poor well head protection often with surface runoff towards the well and accumulation of surface water (pools/ponds) close to or in contact with the well head. Manhole frames (concrete well head protection) used to protect the well head, were often poorly constructed and maintained. Most wells protected by a well house or a concrete well head protection lacked a secure and watertight cap on the top of the casing. The top of the casing was also often terminated below ground. It was suspected that the wells had inadequate or no sealing between the casing and the bedrock, thus causing leakage at the bottom of the casing. To test the latter hypothesis, and to map the extent of insufficient well construction, a second project called Technical quality of bedrock wells was initiated at NGU in 2004 (Storrø *et al.*, 2006). A video camera was acquired in order to inspect the upper 20 m of the wells. This chapter presents the results from this project.

17.2   DATA COLLECTION AND METHOD

The National Well Database, which is part of the National Groundwater Database, is administrated by NGU. Since 1997 well drillers have been obliged by regulations in the Water resources legislation (Olje – og energidepartementet, 2000) to report all drilled groundwater wells to the database. In 2004 it was decided to do a video inspection of around 250 bedrock wells, which at the time corresponded to 1% of the total number of bedrock wells in the database. Today (August 2012), the database contains approximately 60 000 bedrock wells of which >35 000 are used for drinking water.

A colour video camera (Tiny CS3002S from Rico EAB) was acquired in order to inspect the upper 20 m of the wells. The camera is standard equipment used for inspections of water and sewage pipes, and was assembled on a 50 m long cable. Recording was done on a Portable Video Recorder (PVR) from Archos. Both Archos AV380 and Archos AV4100 were used.

The wells were selected with the objective of obtaining a geographically even distribution of wells. The logged wells represent 13 counties and 37 municipalities (Figure 17.1a). To minimize travel costs between 5 and 30 wells were logged in each municipality. In total 270 wells were visited, of which 213 wells were accessible with a video camera. Both private wells and wells owned by waterworks were filmed. Ownership distribution between private wells and waterworks is shown in Figure 17.1b and the area of application for the private wells is shown in Figure 17.1c. Most wells

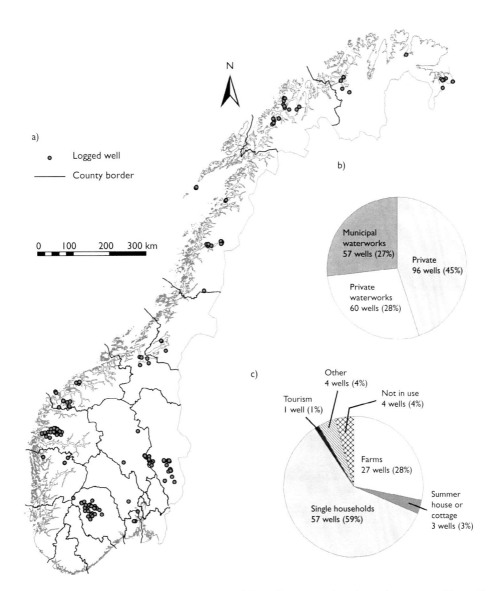

Figure 17.1 a) Geographical distribution of the 213 wells inspected with a video camera. The wells represent 13 counties and 37 municipalities. b) The wells are private (each well supply only one household, farm, cottage etc.) or belong to private or municipal waterworks supplying several households, hotels, schools etc. c) The area of application for the private wells.

were chosen from the National Well Database, the rest were part of the PhD-study by Gaut (2005). During inspection, well installations (e.g. pump, raising main) remained in the well. To facilitate the camera inspection, wells with a diameter of 110 mm or more were selected, thus most wells are drilled after 1985. Likewise, energy wells (used for ground source heating and cooling) and inclined wells were not included.

In the field, the vicinity of the wells was filmed to document the land use and well protection. All 213 wells were logged down below the casing and into bedrock. Most wells (90%) were filmed to more than 10 m depth below ground level (Figure 17.2). The wells could not be filmed deeper than the cable length of 50 m. One well was already logged by the owner down to about 70 m.

Field work was conducted in the period 2004 and 2005 and during different seasons. Data from each well was compiled in an Excel based project-database and logs based on the video films were created.

17.3   RESULTS

Altogether 270 wells were visited, of which 213 wells were filmed with the video camera. Filming in the remaining 57 wells was not possible due to a range of different reasons, including small well diameter (<110 mm), large well installations, additional inner casing, or the well owner did not want the well filmed or the well was buried or otherwise difficult to reach.

Information recorded at each locality was land use around each well, type of well head protection, condition of the casing and possible problem indicators inside

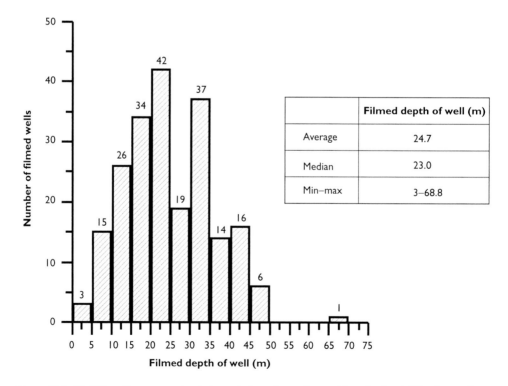

Figure 17.2 All 213 wells are examined down below the casing and into bedrock. The figure shows depth of inspection in intervals of 5 m (0–5, 5–10 etc.). The well inspected to 68.8 m was filmed by the owner with different equipment.

the well. The term possible problem indicators refers to distinct, visually observable, characteristics that may imply, or cause, reduced water quality (or risk thereof) in the well.

17.3.1 Surrounding land use

Land use within 50–100 m of each well was described for 241 wells and often more than one kind of land use (e.g. arable land and cattle) was registered. A total of 185 (77%) wells were located close (< about 50 m) to houses or farms, but only 4 of these were located in densely built-up areas. The study by Gaut (2005) showed that wells situated less than 100 m from farm land (e.g. grazing land and cultivation of grain crops) are more susceptible to microbiological contamination than other wells. In this study, 40 wells (17%) were located on grazing land, while an additional 12 wells (5%) had grazing animals within 100 m. Manure is known to be spread regularly in the vicinity of 10 wells (4%), whereas grain crops or grass were cultivated close to 55 wells (23%). Most wells were located in the vicinity of wooded areas or other types of non-agricultural terrain, but only 44 wells (18%) had no grazing animals, houses or farm land within 100 m of the well location.

17.3.2 Well head protection

Well head protection is installed to protect the borehole and its installations against intruders (animals and humans), but should also ensure easy access to the well when needed. In this study protection was evaluated for 240 wells; it normally consisted of a well house or a concrete manhole frame. Observations from this and other studies (Gaut *et al.*, 2007) show that precautions against intrusion of surface water are not prioritised when the well head protection is constructed. A well house protects 61 wells (26%). Some of the well houses are placed on top of a concrete manhole frame. No protection existed for 72 wells (30%) – this means that, at best, a simple cap is merely placed on the top of the casing. The cap is mostly screwed on to the casing, but not always. The concrete well protections (manhole) used at 107 wells (44%) are of varying quality. They are mostly located below ground when the superficial deposits are deep enough (>1–1.5 m). As for the well houses, not all have a concrete floor. The manhole covers are of varying material (concrete, iron, wood or fibreglass). Holes in the cover causing leakage of water into the manholes were often observed.

As might be expected, no source protection zones had been established for the private wells. There were no fences around the wells and, for the most part, no restrictions existed on land use. This was also the case for many of the wells connected to the waterworks, weather they were public or private.

17.3.3 The casing

The casing is an important part of the well when it comes to protection. In Norway, it is quite common for the casing to be installed by an ODEX-type drilling bit on a down-the-hole hammer rig. The rig drills a short distance, through any superficial deposits, into the bedrock and the ODEX bit is retracted and withdrawn, leaving the casing in the borehole. Today it is recommended that a sealing compound is introduced into the base of the casing, in order to seal it to the rock of the borehole wall. The casing is then pushed or hammered down to close the small gap produced when the ODEX bit is withdrawn. Afterwards drilling continues through the sealed casing at a slightly narrower diameter.

It is recommended that the top of the casing is above ground level. This was the situation for 125 (52%) of the 238 wells where this information was recorded. The casing terminates at ground level for 14 wells (6%) and 99 wells (41.5%) had the casing below ground level. In the latter case, most of the wells were protected by a manhole. For the 139 wells where the top of the casing was at or above ground level, the height of the casing was <0.2 m above ground for 53% of the wells, whereas 12% had the casing from 0.4–1 m above ground level.

The total length of the casing was measured for 205 wells and the results show that 50% of these wells have less than 3 m of casing (Figure 17.3a). It is expected that the casing is emplaced shortly after bedrock is reached. Based on data from the well database on depth to bedrock and measured length of well casing, drilling depth into bedrock was estimated. Figure 17.3b shows that, in about 35% of the wells, casing is emplaced <1 m into bedrock and there are even some wells where the casing is terminated before bedrock is reached.

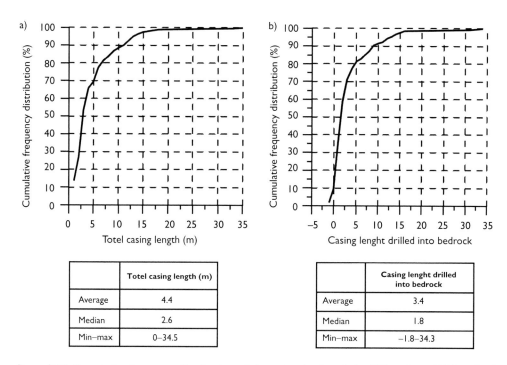

	Totel casing length (m)
Average	4.4
Median	2.6
Min–max	0–34.5

	Casing lenght drilled into bedrock
Average	3.4
Median	1.8
Min–max	−1.8–34.3

Figure 17.3 Cumulative frequency distribution of a) total casing length for 205 wells and b) length of casing drilled into bedrock for 128 wells.

17.3.4 Problem indicators

The concept of problem indicators is used as a description of distinct physical characteristics observed in the inspected wells, and which may imply, or cause, poor water quality. The indicators are defined below, and categories 1–3 are shown in Figure17.4.

1 Leakage of water between the bedrock and the bottom of the casing.
2 Fractures or cavities immediately below the casing.
3 Fractures <10 m below ground level with water entering the well through these fractures.
4 Biofilm and other types of coating on the casing, well installations and the borehole wall.

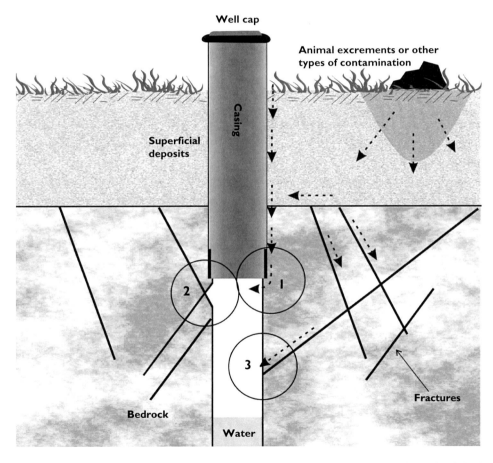

Figure 17.4 The sketch shows problem indicators (numbered 1–3) observed in the wells. 1. Leakage below the casing. If the casing is not pushed down during drilling a gap will be seen here. 2. Cavities and/or fractures observed immediately below the casing, and 3. Fractures observed less than 10 m below surface and water possibly entering the well through these fractures. Water flow is shown by the dotted lines. Figure from www.grunnvann.no.

Due to high water level (above the bottom of the casing) in 63 wells, it was only possible to evaluate the occurrence of leakage between bedrock and the bottom of the casing in 150 wells. Leakage was observed in 60 (40%) of these wells. In about 2/3 of the 60 wells, leakage was small or indistinct, whereas in others water was flowing in. Condensation from the casing wall gathering at the bottom of the casing made evaluation difficult for 5 wells (3%) because it was hard to determine whether the water was from the condensation or caused by leakage. No leakage was observed in 61 (41%) wells, but in six of these there were traces of precipitation that indicate former leakage (see Table 17.1).

Table 17.1 Number of wells exhibiting leakage between bedrock and the bottom of the casing ($n = 150$). High water level made evaluation impossible in 63 wells.

Type of leakage	Number of wells	Percent (%)
Leakage	60	40
Condensation water, often rusty coloured, from the casing makes observation difficult	5	3
Possible leakage, but cannot be evaluated based on the video	24	16
Dry with traces of precipitation	6	4
No leakage	55	37
Total	150	100

Physical sealing between the casing and the bedrock was observed in only 3 of the 213 wells inspected. For 13 wells, the casing was not pushed completely down during construction, causing a gap between the bedrock and the casing. This opening or gap enhances the possibility of water leakage at the bottom of the casing.

Existence of fractures and cavities immediately below the casing were evaluated for 210 wells (No. 2 in Figure 17.4). No visual fractures were observed in 154 wells, whereas fractures or cavities were observed in 39 (19%) wells. Fractures could not be evaluated for the remaining 17 (8%) wells.

Fractures or cavities <10 m below surface were observed in 126 (59%) out of 213 wells. Water inflow was observed in 63 of the wells. At some fractures no water was flowing, but black, white or rust coloured precipitation was observed on the walls.

Forms of coating on the casing, well installations and the borehole wall was common. This is assumed in many cases to be biofilm, but may also be caused by chemical precipitation. An increased amount of biofilm was typically observed in connection with fractures and indicates inflow of water. Within the project framework, it was not possible to classify microbial species or describe any chemical composition. The observations are, therefore, only based on visual examination. Pictures giving examples of the different kinds of biofilm can be seen in the digital report by Storrø et al. (2006, Figure 5.6 therein). What is assumed to be biofilm related is classified in four categories:

- Coating with a distinct starting point, "flowing" downwards along the borehole wall. The same is observed in Norwegian tunnels (Banks & Banks, 1993). Might also be caused by chemical precipitation.

- Pillow like and fluffy coating. Resembles cotton wool except that it is usually iron coloured.
- Coating with a flat surface. Muddy and commonly iron coloured. Might also be caused by chemical precipitation.
- White threads or filaments. One end is fastened to the borehole wall or raising main, whereas the rest is floating in the water.

The presence of chemical precipitation and biofilm was evaluated for 213 wells (Table 17.2). Only 48 (23%) had clean borehole walls. Suspected biofilm was observed in 109 (51%) wells. Pillow-like and fluffy biofilm was the most common biofilm. In several wells the amount of this biofilm caused problems during video recording, because the biofilm was scraped off the borehole walls and was blocking the view of the camera. The same type of biofilm was also often observed only on the casing walls or at the bottom of the casing and not on the bedrock wall. In some wells an increase or decrease in the amount of biofilm occurred at a specific fracture or fracture zone. This was especially noticeable when there was no or very little biofilm in the upper part of the borehole.

Table 17.2 Number of wells with observation of coating (biofilm and/ or chemical precipitation) on the casing, well installations and the borehole wall. Total number of wells is 213.

Type of coating	Number of wells	Percent (%)
No coating. Clean borehole walls	48	23
Chemical precipitation	56	26
Biofilm	23	11
Biofilm and chemical precipitation	86	40
Biofilm and/or chemical precipitation	165	77

In 142 wells (Table 17.2) the coating was interpreted to be caused by chemical precipitation rather than being biofilm related. The mineral precipitation was most often white, grey or rusty coloured, but could also be black or yellow.

It should be pointed out that the bacteria typically causing biofilms in wells and groundwater are non-pathogenic. While some of the biofilm bacteria are common in soils and may conceivably indicate contamination with shallow soil water, several of them are believed to survive in normal, uncontaminated groundwater. The presence of biofilm is thus not an automatic indicator of inferior water quality.

17.4 DISCUSSION

17.4.1 Discussion of results

Most of the wells in this study are located in the vicinity of possible contamination sources such as houses, farmland or grazing animals. Based on the knowledge that wells drilled in crystalline bedrock situated <100 m from these sources are more susceptible to faecal microbiological contamination than other wells (Gaut, 2005), the groundwater from these wells could easily be contaminated if precautions are not taken.

The main focus of the video inspections was to evaluate the existence of water leakage between the well casing and the bedrock. Such water leakages may be microbiologically contaminated, due to the rapid transport between the surface/soil environment and the well. The main purpose of the casing installation is to prevent loose superficial deposits from entering the bedrock well; the casing should thus be installed into solid bedrock (NS 3056, 2012). Consequently the casing length is directly correlated to the thickness of the deposits and the degree of fracturing in the uppermost part of the bedrock. Leakage was observed in 40% of the wells, indicating that sealing between the bottom of the casing and the bedrock was not performed during well construction or that the sealing is insufficient. It is assumed that lack of sealing is the main cause of undesired water leakages, since this has not been a focus in Norway until recent years.

It can be disputed whether any observed leakage from shallow soils or superficial materials is an asset or a problem. Clearly, rapid leakage from the surface or from shallow soils is a potential water quality risk factor. On the other hand, bedrock wells drilled in areas with thick Quaternary deposits are known to yield more water than average (Tiedemann, personal communication). This is because water is stored in the overlying sediments and is infiltrated through fractures or along the casing annulus and into the well. Under such circumstances the leakages could be regarded as an asset. Additionally, a thick protective layer of superficial deposits at the well site and in the recharge area was found by Gaut (2005) to be important in maintaining good microbiological water quality. Several groundwater wells in Norway are drilled in Quaternary deposits where the well screen is located only 3–8 m below ground level. Consequently the abstracted water must be regarded as shallow groundwater. Nevertheless, this water often displays adequate microbiological quality. There is no reason to suppose that this water should be less contaminated by faecal microorganisms than the water infiltrated through the Quaternary deposits down into a bedrock well. However, in Norway, wells located in superficial deposits mainly supply waterworks, and, consequently, are subject to strict drinking water regulations and source protection regimes. This is not the case for private wells in crystalline bedrock.

The results also show that 35% of the wells have a maximum 3 m of casing which is drilled less than 1 m into the bedrock. In addition in 39 wells it was observed cavities and fractures directly below the casing demonstrating that the casing has not been drilled into solid bedrock. Consequently any water leaking into these wells could be a problem and imply the necessity to focus on sealing the bottom of the casing.

In 60% of the wells, different degrees of fracturing in the borehole wall from the bottom of the casing and down to about 10 m below surface were observed. Inflow of water was not always occurring from these fractures, but this can be due to the weather conditions at the time of inspection. At some fractures no water was flowing, but there were black, white or rust coloured precipitation on the well walls. This indicates that water may enter the well during periods with rain and snowmelt, and the fractures represent possible flow channels for contaminated water into the well. A possible solution to prevent inflow of water through fractures close to the surface is to install an inner plastic casing to seal off the fractures.

Biofilm and mineral precipitation are common in Norwegian bedrock wells. A correlation study between the age of the borehole and the amount of biofilm reveals no correlation. In some wells the biofilm is mainly within the casing or in the uppermost part of the well. This is often related to wells where the groundwater level is inside the casing or leakage is observed at the bottom of the casing. The biofilm is in these situations always rust-coloured and may be linked to the corrosion of the steel casing. In other wells the biofilm seems to be connected to fractures or fracture zones and thereby to the hydrogeological conditions in the well. A possible explanation is that the water chemistry either enhances growth or in some cases limits the biofilm. A closer study of the water chemistry and the biofilm may give some answers. Unfortunately this was not possible during the project. Analysis of the biofilm may also reveal whether the biofilm consists only of harmless microorganisms or if human pathogens are present.

17.4.2   Improvement in groundwater protection

Sources of microbiological contamination are described by several authors (e.g., Bitton and Gerba, 1984; Daly, 1985; Macler & Merkle, 2000), and it has been known for many years that contamination occurs in the recharge area or at the well head (Wright, 1995; Macler, 1996; Robertson & Edberg, 1997; Conboy & Goss, 1999; Macler & Merkle, 2000; Korkka-Niemi, 2001; Lilly et al., 2003). As a consequence, guidelines for construction of water supply boreholes have been drawn up in several countries by the geological surveys, environmental agencies, and others (Geological Survey of Ireland, 1979; Wright, 1995; Environment Agency, 2000; IGI, 2007; Commonwealth of Massachusetts, 2008; Sveriges Geologiska Undersökning, 2008; New Hampshire D.E.S., 2010; Scottish Environmental Protection Agency, 2010a; Scottish Environmental Protection Agency, 2010b; Gaut, 2011). Several of these are revisions of older versions. In contrast to the more general groundwater protection regulated through the Safe Drinking Water Act (SDWA) in USA and the Water Framework Directive in the EU, well design and construction are often not regulated by legislation or any legislation only applies to public wells or waterworks. An important incentive has, therefore, been to inform and educate private well owners. The guidelines often include both well design and construction, wellhead completion, and how to decommission abandoned wells (Scottish Environmental Protection Agency, 2010b; Environment Agency, 2012).

In Norway the focus on groundwater through PROVA has led to several seminars and conferences with groundwater as an important topic. The results from the NGU projects Vulnerability of groundwater wells in bedrock to microbiological contamination (Gaut, 2005) and Technical quality of bedrock wells (Storrø et al., 2006) have contributed to increased knowledge and focus on the importance of well location, well construction and protection. This is especially so, since the projects have been able to quantify, prove and visualise the poor state of many wells, a situation which for many years were suspected by Norwegian hydrogeologists.

The possibility of showing videos during presentations has been a great help to visualise the problems with drinking water wells in bedrock. Today the Norwegian Food Safety Authority, well owners and well drillers in Norway understand the importance of proper well construction, and some drilling companies have bought cameras for well inspection. The national associations for well drillers are working towards education (in technical college) for new drillers, which include hydrogeology. One of the two associations has also initiated and paid for a new standard for drilling wells in bedrock (NS 3056, 2012). One of the main goals was to make a standard that can be used for drilling private wells for either water supply or energy purposes.

The recent rapid development of multimedia and the Internet as a common tool, has made it possible to reach both professionals and the public with information about groundwater and wells, and it is now easy to find guidelines. NGU runs the Internet information portal www.grunnvann.no [Groundwater in Norway] which, among other issues, presents detailed instructions on well construction and drinking water protection. Similar pages can be found in other countries hosted by the geological surveys, environmental agencies and others working with groundwater.

17.5   CONCLUSION

Leakage between the well casing and bedrock was observed in 40% of 150 Norwegian bedrock wells. Leakage from overlying Quaternary sediments can be an asset because the well yield is increased, but only if the thickness of the superficial deposits is more than 3 m, the casing is drilled into solid bedrock and the well head is properly protected. In most of the investigated wells, such leakage is more likely a possible cause of contamination. In 35% of 210 wells, the casing is less than 3 m long. Fractures and/or cavities were often observed either immediately below the casing (19% of 210 wells) or less than 10 m below the ground surface (59% of 210 wells). Water flowed into the wells from some of these fractures. In addition poor well head protection was a common problem. Among 240 investigated wells, no protection was observed for 72 wells (30%).

The video inspections showed that biofilm and mineral precipitation is common in Norwegian bedrock wells. This is not related to the age of the well. Biofilm or mineral coating is often linked to rusting steel casing and to fractures in the borehole wall.

Throughout the world, well design and construction are often not regulated by legislation or they apply only to public wells or waterworks. However, internet information portals and guidelines for construction of water supply boreholes have been drawn up in several countries by the geological surveys, environmental agencies, and others to help private well owners. In Norway, the focus on groundwater since the mid 1990s has led to increased awareness among both well drillers and well owners regarding groundwater vulnerability and the importance of proper well construction. A Norwegian standard for drilling private wells in bedrock for either drinking water or energy purposes is published. Results from the two NGU projects covering these topics, and the Internet information portal *Groundwater in Norway* have contributed to the increased awareness.

ACKNOWLEDGEMENT

A special thanks to Frank Sivertsvik, Pål Gundersen and Torbjørn Sørdal at NGU for carrying out parts of the field work.

REFERENCES

Banks D., Banks S.B. & Banks, D. (1993) Groundwater microbiology in Norwegian hard-rock aquifers. In Banks SB & Banks D (eds.) *Hydrogeology of Hard Rocks*, Mem. 24th Congress of International Association of Hydrogeologists, 28th June- 2nd July 1993, Ås (Oslo), Norway, pp 407–418.

Banks D., Røyset O., Strand T., Skarphagen H. (1995a) Radioelement (U, Th, Rn) concentrations in Norwegian bedrock groundwaters. *Environmental Geology* 25, 165–180.

Banks D., Reimann C., Røyset O., Skarphagen H., Sæther O. (1995b) Natural concentrations of major and trace elements in some Norwegian bedrock groundwaters. *Applied Geochemistry* 10, 1–16.

Bank D., Frengstad B., Midtgård Aa. K., Krog J.R., Strand T. (1998) The Chemistry of Norwegian Groundwaters: I. The Distribution of Radon, Major and Minor Elements in 1604 Crystalline Bedrock Groundwaters. *The Science of the Total Environment* 222, 71–91.

Bitton G., Gerba C.P. (1984) Microbial pollutants: Their survival and transport pattern to groundwater. In: Bitton G., Gerba C.P. (editors) *Groundwater pollution microbiology*, pp 65–88. John Wiley & Sons.

Bjorvatn K., Bårdsen A., Thorkildsen A.H., Sand K. (1994) Fluorid i norsk grunnvann – en ukjent helsefaktor. [Fluoride in Norwegian groundwater – an unknown health factor] *Vann* 29, 124–128.

Commonwealth of Massachusetts (2008) *Private well guidelines*. Commonwealth of Massachusetts. 91 pp.

Conboy M.J., Goss M.J. (1999) Contamination of rural drinking water wells by faecal origin bacteria – Survey findings. *Water Quality Research Journal of Canada* 34(2), 281–303.

Daly D. (1985) Groundwater quality and pollution. It affects you. It depends on you. Geological Survey of Ireland, Information Circular 85/1, 25 pp.

Englund J., Myrstad J.A. (1980) Groundwater Chemistry of Some Selected Areas in Southeastern Norway. *Nordic Hydrology* 11, 33–54.

Environment Agency (2012). Good practice for decommissioning redundant boreholes and wells. Environment Agency guidance document LIT 6478/657_12, Environment Agency, Bristol, UK. October 2012.

Environment Agency (2000) Water supply borehole construction and headworks. Guide to good practice. Environment Agency, England & Wales, 8 pp.

Frengstad B. (2002) Groundwater quality of crystalline bedrock aquifers in Norway. Dr. ing. thesis 2002, 53, NTNU, Norway.

Gaut S. (2011) Beskyttelse av grunnvannsanlegg – en veileder. [Protection of waterworks based on groundwater – a guideline] Norges geologiske undersøkelse, 46 pp.

Gaut S. (2005) Factors influencing microbiological quality of groundwater from potable water supply wells in Norwegian crystalline bedrock aquifers. Doctoral thesis 2005:99, NTNU, Norway.

Gaut S., Dagestad A., Brattli B., Storrø G. (2007) Factors influencing the microbiological quality of groundwater in Norwegian bedrock. In: Krásný J., Sharp J.M. (eds) *Groundwater in fractured rocks*. IAH Selected Papers on Hydrogeology 9.

Geological Survey of Ireland (1979) Water Wells. A Guide to the Development of Groundwater for Small Residential and Farm Supplies. Information Circular 79/1, Geological Survey of Ireland, 22 pp.

Hongve D., Weideborg M., Andruchow E., Hansen R. (1994) Landsoversikt – drikkevannskvalitet. Sporrmetaller i vann fra norske vannverk. [National overview – drinking water quality. Trace metals in water from Norwegian waterworks] *Vann* 92, Statens Institutt for Folkehelse, 110 pp.

IGI (2007) Guidelines for drilling wells for private water supplies. IGI Codes & Guidelines.

Korkka-Niemi K. (2001) Cumulative geological, regional and site-specific factors affecting groundwater quality in domestic wells in Finland. Monographs of the Boreal Environment Research, 20, Finnish Environment Institute, 98 pp.

Lilly A., Edwards A.C., McMaster M. (2003) Microbiological risk assessment source protection for private water supplies: Validation study. Macaulay Land Use Research Institute, 53 pp.

Macler B.A., Merkle J.C. (2000) Current knowledge on groundwater microbial pathogens and their control. *Hydrogeology Journal* 8, 29–40.

Macler B.A. (1996) Developing the Ground Water Disinfection Rule. *Journal of American Water Works Association* 88(3), 47–55.

New Hampshire DES (2010) Maintenance of inactive wells and decommissioning of abandoned wells. Environmental Fact Sheet WD-DWGB-1-7.

NS 3056 (2012) Krav til borede brønner i berg til vannforsyning og energiformål. [Requirements for drilled wells in rock for water supply and energy purposes] Standard Norge, 32 pp.

Olje – og energidepartementet (2000) Lov om vassdrag og grunnvann (vannressursloven). [Water resources legislation].

Reimann C., Hall G.E.M., Siewers U., Bjorvatn K., Morland G., Skarphagen H., Strand T. (1996) Radon, fluoride and 62 elements as determined by ICP-MS in 145 Norwegian hard rock groundwater samples. *The Science of the Total Environment* 192, 1–19.

Robertson J.B., Edberg S.C. (1997) Natural Protection of Spring and Well Drinking Water Against Surface Microbial Contamination. I. Hydrogeological Parameters. *Critical Reviews in Microbiology* 23(2), 143–178.

Scottish Environment Protection Agency (2010a) An applicants guide to water supply boreholes. Scottish Environment Protection Agency, Sterling, 11 pp.

Scottish Environment Protection Agency (2010b) Good practice for decommissioning redundant boreholes and wells. Scottish Environment Protection Agency, Sterling, 10 pp.

Storrø G., Gaut S., Sivertsvik F., Gundersen P., Sørdal T., Berg T. (2006) Kvalitet av borebrønner i fjell – inspeksjon av brønnutforming. [Quality of bedrock wells – inspection of well constructions] NGU Rapport 2006.031, Norges geologiske undersøkelse, 30 pp.

Sveriges Geologiska Undersökning (2008) Att borra brunn för energi och vatten – en vägledning. Normförfarande vid utförande av vatten – och energibrunnar. [Well drilling for energy and water wells – a guideline.] Sveriges geologiska undersökning, 36 pp.

Tiedemann, K. personal communication.

Wright G. (1995) Well construction standards. *The GSI Groundwater Newsletter* 27, 2.

Chapter 18

Exploration and characterisation of deep fractured rock aquifers for new groundwater development, an example from New Mexico, USA

T. Neil Blandford[1], Todd Umstot[1], Christopher Wolf[1], Robert Marley[1] & Greg L. Bushner[2]

[1]*Daniel B. Stephens & Associates, Inc., New Mexico, USA*
[2]*Vidler Water Company, Nevada, USA*

ABSTRACT

Increasing demands on water resources throughout the semi-arid south western United States have driven the exploration for groundwater to greater depths. A recently completed project to explore and develop groundwater resources from deep fractured rock aquifers at a site in New Mexico was conducted using geophysical analysis, geologic mapping, isotopic analysis, exploration well zone testing, and production well aquifer testing. The site is composed of a synclinorium bounded by two major fault zones; numerous smaller fault zones and geologic structures are also present. Two deep wells (screened from 412 to 652 m below ground surface [bgs] and 152 to 210 m bgs) were constructed and tested at the site. Aquifer testing of the most productive well indicated a highly productive aquifer of limited extent, with two discernible low-permeability boundaries. These findings are important for the prediction of long-term well yield and pumping effects on adjacent groundwater users and spring flow.

18.1 INTRODUCTION

Over the past decade increasing demands on water resources throughout the semi-arid south western United States, combined with prevailing water laws, have driven the exploration for and development of groundwater to greater depths. The results of field investigation and aquifer testing of deep groundwater resources are presented in this chapter for a site in the semi-arid desert south west near Albuquerque, New Mexico, USA (Figure 18.1). The site location is a ranch within the San Pedro Land Grant of Bernalillo County, immediately east of the Sandia Mountains and about 30 km east of the City of Albuquerque at an elevation of approximately 1980 m above mean sea level. The land surface slopes gently to the north-north-east toward San Pedro Creek and is covered with native grass, piñon, and juniper, typical of high desert environments. The project objective was to investigate, and if possible secure, a viable water supply for future development of the ranch.

For the exploration project, two deep wells (ASE-1 and ASE-2) were constructed and tested at the site. The first well bore was drilled to a depth of nearly 1126 m

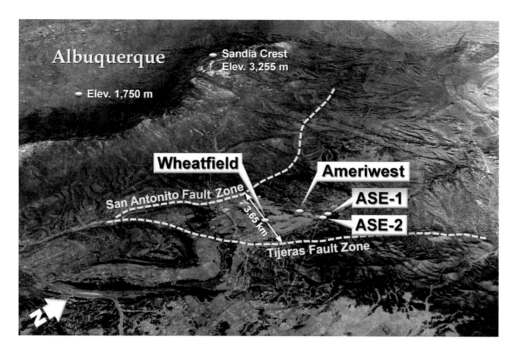

Figure 18.1 Site location. (*See colour plate section, Plate 42*).

below ground surface (bgs), but was completed at a depth of 412 to 652 m bgs in the Abo Formation. The second well was completed nearby from 152 to 210 m bgs and is screened across the San Andres and the Glorieta Formations. Aquifer testing was conducted on the two wells, more extensively so for ASE-2, with multiple observation wells and a monitored spring. Aquifer test data were analysed using a dual porosity analytical model with two low-permeability boundaries that appear to coincide with major fault zones mapped in the region.

Results of the field investigation and aquifer testing indicate that the proposed source of primary water supply, the San Andres-Glorieta aquifer, is bounded on multiple sides. Available groundwater resources at greater depths are insufficient to support the proposed uses. This finding has significant implications for target abstraction rates and the expected nature of drawdown effects on adjacent, existing groundwater users if the water right application before the New Mexico Office of the State Engineer is granted.

18.2　GEOLOGICAL SETTING

The Sandia Mountains have a complex geologic history reflected in the many rock types and geological structures encountered in the area; rocks from Proterozoic to Quaternary ages are present. The geologic units in the area are summarised in Table 18.1. A geological cross-section (Figure 18.2) and block diagram (Figure 18.3)

Table 18.1 Summary of site geological units.

Age	Unit	Summary Description	Thickness at Well ASE-1 (m)
Triassic	Chinle Group (T$_{RC}$)	Dark red shale and feldspathic sandstone	104
	Moenkopi Formation (T$_{Rm}$)	Dark red, micaceous siltstone and sandstones	41
Permian	San Andres Formation (Ps)	Gray, medium – to thick-bedded limestone often with quartz sand grains	62 (combined)
	Glorieta Formation (Pg)	White to pink, well sorted, quartz arenite	
	Yeso Formation (Py)	Orange-red sandstone and siltstone with occasional limestone.	165
	Abo Formation (Pa)	Dark red non-marine sequence of generally fine-grained arkosic conglomerate, sandstone, siltstone, and shale with some thin beds of limestone.	317
Pennsylvanian	Madera Formation (PPm)	Wild Cow member has sequences of arkosic sandstone, sandstone, conglomerate, siltstone, shale and calcarenite.	239
		Los Moyos member is gray limestone, sandstone, and conglomerate.	152
	Sandia Formation (PPs)	Olive green siltstone, sandstone, and conglomerate	39
Proterozoic	Sandia and Cibola granites, undivided (XcYs)	Fine – to coarse-grained granite and monzogranite with potassium feldspar and biotite	Basement – not fully penetrated

illustrate the geological structure and geometry of water-bearing rocks in the study area. The cross-section (Figure 18.2) is coincident with the southern edge of the block diagram in Figure 18.3, which runs east-west through the location of the Wheatfield wells, south of the test wells. Figures 18.2 and 18.3 were constructed on lithological data collected during the drilling of well ASE-1, geological maps (Ferguson *et al.*, 1999b; Read *et al.*, 2000), and a surface geophysical survey. Subsurface exploration at borehole location ASE-1 penetrated the entire stratigraphic sequence, with drilling terminated at 1126 m bgs within intrusive rocks in the Sandia Formation.

Many of the rocks at outcrop and encountered during drilling are jointed and fractured. The entire area has undergone multiple geological events that have broken, faulted, and folded rocks within the study area (Kelley & Northrop, 1975; Karlstrom *et al.*, 1999; Ferguson *et al.*, 1999b). The earliest geologic events date back to the Proterozoic time with ages as old as 1.65 to 0.8 billion years, and these old structural

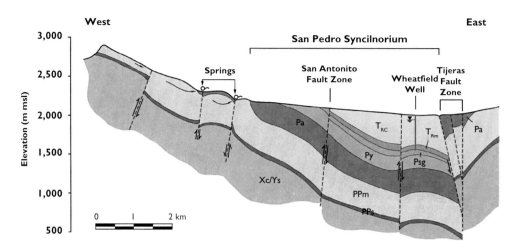

Figure 18.2 West to east geologic cross section, coincident with southern edge of block diagram (Figure 18.3).

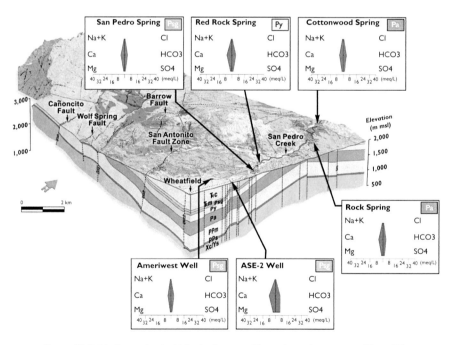

Figure 18.3 Hydrogeological block diagram. (See colour plate section, Plate 43).

systems influenced the expression of subsequent geological structures. The early structures developed along a north east trend, and many of the younger structures follow the same orientation (Figure 18.1).

During the Proterozoic age, contractional forces related to early continent development initiated the rock deformation structures found along the Rio Grande Rift in New Mexico. These forces also created north east to south west structural trends such as the Tijeras Fault zone (Karlstrom et al., 1999). The next major geologic event was formation of the ancestral Rocky Mountains during Mississippian through Permian time, approximately 345 to 225 million years ago. The Pedernal uplift created a mountain range east of the present Sandia Mountains that supplied much of the sediment that constitutes the Madera, Abo, and Yeso Formations.

The Laramide orogeny occurred in the late Cretaceous to early Tertiary. This event reactivated many north east trending faults with a component of dextral strike-slip movement (right lateral fault), including the Tijeras Fault (Figure 18.2). Some of the earlier normal faults experienced reverse slip during this period that caused hundreds of metres of offset.

The most recent tectonic event that created the Sandia Mountains is the formation of the Rio Grande Rift, which began in the Miocene age (Karlstrom et al., 1999). The rift is the result of crustal extension that created a series of north to south trending basins from Colorado through New Mexico and into Texas and Mexico (Hawley, 1996). During formation of the rift, many of the right-slip faults were reactivated and experienced normal thrust (Karlstrom et al., 1999).

Paleozoic and Mesozoic rocks deposited on Proterozoic granites dip to the east along the east side of the Sandia Mountains (Figure 18.2). Offset of the units is evident along the Cañoncito, Wolf Spring, Barrow, San Antonito, and Tijeras Faults (Figures 18.2 and 18.3). The Tijeras Fault zone has a particularly complex geometry related to multiple reactivation events reflecting normal, reverse, and strike-slip movements over geologic time. This fault zone is considered a riedel shear zone represented by conjugate faults developed during strike-slip reactivation (Abbott & Goodwin, 1995; Ferguson et al., 1999a). Within the Tijeras Fault zone, large blocks of rock have been rotated and offset as much as 1600 m (Ferguson et al., 1999a).

Production wells ASE-1 and ASE-2 are completed in a geologic structure named the San Pedro Synclinorium (Ferguson et al., 1999b), which is constrained by the Tijeras and San Antonito Fault zones (Figures 18.1 and 18.2). The San Pedro Synclinorium consists of several broad high-amplitude folds that trend north east and plunge toward the south west (Ferguson et al., 1999b; Kelley & Northrop, 1975). The San Andres and Glorieta Formations are encountered at a depth of about 150 m bgs at the ASE production wells and at about 450 m bgs at the Wheatfield wells (Figure 18.3).

18.3   HYDROLOGICAL SETTING

The hydrology of the Sandia Mountains and the site is controlled by the complex interactions of variable recharge, geological structure, and hydraulic properties of the geologic units. Recharge to groundwater occurs predominantly at higher elevations, with infiltration occurring on exposed outcrops and along arroyos. Groundwater generally moves to the east or north from areas of recharge along the mountain front to areas of discharge such as wells, springs, or adjacent geological units (Bartolino et al., 2010). At some locations on the slopes of the Sandia Mountains, comparison of observed spring flow with recharge rates determined through chloride mass balance suggests that geological structures may channel recharge from upland areas to the north or south rather than downdip to the east (McCoy & Blanchard, 2008).

Predominant aquifers in the study area include the Madera Formation, the Abo and Yeso Formations, and the San Andres and Glorieta formations. The Madera Formation was the initial target aquifer at the beginning of exploration activities, but was found to yield virtually no water at ASE-1. The Abo and Yeso formations can yield sufficient water for domestic wells at many locations, but typically do not yield large quantities of water to wells. Well ASE-1 was completed in the Abo Formation and has limited yield. The San Andres and Glorieta formations can each yield significant quantities of water to wells (e.g., Shafike & Flanigan, 1999). Combined, these formations compose the San Andres-Glorieta aquifer, which is the most productive aquifer in the study area where it occurs. Well ASE-2 is completed in this aquifer unit.

The relationships between rock units and structures and their influence on groundwater flow and chemistry are illustrated in the block diagram (Figure 18.3). Springs tend to occur where faults juxtapose different rock types, or where units are brought to the surface and groundwater discharges, such as at San Pedro and Dragonfly Springs. Rock types and seepage between aquifer units also influence water quality, as shown by the stiff diagrams presented in Figure 18.3. For example, well ASE-2 is completed through the full thickness of the San Andres-Glorieta aquifer, and samples from this well include the effects of upward seepage of water from the underlying Yeso Formation, which has a high gypsum content. The nearby Ameriwest well is also completed in the San Andres-Glorieta aquifer, but does not penetrate the full thickness of the Glorieta sandstone; therefore, water quality from this well has reduced sulfate and overall total dissolved solids concentrations compared to ASE-2. The Ameriwest well has similar water quality to that observed at San Pedro Spring. The direction of groundwater flow is generally to the north (i.e., from the ASE-2 and the Ameriwest wells toward San Pedro Spring) and, therefore, is in the updip direction.

San Pedro Creek (Figure 18.3) is the primary surface water feature in the study area. Surface water runoff occurs in response to precipitation and snowmelt events. Perennial reaches only occur downstream of springs that occur along the creek, the locations of which are controlled by geologic structures that cause aquifer units to be truncated at the surface (Ferguson et al., 1999a).

The largest spring in the area is San Pedro Spring, which discharges at about 230 l/m. The spring occurs where the rocks of the San Andres-Glorieta aquifer outcrop and are for the most part covered by a thin veneer of alluvium at the north end of the synclinorium (Figure 18.3). Discharge from San Pedro Spring forms the first and longest perennial reach (about 3.2 km) that occurs within San Pedro Creek. Red Rock and Dragonfly Springs (identified and named as part of this investigation) also contribute to this perennial reach. Red Rock Spring discharges water from the top of the Yeso Formation that appears to be primarily San Andres-Glorieta aquifer water with a component of Yeso Formation water. Dragonfly Spring occurs near the contact between the Abo and Madera Formations farther to the north, where the Abo Formation is truncated at the surface.

Other smaller springs discharge into San Pedro Creek downstream (north) of the perennial reach formed by San Pedro, Red Rock, and Dragonfly Springs. Rock Spring discharges several feet above the creek channel from outcrops in the transitional contact between the Madera and Abo Formations. Cottonwood Spring appears to discharge from the Abo Formation and alluvium (Bartolino et al., 2010). Surface flow ceases approximately 550 m downstream of Cottonwood Spring due to evapotranspiration and seepage into stream channel alluvium.

Based on young apparent carbon ages relative to groundwater samples from wells and the measurable tritium concentrations (0.69 to 1.09 tritium units) found in San Pedro, Rock, and Cottonwood Springs, localised recharge and shallow groundwater movement along the creek bottom likely affect spring water quality and the estimates of groundwater age, a conclusion consistent with that of other researchers (Bartolino *et al.*, 2010).

18.4   METHODS AND RESULTS

To obtain an appropriation of water under the laws of the state of New Mexico, an analysis must be made of the effects of the proposed pumping on existing users of groundwater and surface water in the basin. This determination was made through development and application of a groundwater flow model, which is not presented in this chapter. However, in order to evaluate aquifer properties for implementation into the model, and to assist with model calibration, aquifer tests were conducted for wells ASE-1 and ASE-2. The ASE-1 test was a single well test that confirmed the limited production capacity of the Abo aquifer at the site. ASE-1 was pumped at 190 l/min for a period of 4.25 days and experienced about 282 m of drawdown. The ASE-2 test was a long-term test that involved multiple observation points.

A step-drawdown test conducted on ASE-2 indicated that the San Andres-Glorieta aquifer is highly productive. Subsequent to the step test, a constant-rate aquifer test was conducted during November 2010 for a period of 7 days at a pumping rate of 3220 l/min. Step testing indicated that well ASE-2 can pump at least twice this rate, but the test rate was selected to significantly perturb the aquifer system while at the same time produce a quantity of discharge that could be managed on-site.

Discharge water was pumped south from ASE-2 to a network of temporary sprinkler systems operated during the test. Since the San Andres-Glorieta aquifer lies more than 150 m below the surface, beneath the low-permeability Chinle Group sediments, water discharged on-site could not affect the observation wells during the pumping or recovery periods. Discharge water was not released to San Pedro Creek due to environmental permitting limitations and, more importantly, so that spring discharge could be monitored during the test without interference from aquifer test discharge water.

In addition to San Pedro Spring, hydraulic head was monitored in the pumping well and four observation wells. Well ASE-1 is about 50 m from pumping well ASE-2, but is completed in the Abo Formation, with the top of the screen about 200 m below the bottom of the screen in ASE-2. The Wheatfield shallow observation well is completed in the Chinle Group that overlies the San Andres-Glorieta aquifer, about 1.8 km south west of the pumping well. As expected, drawdown attributable to the aquifer test was not observed at either of these wells due to the low-permeability sediments that separate the screened intervals of these wells from the San Andres-Glorieta aquifer. The other two observation wells, Ameriwest and Wheatfield-deep, are completed in the San Andres-Glorieta aquifer (Figure 18.3).

The total drawdown (uncorrected for well efficiency) after 7 days of pumping at well ASE-2 was about 9.5 m, indicating that the well has a specific capacity of about 340 l/min/m of drawdown. Figure 18.4 illustrates the observed hydraulic head and spring discharge response during the aquifer test. Nearly immediate responses to pumping at ASE-2 were observed in the Ameriwest and Wheatfield-deep wells and at the San Pedro Spring gauge, followed by nearly immediate recovery trends at the cessation of pumping. Total drawdown at the end of the test was 0.49 m at the Ameriwest well and 0.43 m at the Wheatfield deep well, which is more than twice the distance from the pumping well as is the Ameriwest well (Figure 18.4). San Pedro Spring, located about 1.2 km north of ASE-2, had a reduction in flow up to 68 l/min by the end of the test. Although not discovered until the constant-rate aquifer test data were plotted, the effects of well development and the ASE-2 step test are also evident at the Ameriwest and Wheatfield observation wells, as highlighted on Figure 18.4 for the 3-day period prior to the 7-day test. The test results indicate a highly productive, well-connected San Andres-Glorieta aquifer between the pumping well and the three monitoring points.

The nearly immediate hydraulic response over the significant extent of aquifer could be due to the presence of a karst system in the San Andres Limestone. Although it is likely that solution channels and conduits occur in the San Andres Formation to some extent, the Glorieta Sandstone was observed to be fractured in the drill cuttings. In addition, a temporary test well was installed during the drilling and testing of ASE-1 to evaluate water quality and estimate pumping capacity within the mid-section of the Glorieta Formation. The temporary well had 6 m of screen beginning at

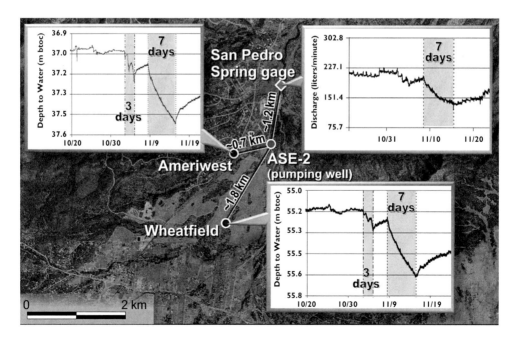

Figure 18.4 Observed water level drawdown and San Pedro Spring discharge during ASE-2 aquifer test.

200 m bgs and 18 m of gravel pack beginning at 197 m bgs. This zone was isolated using a bentonite seal and pumped for more than 500 minutes at two different rates of 340 and 530 l/min, and the observed drawdown indicated a specific capacity of about 200 to 130 l/min/m of drawdown. Compared to the observed specific capacity of 340 l/min/m of drawdown at ASE-2, which is screened across the full San Andres and Glorieta Formation thickness, it is evident that the fractured sandstone of the Glorieta Formation likely contributes approximately half of the ASE-2 well yield, and possibly more due to the limited size and screen length of the test well.

18.5   ANALYSIS

The data from the aquifer tests were analysed using the dual-porosity fractured rock solution developed by Moench (1984) for the analysis of aquifer test data in dual porosity media. The conceptual model for this solution is that the fractures are highly permeable but have limited storage capacity and, therefore, account for the rapid transmission of pumping effects. Conversely, the matrix blocks are characterised by higher storage capacity but low hydraulic conductivity and thereby slowly release water due to the drawdown in the surrounding fractures.

The analytical code AQTESOLV (Duffield, 2007) was used to apply the Moench solution to the observed data; AQTESOLV allows for the simultaneous fit to both drawdown and recovery data at multiple wells. For application of the Moench solution, the assumption was made that the pumping and observation wells completely penetrate the productive portion of the San Andres-Glorieta aquifer. This is true for the pumping well, but the Ameriwest and Wheatfield-deep monitor wells penetrate only the upper portion of the aquifer consisting of San Andres Limestone with some interbedded Glorieta Sandstone. Since both formations are productive and contain no significant low-permeability units, the assumption that the hydraulic head is virtually the same within each unit is reasonable. The aquifer parameters estimated by the Moench solution include the fracture hydraulic conductivity (K), fracture specific storage (S_s), matrix hydraulic conductivity (K'), and matrix specific storage (S_s').

Approximately 1080 minutes into the test, the pump was shut down for about 90 minutes due to mechanical difficulties. When the test was restarted, the observed drawdown and water level trends for the pumping well were different than those observed during the earlier portion of the test. The observed water level response appears to be real but is unexplained. Transducer levels were double-checked and confirmed with manual measurements regularly, and the pumping rate was monitored with two flow meters. The limited shut-down did not significantly affect the observed water levels in the observation wells; only a small noticeable response was observed at the Ameriwest observation well.

Because of the less than ideal test conditions that occurred at the pumping well, the decision was made to conduct the aquifer test analysis by attempting to obtain the best-possible quantitative fit to the observation well data and to treat the observed drawdown at the pumping well, corrected for well efficiency, qualitatively. Analysis of the ASE-2 step test data using the method of Bierschenk (1963) indicates a well efficiency of approximately 30% for the test pumping rate, which indicates a theoretical drawdown at the pumping well of about 3 m, based on the uncorrected drawdown measured in ASE-2 during the test of about 9 m. The drawdown data for ASE-2, corrected for well efficiency, are plotted in Figures 18.5 and 18.6.

Although the water level in the pumping well had fully recovered prior to initiation of the constant rate test, the same was not true for the observation wells (Figure 18.4). Even though the analysis considers the uncorrected drawdown data for the observa-

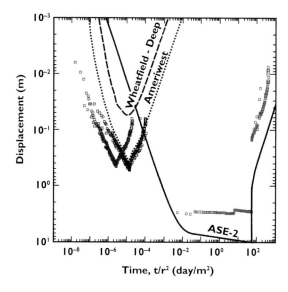

Figure 18.5 Moench aquifer test solution without aquifer boundaries; match to Ameriwest observation well only.

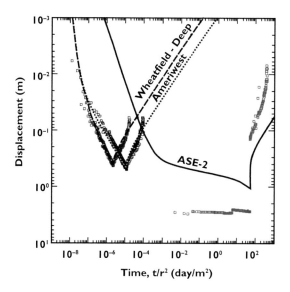

Figure 18.6 Moench aquifer test solution with parallel no-flow boundaries; match to Ameriwest and Wheatfield-deep observation wells.

tion wells, subsequent analysis where the observed drawdown is corrected for residual drawdown from the ASE-2 step test did not significantly change the results. Likewise barometric effects are small relative to the observed drawdown and do not significantly change the observed drawdown behaviour or results of the analysis.

Initially, the Moench solution with no aquifer boundaries was used to estimate aquifer properties for the Ameriwest and Wheatfield-deep observation wells individually. This approach led to good matches to each of the observation wells separately, as illustrated for the Ameriwest well in Figure 18.5. The parameters estimated from the individual fits were similar for both cases with the exception of the matrix hydraulic conductivity. The fracture hydraulic conductivity was about 9 m/d, the fracture specific storage was about 3×10^{-6} 1/m, and the matrix specific storage ranged from about 4×10^{-5} to 2×10^{-4} 1/m. The matrix hydraulic conductivity was low for both wells. Except for the fracture hydraulic conductivity and fracture specific storage, the parameters are correlated in the individual analysis, indicating that the aquifer test solution for those parameters is not unique and various combinations of parameter values may yield a similar fit. In both cases, the simulated drawdown at the pumping well ASE-2 is overestimated by about 7 m.

In order to improve upon the single observation well solutions, additional simulations were conducted considering both observation wells concurrently and introducing no-flow boundaries into the parameter identification process. AQTESOLV allows for the addition of up to four boundary conditions that are either parallel or perpendicular to each other. Zones of limited permeability were expected for the Tijeras Fault zone east of the site (Bartolino *et al.*, 2010; Drakos *et al.*, 1999), the vicinity of San Pedro Spring north of the site where the San Andres-Glorieta aquifer pinches out at the northern extent of the San Pedro synclinorium, and west of the site in the vicinity of the San Antonito Fault zone. A summary of several of the aquifer test solutions considered is provided in Table 18.2 in terms of the root mean squared error (RMS) of the fit between the simulated and observed drawdown at the Ameriwest and Wheatfield-deep observation wells.

The best simultaneous match to both observation wells was obtained using two parallel no-flow boundaries in AQTESOLV (Table 18.2). One boundary was located approximately concurrent with the Tijeras Fault zone, and the second boundary was placed in the vicinity of the San Antonito Fault zone. The two boundaries were placed approximately 3.65 km apart in the analysis, which is the approximate distance between the Tijeras and San Antonito Faults in the vicinity of the Wheatfield wells (Figure 18.1). Several alternative smaller distances between the two boundaries were considered in the sensitivity analysis, but the final distance led to the best fit between

Table 18.2 Summary of aquifer test simulation scenarios.

Scenario	RMS (m)
No-flow boundary coincident with Tijeras Fault zone only	0.08
No flow boundary coincident with Tijeras Fault zone and northern pinch-out of San Andres-Glorieta aquifer (perpendicular boundaries)	0.07
No-flow boundaries approximately coincident with Tijeras Fault zone and San Antonito Fault (parallel boundaries)	0.03

the simulated and observed drawdown data, and distances less than about 3 km led to substantial degradation of the RMS.

Due to mathematical limitations of incorporating boundary conditions into the analytical solution approach, the orientation of the San Antonito Fault zone was assumed to be parallel to that of the Tijeras Fault zone; this assumption is reasonable in the vicinity of the Wheatfield-deep observation well in the southern part of the study area, but the trend of the San Antonito Fault zone diverges sharply to the north west of the region (Figure 18.1). In addition, in the northern portion of the study area the San Andres-Glorieta aquifer outcrops between the San Antonito Fault zone and well ASE-2. These conditions are addressed in numerical simulations not presented here.

The fit between the observed data and the Moench solution was insensitive to the matrix hydraulic conductivity, so the matrix hydraulic conductivity was fixed at 3×10^{-7} m/d based on the results of the individual well analyses conducted previously. The estimated fracture hydraulic conductivity, fracture specific storage, and matrix specific storage for this scenario were estimated to be about 160 m/d, 2.1×10^{-5} 1/m, and 3.3×10^{-6} 1/m, respectively (Figure 18.6). The fracture and matrix specific storage parameters have a high inverse correlation; none of the other parameter pairs had a strong correlation. For the best-fit case for the two observation wells presented in Figure 18.6, the simulated drawdown at the pumping well is about 2 m less than the observed drawdown corrected for well efficiency, and is therefore more accurate than the solution that considers a single observation well (Figure 18.5).

18.6   DISCUSSION AND CONCLUSIONS

The first exploration boring constructed at the project site was advanced to 1126 m in depth. Extensive zone testing, water quality sampling, and borehole geophysics indicated that the initial target aquifer, the Madera, had very low production potential. The first well (ASE-1) was completed in the Abo Formation from 412 to 652 m bgs, with a limited yield of approximately 190 l/min with about 282 m of drawdown. Additional production capacity could have been obtained by screening multiple aquifer units, but regulations of the state of New Mexico prohibit the completion of wells across multiple aquifer units.

Based on observations made during the drilling of ASE-1, well ASE-2 was constructed and screened across the shallower (152 to 210 m) San Andres-Glorieta aquifer. Aquifer testing of the ASE-2 well indicated a highly permeable aquifer with efficient hydraulic connection through fracture networks (likely enhanced by solution activity within carbonate strata) with high hydraulic conductivity and low specific storage; nearly instantaneous responses to pumping were observed at the Wheatfield-deep well 1.8 km south of the pumping well, the Ameriwest well 0.7 km west of the pumping well, and San Pedro Spring 1.2 km north of the pumping well. The southern observation well (Wheatfield-deep) is downdip, where the aquifer is about 215 m deeper than the bottom of screen in the pumping well, and San Pedro Spring is updip, about 123 m higher in elevation than the top of the screen in the pumping well. The nearly immediate response to pumping in all locations indicates that the highly permeable aquifer extends significant distances from the pumping well. This result was not necessarily anticipated, particularly for the Wheatfield-deep well, where smaller primary and secondary permeability was expected due to the greater formation depth and less solution enhancement of permeability since the aquifer is more remote from near-surface recharge sources.

Results of the aquifer testing also indicate that although the aquifer is highly permeable and well-connected between the pumping well and the observation points, it is bounded by zones of sufficiently low hydraulic conductivity that they could be reasonably considered no-flow boundaries in the aquifer test simulations. The eastern no-flow boundary corresponds to the Tijeras Fault zone, which is believed to have low permeability due to the rotated and disjointed nature of the rock mass within the fault zone. Although Ferguson *et al.* (1999a) state that rocks within about 90 m of the Tijeras Fault zone have been highly fractured, they also describe the fault zone as a 'tectonic melange' with blocks of rock that have been translated into position and 'rotated significantly' due to movement along the fault. Fractured rocks that do occur are blocks with variable strikes and dips related to rotations during faulting and displacement along minor faults within the fault zone. Fractures, even if they are open, exist within rotated and displaced blocks that disrupt fracture continuity. In addition, fractures and bedding planes within the fault zone were observed in the field to contain variable degrees of filling by fault gouge or minerals such as carbonates that would further decrease permeability. The low-permeability nature of the Tijeras Fault zone has also been identified by Bartolino *et al.* (2010) based on water level and water quality information.

Inclusion of the San Antonito Fault zone as a hydraulic barrier was required to reasonably match the observed drawdown at the observation wells. The San Antonito Fault zone is not as large as the Tijeras Fault zone but it does consist of multiple faults and, similarly to the Tijeras Fault zone, exhibits significant strike-slip movement in addition to variable amounts of vertical offset along its length (Ferguson *et al.*, 1999a, 1999b). Although the San Andres-Glorieta aquifer is known to pinch out north of ASE-2 in the vicinity of San Pedro Spring, consideration of this region as a no-flow boundary in the aquifer test analysis did not significantly improve the parameter estimation results.

Testing conducted for a water right permit in New Mexico is at the discretion of the applicant, so the period of time for which aquifer tests are conducted and the selected pumping rate are highly variable. Frequently, constant rate tests may be conducted for periods of 24 or 72 hours, at abstraction rates similar to or sometimes less than the expected long-term rate. One reason for this is the New Mexico regulation that groundwater pumping from an exploratory well (a well completed for exploratory purposes that does not have an approved water right) cannot be conducted for more than 10 cumulative days. The purpose of this regulation is to conserve water and ensure that water is not diverted illegally. If the test had been conducted for only 24 hours, only one boundary condition would have been apparent from the observed drawdown response; a 72-hour test would have been sufficient to detect the second aquifer boundary. If the pumping rate had been equal to the requested abstraction rate (rather than nearly double the requested rate), a 72-hour test would also have been insufficient to conclusively detect the second boundary.

For each of the three wells, the simulated rate of recovery of hydraulic head is less than the observed rate of recovery, particularly at later times. One possible explanation for this behaviour is that vertical seepage of groundwater from adjacent geological units, not considered in the Moench analysis, may exhibit an observable effect on the hydraulic heads, particularly later in the recovery period. In addition, for small values of matrix hydraulic conductivity and specific storage relative to the fracture properties, the Moench (1984) solution approaches the same conditions approximated by the Theis (1935) solution, indicating that the aquifer system considered might be reasonably simulated as an equivalent porous media. These issues may be examined in future work.

The bounded nature of the San Andres-Glorieta aquifer in this region is important for the consideration of the future effects of pumping on adjacent users. This is because drawdown from pumping ASE-2 will be magnified between the faults, and will be more limited in regions across the faults, than would otherwise occur for the case of an unbounded aquifer. Given the large number of domestic wells adjacent to the ranch property, estimation of the potential drawdown effects on these wells is an important aspect of the applicant's ability to obtain a permit for a new appropriation of groundwater. At the time of writing, the request for an appropriation of groundwater in the amount of an average abstraction rate of 1683 l/min (slightly more than half of the abstraction rate at which the aquifer test was conducted) is being considered by the New Mexico Office of the State Engineer. The hydrogeological framework of the San Andres-Glorieta aquifer as determined, in part, through the analysis documented in this paper will be one consideration in the State Engineer's decision. The aquifer testing and other characterisation activities, along with additional analyses not presented in this paper, substantiate the long-term viability of the proposed abstraction.

REFERENCES

Abbott J.C., Goodwin L.B. (1995) A spectacular exposure of the Tijeras Fault, with evidence for quaternary motion. In: Bauer P.W., Kues B.S., Dunbar N.W., *et al.* (editors) *Geology of the Santa Fe region, New Mexico*, New Mexico Geological Society 46th Annual Field Conference, September 27–30, 1995, pp.117–126.

Bartolino J.R., Anderholm S.K. & Myers N.C. (2010) Groundwater resources of the East Mountain area, Bernalillo, Sandoval, Santa Fe, and Torrance Counties, New Mexico, 2005. USGS Scientific Investigations Report 2009–5204.

Bierschenk W.H. (1963) Determining well efficiency by multiple step-drawdown tests. *International Association of Science Hydrology* 64, 493–507.

Drakos P., Lazarus J. & Jetter S. (1999) Hydrogeologic characterization of fractured Abo and Madera Formation aquifers, hydrocarbon contamination, and transport along the Zuzax Fault, Tijeras Canyon, New Mexico. In: Pazzaglia FJ, Lucas SG (eds) *Albuquerque Geology, New Mexico Geological Society 50th Annual Field Conference*, September 22–25, 1999, pp. 419–424.

Duffield G.M. (2007) AQTESOLV for Windows, Version 4.50.002, Professional. Developed by HydroSOLVE, Inc. for ARCADIS Geraghty & Miller, Inc.

Ferguson C.A., Timmons J.M., Pazzaglia F.J., Karlstrom, K./E.,Osburn, G.R., & Bauer, P.W. (1999a) Geology of Sandia Park quadrangle, Bernalillo and Sandoval Counties, New Mexico. Open-File Report GM-1, New Mexico Bureau of Mines & Mineral Resources, Socorro, New Mexico.

Ferguson F.A., Osburn G.R. & Allen B.D. (1999b) Geology of the San Pedro 7.5-minute quadrangle. Open-File Geologic Map 29, New Mexico Bureau of Geology and Mineral Resources, Socorro, New Mexico.

Hawley J.W. (1996) Hydrogeologic framework of potential recharge areas in the Albuquerque Basin, Central Valencia County, New Mexico. In: Hawley J.W., Whitworth T.M. (editors) *Hydrogeology of potential recharge areas for the basin and valley-fill aquifer systems, and hydrogeochemical modelling of proposed artificial recharge of the Upper Santa Fe Aquifer, northern Albuquerque Basin, New Mexico.* Open-file Report 402-D, New Mexico Bureau of Mines and Mineral Resources, Socorro, New Mexico, 68p.

Karlstrom K.E., Cather S.M., Kelley S.A., Heizler,M.T., Pazzaglia, F.J. & M. Roy. (1999) Sandia Mountains and Rio Grande Rift: Ancestry of structures and history of deformation. In: Pazzaglia F.J., Lucas S.G. (editors) *Albuquerque Geology, New Mexico Geological Society Fiftieth Annual Field Conference,* September 22–25, 1999, pp. 155–166.

Kelley V.C. & Northrop S.A. (1975). *Geology of the Sandia Mountains and vicinity, New Mexico.* Memoir 29, New Mexico Bureau of Mines and Mineral Resources.

McCoy K.J. & Blanchard P.J. (2008) Precipitation, ground-water hydrology, and recharge along the eastern slopes of the Sandia Mountains, Bernalillo County, New Mexico. USGS Scientific Investigations Report 2008–5179.

Moench A.F. (1984) Double-porosity models for a fissured groundwater reservoir with fracture skin. *Water Resources Research* 20(7), 831–846.

Read A.S., Karlstrom K.E., Connell, S.K., Ferguson, C.A., Ilg, B., Osburn, G.R.,Van Hart, D. & F.J. Pazzaglia (2000) Geology of Sandia Crest quadrangle, Bernalillo and Sandoval Counties, New Mexico. Open-File Geologic Map 6, New Mexico Bureau of-Geology and Mineral Resources, Socorro, New Mexico.

Shafike N.G. &Flanigan K.G. (1999). Hydrologic modeling of the Estancia Basin, New Mexico. In: Pazzaglia FJ & Lucas SG (eds) Albuquerque Geology, New Mexico Geological Society 50th Annual Field Conference, September 22–25, 1999.

Theis C.V. (1935) The relation between the lowering of the piezometric surface and the rate and duration of discharge of a well using groundwater storage. *American Geophysical Union Transactions* 16, 519–524.

Chapter 19

Use of several different methods for characterising a fractured rock aquifer, case study Kempfield, New South Wales, Australia

Katarina David[1], Tingting Liu[2] & Vladimir David[3]
[1]KD, Toongabbie, NSW, Australia
[2]Heritage Computing, Hornsby, NSW, Australia
[3]Argent Minerals, North Sydney, NSW, Australia

ABSTRACT

An integrated interpretation of hydraulic data, test pumping and hydrogeochemical data demonstrates the value of combining several different methods for improved characterisation of a fractured-rock aquifer and assessment of recharge in a temperate climate environment. These methods were applied at the proposed open cut barite-silver mine, in Kempfield, central NSW, Australia. The test pumping results, hydraulic data, assessment of recharge rates, baseflow and geochemical analysis guided the construction of three-dimensional numerical model of groundwater flow to understand the groundwater system. There are no published groundwater studies in this area; hence the analysis in this paper is based on the data collected mainly from the site investigations. The integrated field study of structural pattern and hydraulic tests shows that structural lineaments and fracture porosity are important features that affect groundwater flow in a fractured rock aquifer. The recharge rates have been assessed indirectly using both chloride mass balance and hydrograph analysis, with results favoring the chloride mass balance method. Groundwater-surface water connectivity is limited with an estimated baseflow contribution of 1–7% of streamflow.

19.1 INTRODUCTION

Fractured rock aquifers generally display higher heterogeneity than other aquifers. This has important implications on the conceptualisation of such systems. Due to variability in interconnected primary porosity of the rock formation and irregular occurrence and variability in the fractures, groundwater flow, aquifer characteristics and yield in a fractured rock environment are very difficult to analyse and predict.

The hydraulic properties of fractured rock aquifers may differ significantly within a small area. This heterogeneity is a result of the contrast in hydraulic properties between the high permeability fractures and the low permeability rock matrix and the potential variability in connectivity within fracture systems. The interaction between the flow in a fractured rock aquifer and surface water can also be complicated by the nature of vertical fracture interconnection (Novakowski *et al.*, 2000). In fractured rock aquifers yield is largely controlled by the secondary porosity, a function of the fracturing. Higher yielding groundwater zones are, therefore, likely to occur in the vicinity of major geological structures due to higher fracture density.

This chapter describes a hydrogeological study of fractured meta-sediments and volcanoclastics with the study focused on the development of a conceptual fractured rock model. To develop a conceptual model, it is useful to combine different methods to understand the fracture system behaviour. Structural studies, test pumping data, hydraulic head analysis of recharge, hydrogeochemistry and baseflow analysis are combined to develop a conceptual understanding of groundwater flow beneath the site. This conceptual model is further tested with a numerical groundwater flow model to check its validity. The main objectives of this project were to understand the groundwater system, importance of structural features and connectivity between surface and groundwater systems.

Improved understanding of the interactions between surface and the fractured groundwater systems is of particular importance in the study area due to the ephemeral nature of the creeks, increasing competition for water resources in this catchment, possible change use pattern due to proposed mine development and ecological value of Rocky Bridge Creek as the only breeding ground for the Maquarie Perch.

19.2   STUDY AREA

The Kempfield Project is located 50 km to the west of Bathurst in central west New South Wales, Australia (Figure 19.1). The area of interest, where detailed hydrogeological studies were undertaken, is approximately 2 km east to west and 3 km north to south. The topography in the study area is undulating with gentle slopes. The broad valleys and creeks seen in the area are generally aligned along the fault zones. Elevations in the project area range from 650 m above sea level in the central area to 850 m above sea level to the north. The south-eastern part is densely forested while the area to the west and north is cleared for stock grazing.

The Rocky Bridge Creek crosses the study area from northeast to southwest. The creek has its source at elevations of approximately 1050 m Australian Height Datum (mAHD) in the northeast of the area, and flows generally to the west and southwest toward the Abercrombie River which feeds the Wyangala reservoir. This reservoir supplies water for towns and agricultural irrigation in the area. The study area has a temperate climate (BoM website) with an average annual precipitation of 851 mm and mean annual evaporation of 1350 mm recorded at the Trunkey Creek and Bathurst meteorological station, respectively.

19.2.1   Geology and structure of the study area

Geologically, the Kempfield barite-silver (volcanic hosted massive sulphide – VHMS) deposit is located in the Hill End Trough, one of the several intracratonic basins developed during Silurian–Devonian periods in the eastern province of the Lachlan Orogen, Eastern Australia (Colquhoun et al., 1996; Colquhoun et al., 1997). The Kempfield deposit is hosted in felsic volcanic rock sequence known as the Kangaloolah Volcanics. These have been in-faulted and folded into Middle Ordovician meta-sediments, along the Copperhannia Thrust, and comprise a complex sequence of mainly chlorite schist,

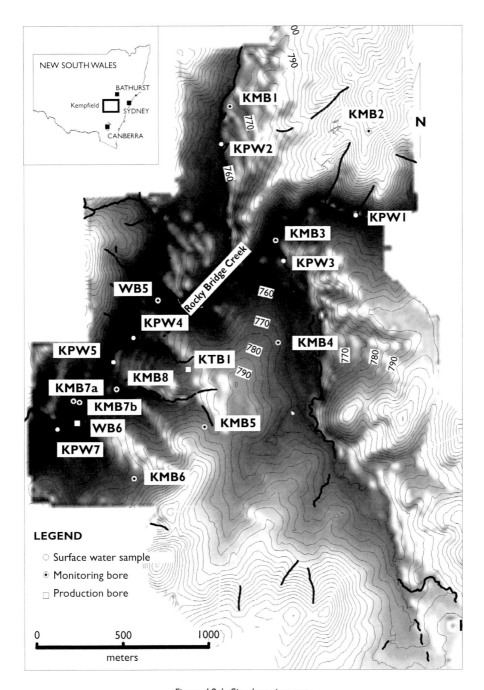

Figure 19.1 Site location map.

siltstone and quartzite (Timms & David, 2011). This felsic volcanic derived sequence of sediments and minor volcanics can be sub-divided into three units: a sequence of fine-grained and quartz-phyllite tuffaceous rocks, volcanoclastic sedimentary rocks which host silver- barite and lead-zinc mineralisation and unaltered siltstone (barren of mineralisation) (Figure 19.2).

The fresh rocks at Kempfield are metamorphosed to lower green schist facies reflected through the different type of schists (Timms & David, 2011). Fresh schist

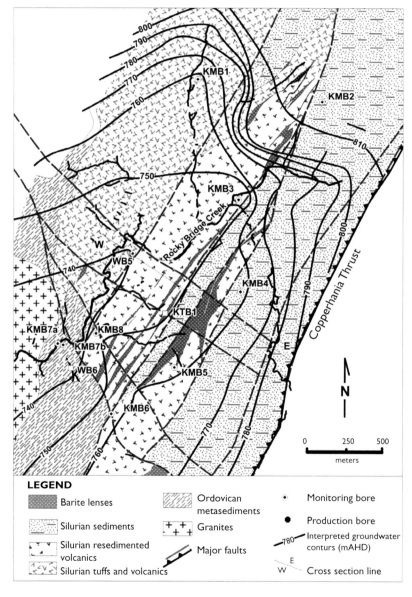

Figure 19.2 Interpreted geology map modified after Timms & David (2011) with potentiometric contours.

gradually changes in the the transition zone (partially weathered 2 to 5 m thick), while the weathered material forms the capping layer. The base of weathering is shown on Figure 19.3.

Structurally, the area is characterised by the orthogonal faulting pattern with sub-vertical structural features extending in the north west-south east and north east–south west direction. The sequence is interpreted as a steeply dipping volcanoclastic sequence. The dominant structural style is normal faulting along the north east trending easterly steeply dipping faults of the tight isoclinal folds (Figures 19.2 and 19.3). The current conceptual understanding suggests that fracture zones that have highest groundwater yield are a result of alignment along sub-vertical structural lineaments.

19.2.2 Hydrogeological characteristics

Three aquifer systems are defined in the study area (Figure 19.3): a fractured (deep) rock aquifer, a weathered (shallow) bedrock aquifer and an alluvial-colluvial Quaternary aquifer, which is poorly developed and associated with Rocky Bridge Creek.

The fractured rock aquifer system is present beneath the entire site. It comprises Silurian sediments, resedimented volcanics, tuffs and barite lenses and Ordovician metasediments (Figure 19.2). These are characterised by secondary porosity as a result of fracturing and display relatively high permeability. Within the study area the fractured rock aquifer is covered by unconfined to semiconfined weathered rock aquifer. This aquifer comprises both partially weathered (transition zone) and completely weathered rock. The bottom of the weathered rock aquifer is defined by the base of weathering and its thickness varies from 5 m–50 m (Figure 19.3). The thickness of this aquifer is dependent of parent geology and hydrothermal alteration rather than elevation. Therefore, shallow weathering is present over the barite lenses whilst deeper weathering is seen along the structural features (Figure 19.3). The weathered rock aquifer is partially saturated to unsaturated, as confirmed by the exploration drilling programme. The alluvial/colluvial system is localised and present only in the vicinity of

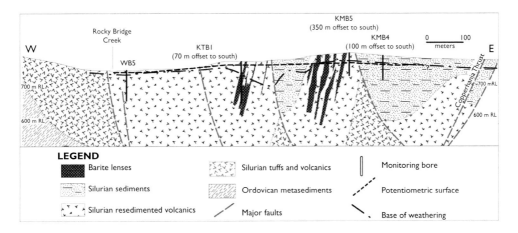

Figure 19.3 Geology west-east cross section.

the water courses. Thicknesses generally between 1 m–2 m are seen. No groundwater bores are installed in this aquifer unit; however its presence and distribution has been confirmed from field inspection and drillers logs.

High permeability groundwater zones occur locally in the vicinity of major geological faults and joint sets due to the increased concentration fracture density and connectivity in these areas. This trend is illustrated through mapping of available bore yield data within 25 km radius of the site (New South Wales Office of Water Database). The water bearing fractures in this geological terrain are generally associated with valley floors, whilst ridgelines are characterised by low permeability rocks and closed fracturing. This is the result of erosional processes increasing permeability along pre-existing structural weaknesses in the valley areas. Away from fault zones, erosion is less pronounced resulting in the presence of resistant and structurally intact features along ridgelines.

The main tectonic structures in terms of hydrogelogy are vertical to subvertical faults, extending in the north west-south east and north east-south west direction (Figures 19.2 and 19.3). These faults affect the groundwater system to some degree, this is demonstrated by the higher groundwater yields that are found where these fractures intersect.

In the study area, the potentiometric surface in the fractured rock aquifer ranges from 40 m below ground surface at the higher elevations to above ground (artesian flow) in the low lying areas. The groundwater flow direction generally follows the topography. There appears to be a groundwater divide to the east of the site aligned roughly north-south along the Copperhania Thrust (Figure 19.2).

19.3 METHODS OF INVESTIGATION

A structural-geological map of the study area was made using available geological data and revised with additional borehole data from recent exploration drilling. This mapping was used to aid the development of a 3D geological framework which was further complemented by aerial photography and borehole log data (Timms & David, 2011). Review of the Rocky Bridge Creek alignment indicates it is controlled by tectonics.

Groundwater levels were measured in monitoring bores (Figure 19.1, Table 19.1) on a monthly basis over a one year period to assess the response to rainfall events and to improve understanding of aquifer recharge. Vertical hydraulic gradients were analysed to understand the connectivity between the shallow and the deep aquifers. A groundwater map was constructed and two pumping tests were performed in areas adjacent to the faults to supplement the existing hydrogeological data and test the aquifer characteristics. In addition, chemical analyses of surface and groundwater samples were undertaken to classify the waters and recharge sources. Recharge to the aquifer was calculated using chloride mass balance (CMB) method and by analysing the groundwater level response to rainfall. Steamflow data from the automated gauging station at Rocky Bridge Creek were used to analyse the baseflow contribution. The results were used to develop a conceptual and numerical model to test the influence of fracturing on flow surface water-groundwater connectivity.

Table 19.1 Summary of groundwater bores.

Site id	Easting MGA	Northing MGA	Screen (m)	Elevation (mAHD)	Aquifer
KMB1	708711	6259714	73–76	768.5	Fractured rock
KMB2	709440	6259578	83–86	862	Fractured rock
KMB3	708951	6259012.7	82–85	750	Fractured rock
KMB4	708964	6258475	82–85	779	Fractured rock
KMB5	708571	6258031	72.3–75.3	787	Fractured rock
KMB6	708201	6257765	72.3–75.3	768.5	Fractured rock
KMB7A	707911	6258149	3.5–6.5	735	Weathered bedrock
KMB7B	707913	6258153	67–70	735	Fractured rock
KMB8	708105	6258232	Open hole to 210 m	742	Fractured rock
KTB1	708524	6258463	84–138	771.0	Fractured rock
WB5	708207	6258532	19–22	745.0	Weathered bedrock
WB6	707927	6258099	84–90	743.0	Fractured rock

19.4   RESULTS

19.4.1 Groundwater levels

The groundwater levels are relatively deep on the slopes, between 20 m and 50 m below ground level. In topographically low areas, the water level is shallow at ground surface or under artesian pressure. Interpreted groundwater potentiometric contours (Figure 19.2) indicate that in the north east the flow is to the west along the Rocky Bridge Creek and further south, it is to the north east towards the creek. The groundwater heads vary from 800 mAHD in the topographically elevated areas to 730 mAHD close to the creek, where the aquifer is artesian and the untapped groundwater bores flow. In the low-lying areas, the measured head is between 1 m and 7 m (KMB7B and KMB8 respectively) above surface and several springs feed small wetland systems. The artesian flow and presence of springs are the result of the steep topography, change in slope, and presence of an impermeable layer along which flow occurs. The potentiometric map (Figure 19.2) was constructed using data from a small number of boreholes; the equipotential lines indicate that groundwater recharges the Creek.

Hydrographs for boreholes located along the transect perpendicular to potentiometric lines (KMB5, KTB1 and KMB7B) indicate that the response to recharge is variable (Figure 19.4). KMB7B shows that a 1 m rise following significant period of above average rainfall, KTB1 shows a 1.8 m rise, while KMB5 records a decline in head. The groundwater flow direction may not be perpendicular to potentiometric lines, but rather at an angle, as expected in an anisotropic environment.

The difference in head (approximately 2 m) between shallow and deep piezometers KMB7 A and KMB7B indicates that there may be limited hydraulic connection between the two aquifers.

19.4.2 Recharge

Relatively little research has been done on interpretation of hydrograph response in fractured-rock aquifers (Healy & Cook, 2002; Rancic *et al.*, 2009; Cook, 2003; Simmons *et al.*, 1999), however, an attempt is made to analyse the hydrographs based on the monthly monitoring data. The response to rainfall is variable in different bores installed around the site (Figure19.4), with average fluctuations of 1.7 m, a minimum of 0.6 m (KMB7B) and a maximum of 3 m (KMB1). The exception is artesian bore KMB8, which appears to have been impacted by nearby exploration drilling during August and September 2011 which reduced its pressure. The value of residual rainfall mass (BoM, Station No. 063061, BoM, 2013) has been plotted on Figure 19.4 for comparison. Rainfall residual mass represents the cumulative sum of the residuals between actual monthly rainfall and long term mean monthly rainfall. The algorithm represents long term trends using monthly rainfall information. A positive slope of the curve indicates that the period is characterised by the rainfall being greater than the long-term mean, whilst a falling trend indicates the opposite (Rancic *et al.*, 2009). Where rainfall is a significant contributor to groundwater recharge, measured groundwater hydrographs are expected to follow the shape of the rainfall residual mass curve. As water level readings were recorded, the representation of rainfall as residual mass to show the general trends was considered more suitable than the use of daily rainfall data.

The potentiometric response to the average rainfall events (December 2011 to March 2012) is low, indicating that the water level in the fractures and the matrix respond together. The weathered rock aquifer locally has a high clay percentage, hence it inhibits vertical flow of water resulting in a smooth hydrograph. The potentiometric

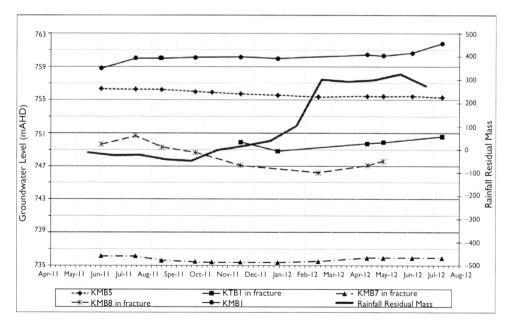

Figure 19.4 Groundwater hydrographs and rainfall residual mass.

surface in KTB1, KMB1 and KMB7 does not show significant response to rainfall recharge throughout the year, with the exception of a high rainfall period experienced during February 2012. The response to these events in the boreholes has a time lag of about a month. Following an extensive period of above average rainfall events between December 2011 and March 2012, it appears that the declining ground-water trend is reversed to become a rising trend with 1 m to 2 m rise over a period of several months. Generally, in arid inland areas in New South Wales higher rainfall events in summer tend to be offset by high evapotranspiration rates and, therefore, a decline in groundwater levels. However, there is limited potential for evapotranspira-tion at the Kempfield site due to the presence of low bush and shallow rooted grass across the site and the relatively deep groundwater levels. In line with groundwater rises, the decline in groundwater level is slow as a result of slow drainage from the matrix.

The aquifer matrix may, therefore, provide baseflow to streams for a long period of time after the water table has declined (Price *et al.*, 2000). This is also reflected in continuous artesian flow and slow loss of pressure in KMB8 (artesian bore) as a result of slow natural discharge (Figure 19.4). It is typical in such aquifers that recharge through a low porosity fractured-rock aquifer causes large magnitude water-table fluctuations (Cook, 2003) however this is not seen in hydrographs possibly due to the presence of the weathered rock aquifer above. The unsaturated hydraulic character-istics, thickness of the weathered zone and deep potentiometric surface on the slopes control the magnitude and timing of these small fluctuations. Hydrograph variations for piezometers installed in fractured rocks often provide a poor record of water-level variations within the aquifer itself, in particular where the storativity of the aquifer is very low relative to the storativity of the piezometer (Healey & Cook, 2002). This can occur as a result of attenuation and large storage capacity of the bore (50 mm PVC) (Simmons *et al.*, 1999).

Recharge calculated using the head fluctuation method at the Kempfield site, gives a value of around 4%–6%, assuming storage of 0.01 (representative of frac-tures and matrix). Thus calculated recharge was obtained by the modifying water table fluctuation (WTF) method and using cumulative rainfall during a wet period rather than single rainfall event (Moon *et al.*, 2004). Calculation of recharge using field storativity data (0,004) from observation bore KMB7B during WB6 pumping test (Figure 19.8), gives extremely low <0.5% rainfall recharge. Due to the complex-ity of the fractured rock system, and difficulty in interpreting the water level rise, the application of this method is questionable. The head fluctuation method is best applied to a shallow water table and an unconfined aquifer that displays sharp water level rises (Crosbie *et al.*, 2010) and where a minimal time lag exists (Healey & Cook, 2002).

Steady state chloride mass balance (CMB) method has been used extensively to estimate the recharge rates in the Australia (Crosbie *et al.*, 2010). The method requires the following data input: estimates of annual precipitation, total chloride (Cl) input and groundwater concentrations (Gee *et al.*, 2005) as follows:

$$R = \frac{c_p P}{c_r} \tag{1}$$

where R = the recharge rate; C_p = Cl concentration in precipitation; P = annual precipitation; and C_r = Cl concentration in groundwater recharge.

Across the Kempfield site the average Cl concentration in groundwater samples is 41.9 mg/l, with average annual precipitation of 851 mm and average rainfall Cl concentration of 1 mg/l (Turner *et al.*, 1996). The groundwater recharge has been calculated at 20 mm/year or 2% of rainfall. Since the runoff component in this catchment is significant (12% of rainfall), the method has been corrected for runoff component. The average Cl concentration in surface water has been assumed to represent runoff concentration (5 mg/l) and the groundwater recharge calculated is 8 mm/year or 1% of rainfall. Although the uncertainty associated with accounting for runoff is small compared to the uncertainty in chloride deposition (Crosbie *et al.*, 2010), in this case the results are close. Therefore, the CMB method appears to be the most suitable method for recharge calculation in this catchment.

19.4.3 Hydraulic testing

19.4.3.1 *Slug testing*

Hydraulic conductivity of the fractured rock aquifer across the site varies from 0.02 m/d to 0.56 m/d as shown in Table 19.2 (Argent minerals, 2013). The values presented were obtained by analysing hydraulic (slug) testing data from monitoring bores using the method by Hvorslev (1951). The Hvorslev method was used as it can be applied to both confined and unconfined aquifers, although the Bower and Rice (1976) method gave similar results and could be applied as the distance from the screen to standing water level is sufficiently great (Brown *et al.*, 1995).

19.4.3.2 *Pumping tests*

Pumping tests may not provide a good indication of the permeability of the aquifer in the fractured rock environment due to the interpretation assuming simplified radial flow and the permeability being spatially variable (Cook, 2003). Nevertheless, these data can give an overall bulk value of permeability which can be used to constrain the conceptual groundwater model.

Two pumping tests (Argent minerals, 2013) were undertaken to assess the hydraulic properties of the aquifer, the role of structural boundaries and the extent of the aquifer. The first test was undertaken in 130 m deep production bore KTB1 which was pumped for 72 hours. The results indicate continuous drawdown and dewatering from the storage within fractures. Two step tests were undertaken prior to setting the constant rate test at 4.8 l/s. This rate was selected as the last two rates had the same drawdown slope (Figure 19.5). The transmissivity calculated from the pump-

Table 19.2 Hydraulic test (slug) results.

Bore	KMB1	KMB2	KMB3	KMB4	KMB5	KMB7B
K (m/day)	0.16	0.02	0.56	0.4	0.25	0.18

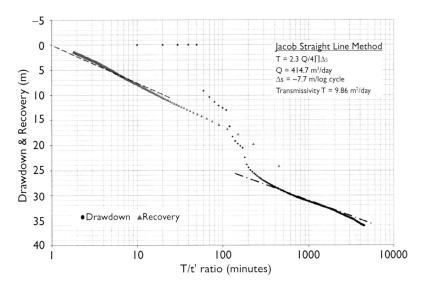

Figure 19.5 72 hour pumping test at KTB1.

ing test using the method by Jacob and Cooper (1946) was 9 m²/day. No recharge or discharge boundaries were intercepted during the pumping period and no drawdown occurred in the closest monitoring bores KMB5 (located 200 m up gradient) and KMB7B (500 m down gradient).

The second 48 hour pumping test was undertaken in the 76 m deep WB6 bore located close (20 m distance) to the Rocky Bridge Creek (Figure19.6). Monitoring was undertaken in shallow and deep monitoring bores at 56 m, 250 m and 600 m distance from the test bore. The constant pumping rate was set at 8 l/s, and recovery monitored following the test. Pumping induced a maximum drawdown of 34 m in the production bore, 11.05 m in deep KMB7B and 0.99 m in shallow KMB7 A, while no drawdown was recorded in test bore KTB1 (Figure 19.7). Rocky Bridge Creek was monitored visually using the gauge in the creek, no change was recorded. The pumping test results for WB6 indicate initial steepening of the curve as a result of the effect of release from matrix storage as the cone of depression has not yet been affected by subsequent recharge boundary. The flattening of the curve occurs at around 19 m drawdown level (Figure 19.7) with possible recharge from a fracture zone at that depth. Below 20 m depth, the rate of drawdown increases, suggesting that the main aquifer inflow zone is between 18 mbgl–20 mbgl with only minor inflow below this zone. Steepening of the curve after 1000 minutes indicates the existence of a negative boundary effect possibly due to the presence of an impermeable boundary. KMB7B shows a very similar response to the pumping bore, indicating that they are both in the same fractured rock system and that KMB7B is hydraulically connected to the pumping bore (Figure 19.7). KMB7 A, installed in the weathered rock aquifer, shows a pressure response with an overall decrease of 0.9 m. Although the creek is located 20 m from the pumped bore, the test results

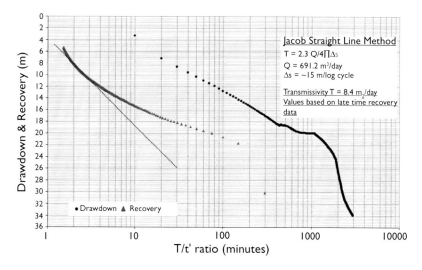

Figure 19.6 48 hour pumping test WB6.

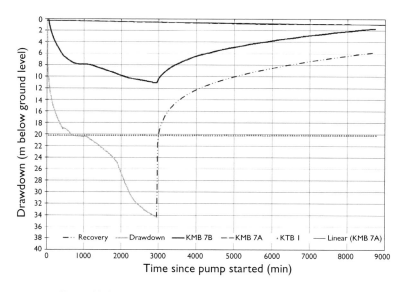

Figure 19.7 Response to pumping WB6 in monitoring bores.

indicate that a recharge boundary was not reached during the pumping test. This implies that limited connection may exist between the fractured rock groundwater and surface water within the area of influence. The Jacob straight line method was used to analyse the response in KMB7B due to pumping in WB6. The results are given in Figure 19.8, and show the calculated transmissivity of 17.3 m²/day and storativity of 4×10^{-4}.

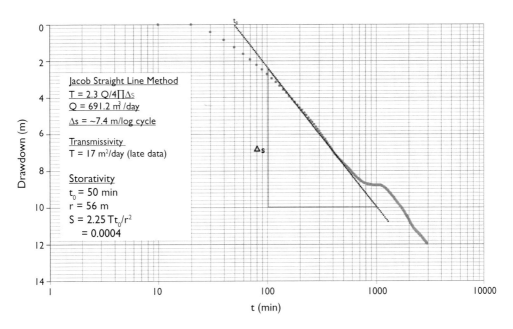

Figure 19.8 Response in bore KTB7B to 48 hour pumping in WB6.

19.4.4 Baseflow

The gauging station was established on Rocky Bridge Creek in 2011 about 300 m downstream of the site. A good daily data set is available during the period from December 2011 to February 2012. This dataset was analysed by using an automated baseflow separation technique and applying the Lyne and Hollick Filter (Nathan & McMahon, 1990) to separate baseflow from streamflow. Using 0.92 alpha filter, a reasonable baseflow separation was achieved, with the results indicating that the baseflow percentage of the average stream flow in the Rocky Bridge Creek is over 1% (10 m³/day) (Figure 19.9). A filter parameter of 0.92 was considered appropriate based on the Nathan & McMahon (1990) case studies on a number of locations in southern Australia (Engineers Australia, 2009).

19.4.5 Hydrogeochemistry

The hydrogeochemistry was determined for groundwater in shallow and deep hydrostratigraphic units, as well as surface water. The regional characterisation of groundwater hydrogeochemistry, was carried out during 2010–2011, and consisted of groundwater sampling from eight piezometers screened in the bottom 3 m in the fractured and weathered rock aquifer and from two test bores which are screened at the bottom 40 m of the bore. Surface water samples were collected from the Rocky Bridge Creek and its tributaries (Figure 19.1). The composition of samples is shown in Table 19.3 and in Figures 19.10 and 19.11. Results indicate the highly variable characteristics of the fractured rock aquifer and distinct differences within the surface water samples.

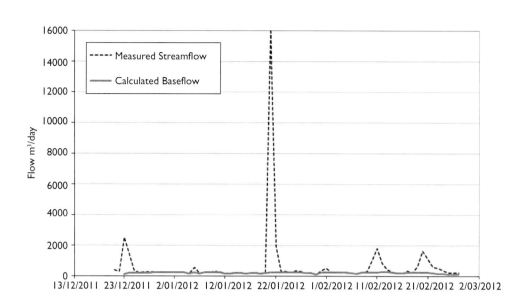

Figure 19.9 Baseflow separation using Lyne and Hollick method applied to daily data for Rocky Bridge Creek.

Table 19.3 Surface and groundwater chemistry (July 2012).

		pH (units)	EC (μS/cm)	Na (mg/L)	K (mg/L)	Ca (mg/L)	Mg (mg/L)	Fe (mg/L)	CO₃ (mg/L)	HCO₃ (mg/L)	SO₄ (mg/L)	Cl (mg/L)
Groundwater samples												
KMB1		8.02	496	105	4	9	4	51.2	0.001	235	12	15
KMB2		7.57	276	15	4	28	6	0.31	0.001	117	6	11
KMB3	Mineralised zone	6.05	144	11	5	7	3	0.11	0.001	31	26	7
KMB4		7.02	218	10	4	22	6	0.14	0.001	85	1	12
KMB5	Mineralised zone	7.28	265	12	6	23	10	0.09	0.001	86	23	13
KMB6	Mineralised zone	7.67	461	26	6	25	27	15.4	0.001	144	51	24
KMB7A		7.36	1950	483	10	22	31	15.8	0.001	1090	0.1	50
KMB7B		7.3	1230	157	12	39	44	0.34	0.001	591	3	50
KMB8	Mineralised zone	6.32	1270	30	12	51	109	8.01	0.001	125	497	23
KTB1	Mineralised zone	7.2	559	26	9	23	35	0.17	0.001	107	115	36
Surface water samples												
KPW1	Upstream	8.8	308.0	14.0	2.0	29.0	11.0	0.1	1.0	120.0	0.1	18.0
KPW3	Upstream	8.3	93.0	9.0	7.0	9.0	6.0	0.0	1.0	22.0	0.2	2.0
KPW4	Spring	8.1	648.0	13.0	10.0	15.0	49.0	0.4	1.0	110.0	185.0	33.0
KPW5	Upstream	8.3	220.0	47.0	11.0	20.0	12.0	0.0	1.0	25.0	1.0	5.0
KPW6	Downstream	8.2	278.0	18.0	4.0	20.0	15.0	0.8	1.0	136.0	14.0	28.0
KPW7	Downstream	7.6	420.0	17.0	2.0	23.0	28.0	3.2	1.0	105.0	49.0	16.0

The surface water was slightly alkaline, while the groundwater was neutral to alkaline, and where the aquifer is mineralised it is slightly acidic. Surface water quality is fresh with an average conductivity of 231 µS/cm upstream and 307 uS/cm downstream in the Rocky Bridge Creek (Table 19.3 and Figure 19.10). The tributaries were generally found to be fresher than the main creek. Groundwater within the mineralised area was brackish with average values of up to 1500 µS/cm, while in other non-mineralised bores groundwater was fresh (up to 500 µS/cm). The mineralised groundwater occurs in the Silurian mineralised (barite-sulphide) resedimented volcanics bounded with north east trending normal faults (Figure 19.2).

Upstream surface water samples (outside of the mineralised zone) are Ca-HCO$_3$ type waters, while downstream they are typically Mg-Ca-HCO$_3$ type waters, both the result of weathering of volcanics and surface runoff as can be seen on the Piper diagram (Figure 19.10). Groundwater from the mineralised zone tends to be Na+Mg- SO$_4$ dominated while elsewhere within the fractured rock aquifer the water is Na- HCO$_3$ and Ca+Mg- HCO$_3$ dominated. Sulphate dominated waters are the result of barite dissolution and this signature is not present in groundwater outside the mineralised zone as presented on Schoeller diagram (Figure 19.11). Groundwater in artesian bore KMB7B in Ordovician metasediments down gradient from the mineralised zone and on the southern side of the north west-south east trending fault is also mineralized, however, it is of a Na-HCO$_3$ type. This implies that a north west-south east trending

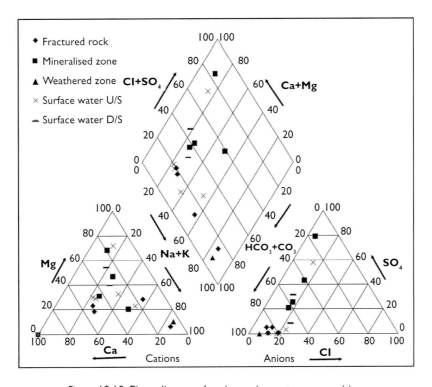

Figure 19.10 Piper diagram showing major water composition.

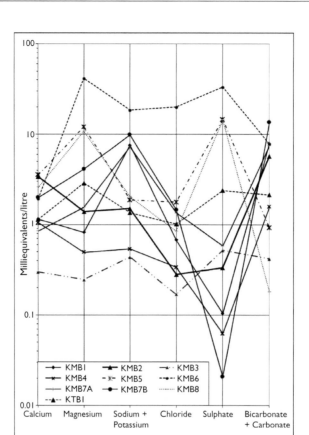

Figure 19.11 Schoeller diagram showing groundwater composition.

fault located close to KTB7B acts as an impediment to flow with distinctly different types of water found on either side of this lineament.

19.4.6   Conceptual groundwater model

The conceptual understanding supports the presence of three generalised groundwater systems:

- Poorly developed alluvium associated with Rocky Bridge Creek;
- Weathered rock of variable thickness; and
- Fractured rock.

Recharge to the groundwater systems occurs via rainfall and runoff infiltration directly and as seepage through the weathered aquifer into the fractured aquifer. Although groundwater levels are maintained by rainfall infiltration with a significant delayed response, they are controlled by topography and geology. The discharge from the local groundwater system occurs to a limited extent as baseflow, seepage at the break of slope, and to a lesser extent as loss by evapotranspiration in the low lying areas where the water table is near the ground surface (generally within 2 m to 3 m of ground level).

The geology structures do not transmit groundwater at shallow depths, which is evident in the hydrograph responses to rainfall. The behaviour of faults and structural features at greater depth is important both as barriers and as preferential flow paths as was confirmed by the difference in geochemical signature of groundwater up gradient and down gradient from the north west-south east trending fault on the contact between Silurian volcanics and Ordovician metasediments. It is not entirely clear how each fault zone behaves, however, the highest yield is obtained from the bores located on the intersection of structural lineaments, noted during this investigation programme and based on data from registered bores in the area (New South Wales Office of Water Database). The only stress to the system is represented by the 72 hour pumping test.

19.4.7   Numerical groundwater model

A numerical model was constructed to test the conceptual model, the validity of the fracture porosity assumption related to the pumping test, the geochemistry and the results obtained in the field, in order to replicate the natural environment.

A 3-Dimensional finite difference model was setup as equivalent porous media using the MODFLOW code (McDonald & Harbaugh, 1988) with flow calculations undertaken utilising the MODFLOW SURFACT (Version 3) code to allow for both saturated and unsaturated flow conditions. The modelling was undertaken using the Groundwater Vistas (Version 5.16) software package (ESI, 2006). The model covers an area of 14.1 km × 15.1 km (212.9 km²), with a variable grid size ranging from 20 m × 20 m in the study area and increasing gradually up to 100 m × 100 m near the outer model boundary. The model comprises four layers built to represent the hydrogeological units with their details and calibrated parameters shown in Table 19.4. The conceptual model boundary can only be defined with certainty on the eastern side of the project area where the Copperhania Thrust Fault extends as a north-south trending feature. A general head boundary (GHB) is set at the western edge of the model domain in layer 2 to 4, elsewhere the boundaries are set as no flow. The head value is set at 1 m below the top of layer and conductance is 1000 m²/d.

Average long-term annual rainfall data for the area were used for the steady state model calibration. Three rainfall recharge zones are defined in the model: creeks with

Table 19.4 Numerical model layers and calibrated parameters.

Layer	Hydrogeological units	Default layer thickness (m)	Horizontal hydraulic conductivity K_h (m/d)	Vertical hydraulic conductivity K_v (m/d)	Sy (%)	Ss (1/m)
1	Creeks Alluvium and Regolith	5	0.3 to 7	8.8×10^{-3} to 1.1×10^{-2}	2–10	0.001–0.005
2	Weathered Rock	15	2.2×10^{-3} to 2.3	4.5×10^{-5} to 2×10^{-2}	0.1	0.0005
3	Transition zone	15	7.8×10^{-6} to 7.4×10^{-3}	2.1×10^{-5} to 9.4×10^{-4}	0.1	0.0005
4	Fractured Rock	250	1×10^{-6} to 0.2	1.6×10^{-5} to 6×10^{-4}	0.1	0.0005

an infiltration coefficient rate of 85 mm/year, the State Forest area 0.5 mm/year, and 34 mm/year rainfall elsewhere. Evapotranspiration has been included in the model using the Evapotranspiration (EVT) package of MODFLOW. The EVT parameter values adopted were a constant rate of 600 mm/year with an extinction depth of 1 m. Surface drainage lines have been represented using either the River (RIV) or Drain (DRN) packages of the MODFLOW software. Rocky Bridge Creek is a perennial creek represented by the RIV package to allow leakage from the creek and baseflow contribution. The river stage elevations were set to 1 m below the creek bank elevation, and river bed levels set to 0.2 m below the stage in the creek. Where ephemeral streams are present within the area, these have been represented within the model as drain cells. Drain cell elevations were set to 2 m below the surface elevations. The river and drain conductance parameter was calibrated to 25 m²/day; this parameter controls the exchange between surface and groundwater.

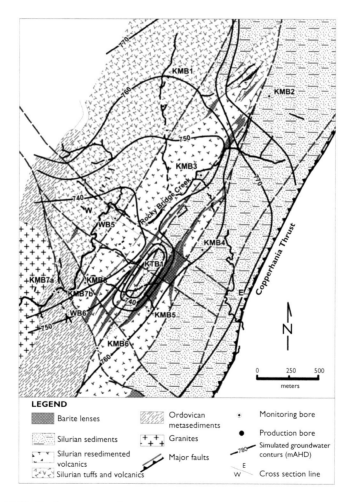

Figure 19.12 Geology map with simulated groundwater contours as a result of the 72 hour pumping test.

Steady state calibration was undertaken using heads and baseflow separation analysis data. Steady state baseflow contribution to the creek is predicted at around 7% (80 m³/day) of the average flow as recorded at the gauging station just downstream of the site. Baseflow calculated using the automated technique is lower. This is most likely due to the short period of wet weather data which may not be representative of all seasons and also due to the choice of alpha filter. Nevertheless, the model achieved a calibration of less that 3% SRMS which is within the acceptable range defined in groundwater modelling guidelines (Barnett *et al.*, 2012). The transient model was then matched to the results of the 72 hour pumping test, to see how the system behaves under stress (Figure 19.12). The potentiometric contours agree well with the results of the pumping test indicating that the hydraulic properties are well defined within the zone of influence of the pumping test.

19.5 SUMMARY

Several different methods were used to characterise the fractured rock aquifer in volcanics and metasediments and the results compared. An approach to combine hydrographs, pumping test data, hydrochemistry, baseflow separation analysis and numerical modelling has yielded a better understanding of this complex aquifer. The groundwater contour maps and the response in hydrographs indicate that the flow in this fractured rock aquifer is from elevated ridgelines to low lying Rocky Bridge Creek valley and the response to recharge is variable. The influence of structures on flow can be inferred from geochemical data. The geochemical signature of up gradient groundwater in Silurian resedimented volcanics (mineralised zone) differs from the signature of that in down gradient non-mineralised Ordovician sediments across the north west-south east trending fault line. The groundwater recharge in the study area is low; 1–2% of rainfall precipitation calculated using the CMB method, which is considered to be more accurate than the WTF method. This is within the range of groundwater recharge typical for other projects in SE Australia (NATIONAL WATER COMMISSION, 2005; Leaney *et al.*, 2011), and similar to that obtained during the numerical model calibration. The results of two pumping tests indicate that the system is a typical fractured rock aquifer, where water is sourced from fractures as well as matrix porosity. Geological structures may act as barriers or conduits to flow, and at present this is not well understood. However, the highest groundwater yields are generally seen in bores located at or near to intersections of structural lineaments. The groundwater model based on the conceptualisation of the system, replicates the pumping test results well. Limited groundwater-surface water connectivity is further confirmed by the modelling results. The baseflow contribution to the creek predicted by the model (7% of average flow) was higher than that calculated using baseflow separation analysis (1%). However, uncertainty in the baseflow contribution and high variability in hydraulic conductivity and connectivity of fractures may be partially resolved by running sensitivity analysis to assess the range of flow for different scenarios.

Understanding the influence of fractures is important in planning field characterisation and consequently future development, dewatering and any associated water losses and changes in quality in fractured rock aquifers.

ACKNOWLEDGMENT

The authors would like to thank Argent Minerals for allowing the use of their data, information and for constructive discussions during the preparation of this paper. The authors would also like to acknowledge the reviewers who provided valuable comments.

REFERENCES

Argent Minerals (2013) Environmental Impact Statement for the Kempfield Project, submitted to Department of Planning and Infrastructure on 8 April 2013.

Barnett B., Townley L.R., Post V., Evans R.E., Hunt R.J., Peeters L., Richardson S., Werner A.D., Knapton A., Boronkay A. (2012) *Australian groundwater modelling guidelines*, Waterlines report, National Water Commission, Canberra.

BoM (2013) URL: http://www.bom.gov.au/; last visited 15 January 2013.

Bouwer H., Rice R.C. (1976) A slug test method for determining hydraulic conductivity of unconfined aquifers with completely or partially penetrating wells, *Water Resources Research* 12(3), 423–428.

Brown D.L., Narasimhan N.T., Demir, Z. (1995) An Evaluation of the Bouwer and Rice Method of Slug Test Analysis, *Water Resources Research* 14 (5), 1239–1246.

Colquhoun G.P., Meakin N.S. & Krynen J.P. (1996) Chesleigh Group. *In*: Meakin N.S. & Morgan E.J. compilers *Dubbo 1:250 000 Geological Sheet SI/55–4, Explanatory Notes 72–77.* Geological Survey of New South Wales, Sydney.

Colquhoun G.P., Meakin N.S., Krynen J.P., Watkins J.J., Yoo E.K., Henderson G.A.M., Jagodzinskie A. (1997) Stratigraphy, structure and mineralisation of the Mudgee 1:100 000 Geological Map Sheet. *Geological Survey of New South Wales Quarterly Notes* 102, 1–14.

Cook P.G. (2003) A guide to regional groundwater flow in fractured rock aquifers, CSIRO Land and Water, Glen Osmond, SA, Australia.

Cooper H.H., Jacob C.E. (1946) A generalized graphical method for evaluating formation constants and summarizing well field history. *American Geophysical Union Transcription* 27, 526–534.

Crosbie R., Jolly I., Leaney F., Petheram C., Wohling D. (2010) Review of Australian Groundwater Recharge Studies. CSIRO: Water for a Healthy Country National Research Flagship.

Engineers Australia (2009) Austalian Rainfall and Runoff Revision Project 7: Baseflow for Catchment Simulation, Stage 1 Report – Volume 1.

ESI, (2006) Groundwater Vistas. Version 5.16 User's Manual

Gee G.W, Zhang Z.F., Tyler S.W., Albright W.H. (2005) Chloride Mass Balance: Cautions in Predicting Increased Recharge Rates, available at www.vadosezonejournal.org. *Vadose Zone Journal* 4, 72–78.

Healy R.W., Cook P.G. (2002) Using groundwater levels to estimate recharge. *Hydrogeology Journal* 10, 1431–2174.

Hvorslev M.J. (1951) Time Lag and Soil Permeability in Ground-Water Observations, *Bulletin Waterways Exploration State Corps of Engineers U.S. Army*, 36, 1–50, Vicksburg, Mississippi.

Leaney F., Crosbie R., O'Grady A., Jolly I., Gow L., Davies P., Wilford J., Kilgour P. (2011) Recharge and discharge estimation in data poor areas: Scientific reference guide. CSIRO: Water for a Healthy Country National Research Flagship.

Moon S.K., Wooa N.C. Lee K.S. (2004) Statistical analysis of hydrographs and water-table fluctuation to estimate groundwater recharge. *Journal of Hydrology* 292, 198–209.

McDonald M.C., Harbaugh A.W. (1988) MODFLOW, A Modular Three-Dimensional Finite Difference Groundwater Flow Model. U.S. Geological Survey, Open File Report, 91–536, Denver.

Nathan R.J., McMahon T.A. (1990) Evaluation of automated techniques for base flow and recession analyses. *Water Resources Research* 26(7), 1465–1473.

National Water Commission (2005) Australian Water Sources URL: http://www.water. gov.au/WaterAvailability/Whatisourtotalwaterresource/GroundwaterRecharge/index. aspx?Menu = Level1_3_1_6, last visited 1 June 2013.

Novakowski K.J., Owtobee P.A., Kryger P.W. (2000) Potential discharge of Lockport groundwater into Twenty Mile Creek downgradient for the CWML site, Smithville, Ontario. Final report to the Smithville Phase IV Bedrock Remediation Programme.

Price M., Low R.G., McCann C. (2000) Mechanisms of water storage and flow in the unsaturated zone of the Chalk aquifer. *Hydrogeology Journal* 233, 54–71.

Rancic A., Salas G., Kathuria A., Ackworth I., Johnstone W., Smithson A., Beale G. (2009) Climatic influence on shallow fractured-rock groundwater systems in the Murray Darling Basin, NSW, Department of Environment and Climate Change, NSW.

Simmons C.T., Hong H., Wye D., Cook P.G., Love A.J. (1999) Signal propagation and periodic response in aquifers: the effect of fractures and signal measurement methods. In: Water 99 Joint Congress, Brisbane, Australia, 727–732.

Timms D., David V. (2011) Kempfield Silver, Barite and Base metal (Pb-Zn) Deposit, Lachlan Orogen, Eastern Australia. Eighth International Mining Conference, Queenstown, New Zealand, 22–24 August 2011.

Turner J., Lambert M., Knott J. (1996) Nutrient Inputs from Rainfall in New South Wales State Forests, Forest research and development division State forests of New South Wales, Sydney, Research paper, 17.

Chapter 20

Main features governing groundwater flow in a fractured Basalt Aquifer System of South-Eastern Australia

Irena Krusic-Hrustanpasic & Frederic Cosme
Golder Associates Pty Ltd., Richmond, Australia

ABSTRACT

The western suburbs of Melbourne, Australia are underlain by a thick sequence of Cainozoic age basalt rocks (Newer Volcanics Province), which resulted from multiple phases of volcanic activity. Historically, most industrial development occurred in these suburbs, which resulted in several contaminant plumes within the area. A hydrogeological conceptual model was developed for one of the industrial sites to understand impacts of the site on groundwater and evaluate remedial options. Three main phases of lava flows were recognised under the study site. The time breaks between the phases were sufficiently long to allow for inter-phase deposits (palaeosols) to develop over extended areas. Within each of the main lava flow phases, a number of individual flows occurred leading to a complex structural profile of the basalt unit. A number of aquifers and sub-aquifers occur through the basalt unit resulting in a complex inter-relationship and lateral movement of contaminants.

20.1 INTRODUCTION

The western suburbs of Melbourne (Figure 20.1) are commonly referred to as the Melbourne Industrial Hub. Due to the proximity of the Central Business District and a number of the waterways that supported water supply requirements, waste water discharge and transport needs, most historical industrial development occurred within this area. The industrial activities started as early as in the middle to late 1800s and included manufacturing of a wide range of products. Some of Melbourne's oldest chemical manufacturing plants are or have been located within this area. Such industrial activities resulted in several contaminant plumes. The plumes included inorganic and organic components as well as dissolved and free phase NAPLs (non-aqueous phase liquids).

Over the last few decades the population of Melbourne increased significantly resulting in a rapid expansion of the city and increased demand for housing close to the Central Business District. The value of land in proximity to the Central Business District soared during the 1990s and 2000s. Consequently, a number of industries moved further away from the Central Business District, where the land values remained lower and many of the old industrial sites have been offered for sale. In

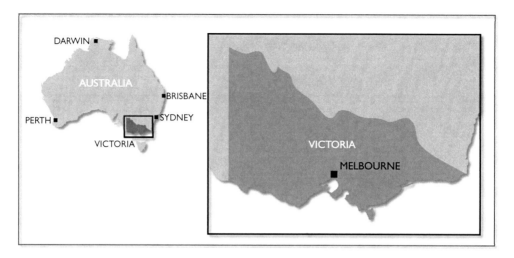

Figure 20.1 Study area general location.

order to make a profit, a planning permit would be applied for from the local Authority to change the land use and rezone the land from industrial to non-industrial uses.

Consistent with the Commonwealth framework, the Victorian legislation (EPA, 1970, The Act) requires that for sites where contamination is suspected based on past land use or being adjacent to known contaminated sites, either a Statement or Certificate of Environmental Audit be issued stating what uses the site is suitable for and any conditions required. These documents would be issued after a statutory environmental audit of a property has been conducted under the Act. The objective of the audit is to understand the risk that the property may pose to elements or segments of the environment and to evaluate the potential for the site to be cleaned up to the extent practical. An Environmental Site Assessment is typically conducted to meet these requirements.

A conceptual site model was developed for one of the industrial sites (the site) and is described in this chapter. The development of the conceptual site model was part of an Environmental Site Assessment conducted for the site in order to understand the groundwater flow and contaminant migration through the hydrostratigraphic units recognised under the site. The objective of the conceptual site model was also to support an evaluation of the site remedial options.

20.2 GEOENVIRONMENT

The site is located about 15 km west of the Central Business District and occupies approximately 150 ha of land. It sits within a landscape unit known as the Werribee Plains, which developed during the late Tertiary to Quaternary age periods. The plain is intersected by creeks and rivers with the valleys of the main creeks/rivers, generally, developed along the edges of the basalt flows.

The landscape of the broader site area is relatively flat to slightly undulating and with typical elevations of the site between RL 48 m AHD[1] and RL 55 m AHD. An extinct volcano rising to over RL 60 m AHD is the dominant topographic feature within the broader area of the site.

The main geological unit within the broader area of the site is the Cainozoic age basalt rock unit. The unit was formed by a sequence of stacked basaltic lava flows formed during episodes of volcanic activity and is part of the Newer Volcanics Province. The Newer Volcanics Province covers an extensive area of about 15 000 km[2], from Central Victoria to South Australia (Figure 20.2).

The formation of the Newer Volcanics Province resulted from prolonged volcanic activity that began about 4.6 million years ago (Ma) and which was part of more extensive volcanic activities associated with the breakup of Gondwana (Hare & Cass, 2005; Hare *et al.*, 2005). The intermittent phases of volcanic activity resulted in a thick sequence of stacked basalt lava flows with a total thickness in excess of 100 m. Over 250 eruption points produced valley flows, which covered a well developed Tertiary palaeo-topography.

The basalt of the Newer Volcanics is the youngest geological unit under the site. The unit is about 50 m thick and was the main hydrostratigraphic unit of interest in relation to the potential site impact on groundwater. The basalt is underlain by the Tertiary age sediments of which the Brighton Group[2] is the youngest unit.

Three main phases of lava flow within the basalt unit were recognised under the site (Figure 20.3). The time breaks between the main phases of volcanic activities were sufficiently long to allow for inter-phase deposits to develop over an extensive area. The inter-phase deposits are referred to as palaeosols. Multiple flows of basaltic lava also appear to have occurred within some of the main phases of volcanic activities. The gaps between these flows appear to have been relatively short, resulting in the

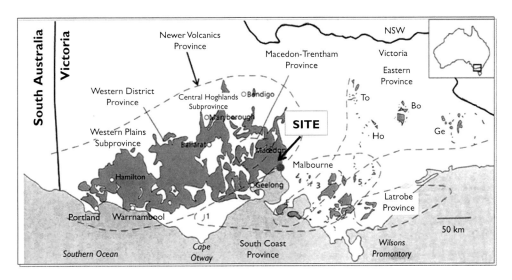

Figure 20.2 Extent of The Newer Volcanics Province.

development of interflow palaeosols of limited and localised in extent. These inter-flow palaeosols are referred to as intra-clays.

A number of hydrostratigraphic units were recognised within the Newer Volcan-ics unit, based on their composition, position in the vertical geological profile, role in groundwater movement and hydraulic properties. Each of the basalt flow phases was considered to be an aquifer (Figure 20.3):

- Upper Basalt Aquifer: Phase 3 basalt flow unit characterised by discontinuous and localised saturation.
- Middle Basalt Aquifer: Phase 2 basalt flow unit inferred to act as an unconfined to confined aquifer.
- Lower Basalt Aquifer: Phase 1 basalt flow unit acting as a confined aquifer.

Based on the site investigations that included hydraulic testing and visual inspec-tion of the rock cores, the following features were identified as the main features controlling groundwater flow within each aquifer:

- Fractured and vesicular basalt characterised by high to medium hydraulic conductivity.
- Horizontal fractured zone characterised by closely spaced sub-vertical fracturing system. The zone is inferred to be highly conductive.
- Massive columnar basalt of a low hydraulic conductivity permitting limited verti-cal flow.
- Extremely to completely weathered basalt characterised by a low hydraulic con-ductivity. Generally, the zone is acting as a hydraulic barrier.

The palaeosols and intra-clays were found to be of a similar composition which varied from in-situ residual basaltic clayey soils to calcrete of a pisolitic

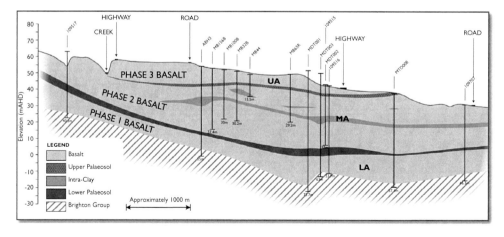

Figure 20.3 Site geological cross section and main aquifers (UA – upper basalt aquifer, MA – middle basalt aquifer, LA – lower basalt aquifer).

Figure 20.4 Palaeosol veining system section view.

texture. Typically, the upper 0.5 m to 1.0 m of the palaesols and to a lesser extent the intra-clays, is thermally altered resembling the red bricks. The zone is characterised by a veining system developed in a polygonal pattern (Figure 20.4). Although appearing hard, the palaeosols samples broke along the veining system under pressure applied by a geological hammer, indicating the presence of weaknesses (defects) along the veining systems. Based on the composition of the palaeosols and presence of these defects, the palaeosols were considered to act as leaky aquitards that impede the vertical movement of groundwater to varying degrees.

20.3   SITE INVESTIGATIONS

Groundwater impacts under the site were identified within the Upper Basalt Aquifer and Middle Basalt Aquifer. The free phase organic compounds and dissolved phase organic and inorganic components were observed in the Upper Basalt Aquifer, while only dissolved phase organic and inorganic components were detected in the Middle Basalt Aquifer. No contamination was detected within the Lower Basalt Aquifer. Detailed groundwater investigation of the Upper Basalt Aquifer and Middle Basalt Aquifer, therefore, were conducted. Emphasis was placed on the better characterisation of the groundwater flow system and vertical migration of the contaminants of interest within the Upper Basalt Aquifer considering that sources of contamination were already identified to be within this aquifer.

The groundwater investigations included:

- installation of 250 monitoring wells of which about 20% were continuously cored
- hydraulic testing,

- installation of passive flow meters to assess contaminant stratification within the wells and to estimate Darcy velocity,
- assessment of the NAPL presence using 'FLUT liners',
- NAPL sampling and characterisation,
- groundwater gauging and sampling (a number of monitoring events were undertaken),
- soil vapor investigation including installation of permanent and temporary vapour probes using gastec technology.

The drilling programme included drilling of about 50 continuously cored boreholes using predominantly diamond core drilling technique. The sonic drilling technique was also implemented at two locations but the quality of the core for the rock assessment was inferior to that obtained by diamond drilling. The vibrations generated by the sonic rig interfered with integrity of the *insitu* fracturing network with a tendency to enlarge and break the core recovered. The objective of coring was to enable visual inspection of the subsurface conditions, better characterisation of rock properties, fracture distribution, position and properties of palaeosols and intra-clays and to assist with installation depths and elevations of the screen intervals in the monitoring wells. Decisions on the monitoring well locations were also guided by the results from the soil vapour investigations.

In general, groundwater monitoring wells were completed with a 3 m long screen interval. A number of wells with a shorter screen targeting particular zones of the aquifers (horizontal fractured zone and massive columnar basalt sub-aquifers), multiple screened wells and paired wells were installed to further develop conceptual site understanding, i.e., further investigate the distribution of saturated zones, vertical hydraulic head distribution and vertical distribution of contaminants.

Hydraulic testing was undertaken at a number of locations. This included two pumping tests and about 20 slug tests. The slug tests were undertaken in the open boreholes and involved testing of selected short intervals within individual boreholes. The test intervals (length and level) were determined based on the visual inspection of the cores and were generally between 1 m and 1.5 m long. The aim of the tests was to assess hydraulic properties of specific aquifer zones (i.e., fractured zone, massive columnar). The pumping tests targeted the upper Middle Basalt Aquifer zone due to thicker and continuous aquifer saturation zone and overall intend to focus remediation efforts within this zone. The pumping tests were, in general, of a short duration, 9 hours and 3 hours, due to issues with the off-site disposal of highly contaminated groundwater.

The results of slug tests indicated hydraulic conductivities of the highly fractured zones to be range from about 5E-04 m/s to 2E-03 m/s, moderately fractured to fractured zones from 1E-07 m/s to 5E-05 m/s and massive columnar zones from less that 1E-10 m/s to 1E-09 m/s. The results of the pumping test indicated hydraulic conductivity of the upper Middle Basalt Aquifer zone to be in order of about 1E-06 m/s.

Additional to the site groundwater investigations, the rock outcrops of the Upper Basalt Aquifer in a nearby quarry and the Middle Basalt Aquifer in a nearby creek valley were inspected to better understand the rock profile and distribution of fracturing system.

20.3.1   Upper Basalt Aquifer

Three main zones were identified within the Upper Basalt Aquifer. Baseed on the visual inspections of the cores, it was indicated that typically the upper 5 m to 8 m of the unit was highly fractured and vesicular, typical of the fractured and vesicular basalt (Figure 20.5). The fractured and vesicular basalt zone was underlain by a 2 m to 3 m thick sub-horizontal fracture zone (horizontal fractured zone) while the lower zone of the aquifer was represented by a 2 m to 5 m thick massive columnar basalt. A very thin zone, generally less than 20 cm thick, characterised by very large vesicles (up to a few centimetres) and cavities was typically encountered at the base of the massive columnar basalt and above the palaeosol that separates the Upper Basalt Aquifer from the Middle Basalt Aquifer.

The groundwater level data indicated localised areas of groundwater saturation within the Upper Basalt Aquifer. The horizontal fractured zone and massive columnar basalt zones were indicated to behave as distinct sub-aquifer units (Figure 20.6).

Due to the closely spaced and interconnected fractures, the horizontal fractured zone was inferred to behave as a single system and, on a site scale, likely to behave similarly to a porous medium aquifer. The horizontal fractured zone sub-aquifer responds to rainfall by shrinking and extending, both vertically and laterally, indicating a low downwards seepage rate to the massive columnar basalt sub-aquifer. The massive columnar basalt sub-aquifer, therefore, was inferred to be a partial barrier to downward groundwater flow from the horizontal fractured zone sub-aquifer. The rate of groundwater seepage through the massive columnar basalt sub-aquifer along with the surface water infiltration rate was considered to be the main factor

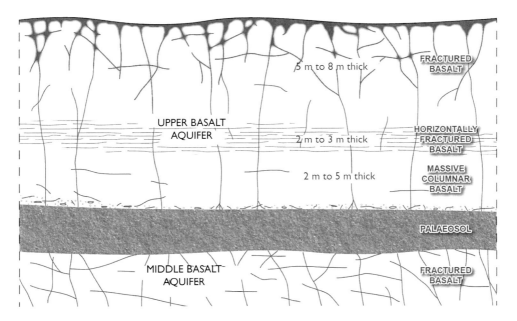

Figure 20.5 Distribution of main zones within the Upper Basalt Aquifer. (*See colour plate section, Plate 44*).

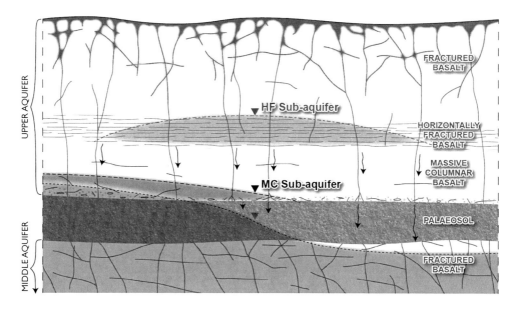

Figure 20.6 Sub-aquifers within the Upper Basalt Aquifer and distribution of groundwater saturation zones. HF – Horizontal fractured zone, MC – Massive columnar basalt). (*See colour plate section, Plate 45*).

controlling the lateral extent of the saturation within the horizontal fractured zone sub-aquifer.

The massive columnar basalt sub-aquifer was inferred to behave as a dual porosity medium with groundwater within this sub-aquifer moving predominantly through widely spaced columnar fractures. Many of the groundwater wells installed within the massive columnar basalt sub-aquifer were dry, although the overlaying horizontal fracture zone sub-aquifer contained groundwater. This sugested a very low permeability of the compact rock mass as it was indicated by results of slug tests. The areas of groundwater saturation within the massive columnar basalt sub-aquifer were relatively small and generally to coincide with the areas where the hydraulic heads within the Middle Basalt Aquifer were above the top of the palaeosol, i.e. the Middle Basalt Aquifer piezometric surface occurs within the massive columnar basalt sub-aquifer zone (see Figure 20.6). In the areas where the hydraulic heads within the Middle Basalt Aquifer were below the palaeosol, the massive columnar basalt sub-aquifers was always found to be dry. It was inferred that the saturation within the massive columnar basalt sub-aquifer is controlled predominantly by the Middle Basalt Aquifer groundwater levels, which preclude or allow for vertical migration of groundwater, rather than the degree of separation provided by the underlying palaeosol. The saturation of the extremely to completely weathered basalt sub-aquifer was not, therefore, directly related to saturation of the horizontal fractured zone sub-aquifer. This suggested that vertical hydraulic conductivity of the massive columnar basalt sub-aquifer is lower than the vertical hydraulic conductivity of the paleosol.

Downwards movement of contaminants and NAPL was inferred to occur through the upper fractured and vesicular basalt zone downwards to the horizontal fractured zone sub-aquifer and then further downwards through the widely spaced columnar fractures of the massive columnar basalt basalt zone (thick lines in Figure 20.7). Due to the low permeability of these fractures, the bulk of the contaminant mass appears to remain trapped within the horizontal fractured zone sub-aquifer with only a small volume moving downwards through the massive columnar basalt sub-aquifer. This suggests that the remediation of the Upper Basalt Aquifer would require significant effort at a high cost with potentially only a small portion of the contaminant mass accessible for removal.

Additionally, the 40-year period since the active release and the nature of the NAPL observed suggested that the NAPL is of a limited mobility and to occur, generally, at residual saturation. It was considered that this is likely to limit the practicability of the use of aggressive remediation to reduce the contaminant mass at the source.

20.3.2 Middle Basalt Aquifer

The Middle Basalt Aquifer is an extensive regional aquifer. The aquifer conditions range from unconfined with the water table within the aquifer to confined conditions with the aquifer hydraulic head within or above the palaeosol. The thickness of the aquifer generally ranges from about 15 m to 30 m. The groundwater flow within this aquifer is governed predominantly by the regional flow conditions.

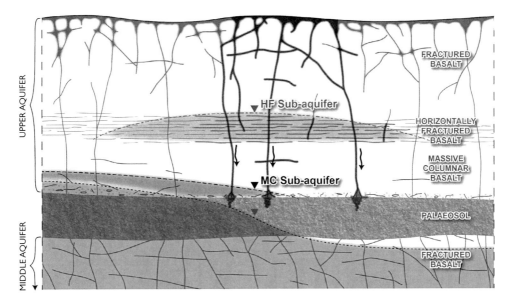

Figure 20.7 Downward movement of contaminants (indicated in thick lines) HF – Horizontal fractured zone, MC – Massive columnar basalt. (*See colour plate section, Plate 46*).

No source of contamination was identified within this aquifer and the contamination within the Upper Basalt Aquifer is considered to be the source of the Middle Basalt Aquifer pollution. The groundwater plume identified within the Middle Basalt Aquifer comprises inorganic and dissolved organic components only, i.e., no free phase contamination was observed.

This aquifer provides the main pathway for contaminat groundwater to reach the receptors of concern (aquatic eco-system) and this is the aquifer most likely to be targeted for groundwater resources. Preventing further migration of the groundwater plume through the Middle Basalt Aquifer, therefore, was considered to be a priority for protection of the beneficial uses of groundwater. Due to the saturated thickness (up to 30 m) of the aquifer and inferred hydraulic properties (generally in order of IE–06 m/s), remediation of this aquifer was also infered to be more practical and with a higher chance for success than remediation of the Upper Basalt Aquifer.

20.4 CONCLUSIONS

For the purpose of environment protection, a conceptual site model was developed for one of the industrial sites in Melbourne. The objective of the conceptual site model was to understand groundwater and contaminant movement through a basalt rock aquifer underlying the site.

The internal structural characteristics of basalt rock are typically complex due to different processes involved during the lava flow(s), cooling period and the weathering of the rock. The groundwater flow and lateral movement of contamination are governed by internal structural features that could limit feasibility of site remedial options. The main structural features identified under the site that was subject of the conceptual site model were horizontal fractured basalt, massive columnar basalt, fractured and vesicular basalt and extremely to completely weathered basalt. Overall, the horizontal fractured zones were identified as highly conductive zones with predominantly horizontal flow, the massive columnar basalt as barriers with limited vertical flow, fractured and vesicular basalt as high to medium conductive zones with no preferential flow direction and extremely to completely weathered basalt as low conductive zones (hydraulic barriers).

Three main aquifers within the basalt unit were distinguished under the site. The Upper Basalt Aquifer and Middle Basalt Aquifer were subject of intense investigation. The results of investigation indicated that the horizontal fractured zone and massive columnar basalt zone within the Upper Basalt Aquifer behave as distinctive sub-aquifers. The variation in the vertical hydraulic conductivity between these two sub-aquifers results in a perched sub-aquifer within the horizontal fractured zone and restriction of the contaminant movement downwards to the massive columnar basalt sub-aquifer and deeper towards the Middle Basalt Aquifer. This resulted in the bulk of the contaminant mass being trapped within the horizontal fractured zone, which is characterised by a relatively small thickness and limited areas of groundwater saturation. Groundwater saturation within the underlying massive columnar basalt sub-aquifer was also of a limited extent and controlled predominantly by the Middle Basalt Aquifer groundwater levels. This, along with the inferred hydraulic properties of the sub-aquifers and localised extent of imobile NAPL, suggested that the remediation of the Upper Basalt Aquifer would require significant effort at a high cost with potentially only a small portion of the mass accessible for removal. The use of aggressive remediation to reduce the contaminant mass at the source, therefore, was infered not to br feasible.

The results of investigation suggested that the remediation efforts should be concentrated onto the Middle Basalt Aquifer. This aquifer provides the main pathway for contaminat groundwater to rech the receptors of concern (aquatic eco-system) and it is the aquifer most likely to be targeted for groundwater resources. Additionally, it was assessed that the clean-up of this aquifer would be more practical and have a higher chance for success.

An overall conclusion of this study was that a comprehensive conceptual model is an essential tool to assist with decision making on remediation and management of contaminated groundwater systems.

ACKNOWLEDGMENT

The authors would like to thank Mr Richard Heath of Golder Associates Pty Ltd for drafting support.

REFERENCES

Birch W.D. (ed.) (2003) *Geology of Victoria*. Geological Society of Australia (Victoria Division), Special Publication 23. 2003.

EPA (1970) (Vic): Environment Protection Act 1970, No 8056 of 1970.

Hare A.G., Cas R.A.F. (2005) Volcanology and evolution of the Werribee Plain Intraplate Basaltic Lava Flow-field, Newer Volcanics Province, Southeast Australia, *Australian Journal of Earth Science* 52, 59–78.

Hare A.G., Cas R.A.F., Musgrave R., Phillips D. (2005) Magnetic and Chemical Stratigraphy for the Werribee Plain Lava Flow-field, Newer Volcanics Province, Southeast Australia, *Australian Journal of Earth Science* 52, 41–57.

Subject index

Author index

Place index

Colour plates

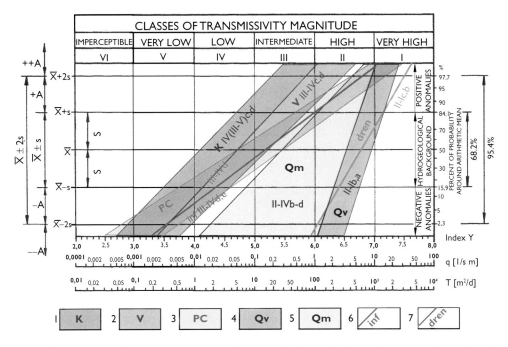

Plate I Distribution of transmissivity values of selected hydrogeological environments in the Czech Republic expressed as cumulative relative frequencies. (*See Fig. 1.1*).

Index Y = index of transmissivity Y = log (10^6 q), q = specific capacity [l/s m], T = coefficient of transmissivity [m^2/d]; $\overline{x}$ = arithmetic mean of statistical samples, s = standard deviation, ++A, +A, –A, – –A = fields of positive and negative anomalies (+A, –A) and extreme anomalies (++A, – –A) outside the interval $\overline{x} \pm$ s of prevailing transmissivity (= hydrogeological background).

Classes of transmissivity magnitude and variation after Krásný (1993a).

Prevailing transmissivity values in different hydrogeological environments expressed as fields or lines of cumulative relative frequencies determined by aquifer tests in hydrogeological boreholes. Represented ranges of transmissivity encompass the greater part of hydrogeological environments in Czech between the lowest values less than 1 m^2/d (field K) and the highest values of more than 1,000 m^2/d (field Qv):

 1 – near-surface aquifer of most of "hard rocks" (hydrogeological massif – K) except of environments represented in the field V (2);
 2 – crystalline limestones and other rocks of hydrogeological massifs with relatively higher prevailing transmissivity – V;
 3 – Permo-Carboniferous basins PC – data from boreholes up to the depths of several tens of meters;
 4 – most of Quaternary fluvial deposits along the main water courses – Qv (rivers Labe and Morava etc.); adjacent field Qm (5) represents Quaternary deposits along smaller water courses and of accumulations at higher terrace benches;
 6, 7 – lines reflect possible considerable differences in transmissivity of sandstones of the Bohemian Cretaceous basin as determined in boreholes located at landform elevations and slopes (hydrogeologically groundwater recharge zones – inf) and in valleys (zones of groundwater discharge – dren).

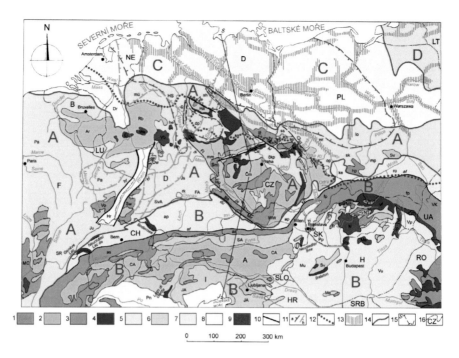

Plate 2 Geological and hydrogeological position of the Czech Republic in the Central Europe. (*See Fig. 1.6*).

Defined hydrogeological megaprovinces: A – Megaprovince of European Variscan units and their platform cover; B – Alpine-Carpathian Megaprovince; C – Megaprovince of Central European Lowland; D – Megaprovince of East-European and North-European (Fennosarmatian) Platform.

1–9 – Main types of hydrogeological environment:

1–3 – *hydrogeological massif:*
  1 – igneous rocks of different age except for young Tertiary and Quaternary volcanic rocks,
  2 – metamorphic rocks and intensively folded preorogenic sediments, mosty of Precambrian and Paleozoic age,
  3 – Mesozoic and Tertiary intensively folded, usually non-carbonate sediments of Alpides (e.g. Outer Carpathians Flysch belt that mostly has a character of hydrogeological massif).

4–8 – *hydrogeological basins:*
  4 – mostly less intensively folded post-orogenic sediments of different age: Upper Paleozoic of the Variscid platform cover, Intracarpathian Paleogene;
  5 – Mesozoic and Tertiary unmetamorphosed and slightly metamorphosed sediments of Variscid platform cover,
  6 – Mesozoic carbonate rocks of Alpides, often forming hydrogeologic basins;
  7 – Neogene sediments of Alpine and Carpathian Foredeeps („Molasse") and intramountain basins of Alpides;
  8 – Quaternary sediments of the Middleeuropean Lowland, of the Upper-Rhein graben and of the Po-Lowland with their Tertiary and Mesozoic basements;

  9 – *young volcanic rocks;*
10 – *important structural elements, faults and fault zones;*
11 – *regions with frequent occurrences of salt deposits and domes* in hydrogeological megaprovinces A and C: a) mostly of Permian age, b) mostly of Triassic, partly also of Jurassic age;
12 – *southernmost limit of the Pleistocene continental ice-sheet;*
13 – *approximate extension of the main overdeepened valleys of Quaternary glaciofluvial origin* („Urstromtäller") in the Middle-Europian Lowland, partially extended to the south into the Megaprovince of European Variscan units;
14 – *boundaries of hydrogeological megaprovinces;*
15 – *line of the section represented in Fig. 1.7.*
16 – *state boundaries and symbols of states.*

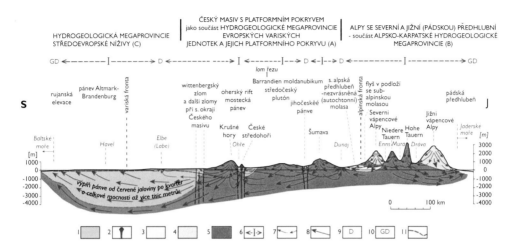

Plate 3 Schematic geological-hydrogeological section through the Central Europe between the Baltic and the Adriatic Seas. (*See Fig. 1.7*).

The section in the north-south direction represents the main geological units and defined hydrogeologic megaprovinces A–C, character of hydrogeological environments, main features of vertical hydrodynamic and hydrochemical zonalities and position of the Bohemian massif in the framework of regional and global groundwater flow between the Baltic and Adriatic seas with the most important zones of groundwater recharge and discharge. Some units are projected into the section. Relatively small thickness of shallow groundwaters with small TDS contents compared with deeper saline groundwaters (brines) can be observed.

A–C: Hydrogeological megaprovinces: A – Megaprovince of European Variscan units and their platform cover, B – Alpine-Carpathian Megaprovince, C – Megaprovince of Central European Lowland.

  1 – Cenozoic, Mesozoic and Upper Paleozoic deposits of Variscan platform cover and of other platform units occurring more to the north,
  2 – Tertiary and Quaternary volcanic rocks at the Ohře (Eger) rift,
  3 – sediments of Alpine Foredeep (Molasse),
  4 – Alpine carbonate rocks,
  5 – igneous, metamorphic and diagenetically lithificated and/or intensively folded Variscan rocks, rocks of other platform units, of Alpine core and of Flysch Belt ("hard rocks")
  6 – zones of prevailing groundwater recharge, with only local groundwater discharge,
  7 – main inferred directions of local to regional groundwater flow,
  8 – main inferred directions of regional to continental/global groundwater flow, somewhere directed out of the line of the section or even with stagnant groundwater,
  9 – important zones of groundwater discharge,
  10 – zones of discharge of continental groundwater flow (Baltic and Adriatic seas),
  11 – inferrred boundary between groundwaters with small TDS contents and deep saline groundwaters (brines).

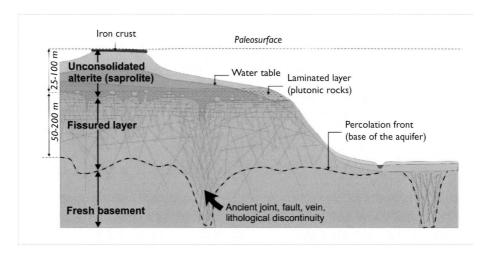

Plate 4 Conceptual model of a partly eroded paleo-weathering profile on hard rocks. (*see Fig. 2.2*).

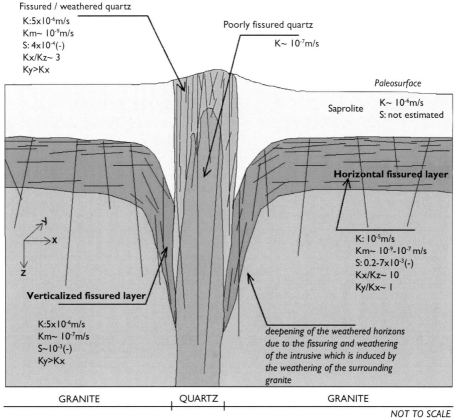

Plate 5 Conceptual hydrodynamic model of a vertical discontinuity in hard rock. (*see Fig. 2.7*).

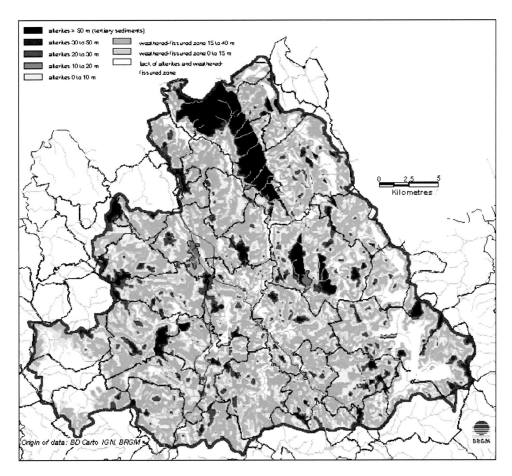

Plate 6 Geological map of the weathering cover on the Truyère, Lozère, France watershed: thickness of the saprolite (increasing thickness from yellow to red and black) and the fissured layer (increasing thickness from blue to green), white: weathering profile totally eroded. (see Fig. 2.14a).

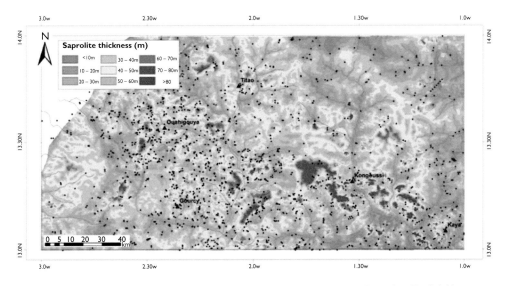

Plate 7 Saprolite-thickness over a ~250 × 100 km area in Burkina Faso. (see Fig. 2.14c).

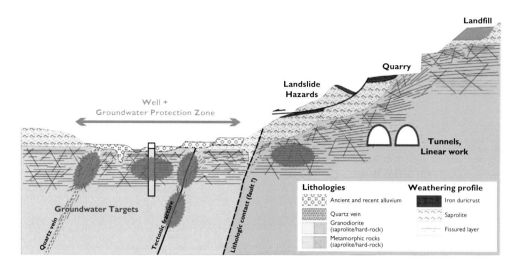

Plate 8 Various uses of a high resolution mapping of the weathering cover at the local scale (see Fig 2.6 for the legends). (see *Fig. 2.15*).

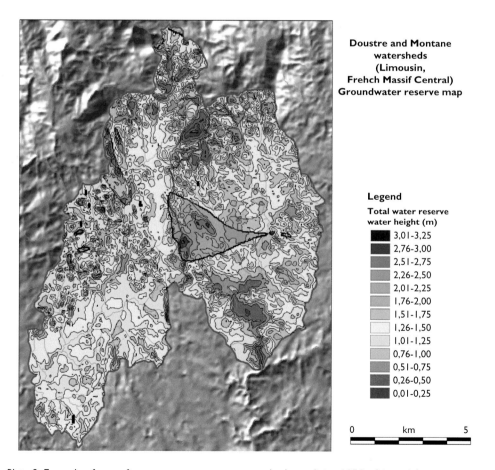

Plate 9 Example of map of water reserve on two watersheds totalizing 115 km² in mainly granitic context, Corrèze, French Massif Central. The triangular yellow, green to blue area in the middle of the figure corresponds to a biotite granite with higher groundwater reserves. The other lithological contours do not generate an enough groundwater reserve contrast to be easily identified on this map. (see *Fig. 2.19*).

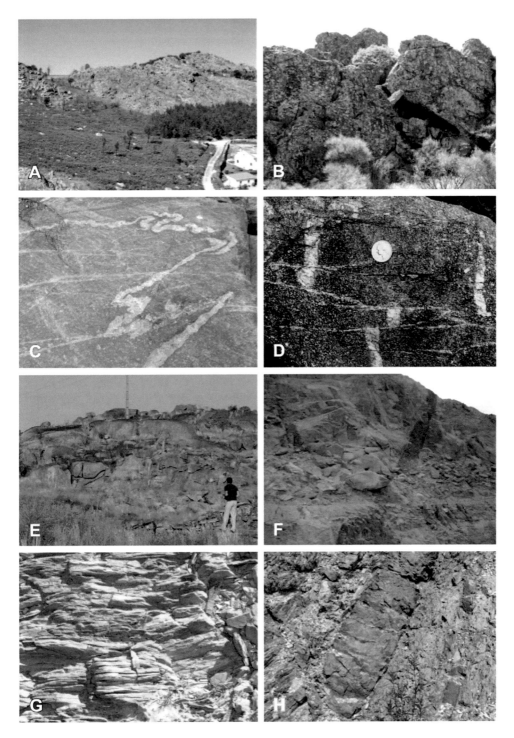

Plate 10 A-B: Quartzites, Castelo de Vide, Portugal; C-D: Migmatites, Helsinki, Finland; E: Granites, Portalegre, Portugal; F: Metamorphic rocks cut by a volcanic dyke, Aqaba, Jordan; G-H: Shales, Mértola, Portugal, with the intercalation of a quartzite layer in case H. (*see Fig. 4.5*).

Plate 11 Excavation for geotechnical purposes, important for the selecting drilling sites (the objective was to create an industrial area in this area and both studies were occurring at the same time). Figures B, C and D show more favourable features for water than Figure A (fractures filled with highly fractured quartz veins). (*see Fig. 4.6*).

Plate 12 Area in metamorphic rocks (shales, schists, greywackes) in Granja, Mourão (Portugal). With such quantity, position and fracturing of quartz veins, it would be a good place for drilling. In all the other visible cuts in this village the quartz veins are absent. (*see Fig. 4.7*).

Plate 13 Example of a well in metamorphic rocks (shales, schists, greywackes) in central Portugal (area of Nisa). Quartz veins and fractures cut the rock in this specific position, where there is a productive well (yield of 8 l/s) which supplied a population of 70 inhabitants (the well is no more in use for water supply). (see *fig. 4.8*).

Plate 14 Drilling in hard rocks (A). Steps during the construction of the well: drill cuttings taken every 3 m (B), a section showing material which indicates a probable fracture (C), close analysis of the cuttings directly picked from the well during drilling (D), or after washing the cuttings with water to clean away dust (E), and observing the flow coming from the well during drilling, using compressed air directed to the bottom of the well (F). (see *Fig. 4.12*).

Plate 15 A granitoid outcrop showing the discontinuity network of the fissured layer. (see *Fig. 5.3*).

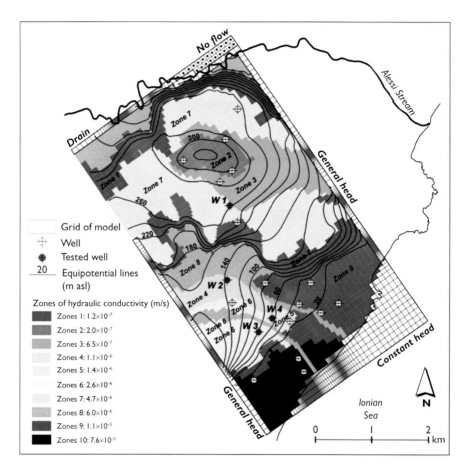

Plate 16 Model area, boundary conditions and results of the steady-state calibrated model showing the hydraulic conductivity zones and potentiometric contour map. (*see Fig. 5.6*).

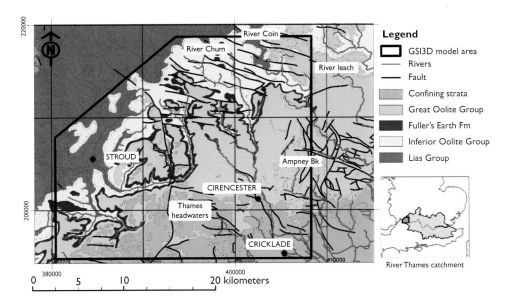

Plate 17 Location of the study area showing extent of the Cotswolds 3D geological model and 250k mapped geology with location of faults and drainage network. The position of the River Thames catchment (grey) and study area (green) is indicated on the smaller location map. Contains Ordnance Survey data © Crown Copyright. (see Fig. 6.1).

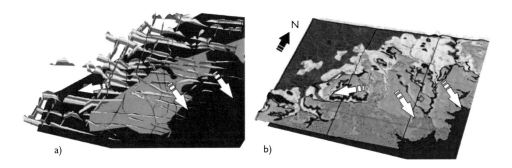

Plate 18 Outputs from the GSI3D model showing a) the geological cross-sections with maximum groundwater level surfaces for the Inferior Oolite aquifer (dark blue) and the Great Oolite aquifer (light blue), and; b) the mapped geology at surface (1:250 000 scale) draped over the digital terrain model along with the groundwater level surface to highlight where artesian groundwater levels exist in the Inferior Oolite aquifer under maximum groundwater level conditions (dark blue). White arrows highlight the regional groundwater flow direction; note the catchment divide between the River Thames to the east and the River Severn to the west. Contains Ordnance Survey data © Crown Copyright. (see Fig. 6.2).

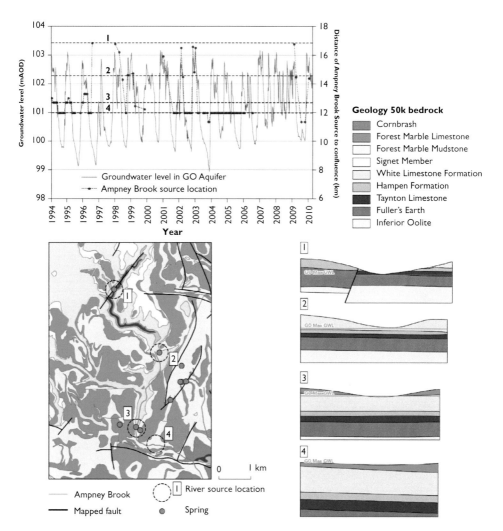

Plate 19 Stream head migration for the Ampney Brook during 1994–2010 is shown along with ground-water level variations within a Great Oolite observation well. Four locations to which the Ampney Brook migrates to are identified. The surface and sub-surface geology at each of these locations is shown. Groundwater level and river source location data © Environment Agency copyright and/or database rights 2012. All rights reserved. (*see Fig. 6.3*).

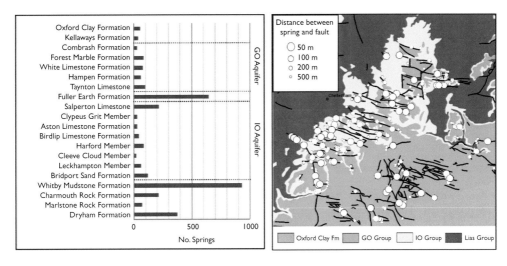

Plate 20 The number of springs associated with each of the geological unit mapped at surface (1:50 000), and; the extent to which springs are associated with mapped fault. (see Fig. 6.4).

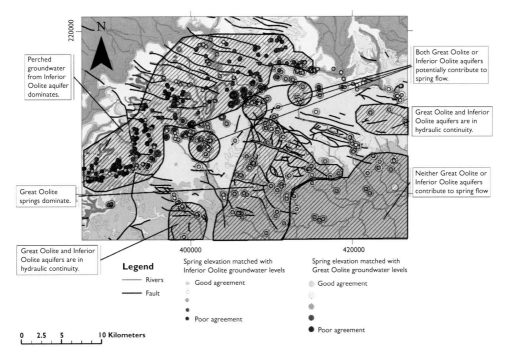

Plate 21 Hydrogeological domains for the Cotswold aquifers derived using the relationship between spring elevation and groundwater levels within the underlying Great Oolite and Inferior Oolite aquifer as an indicator of groundwater discharge processes and aquifer inter-connectivity. Contains Ordnance Survey data © Crown Copyright. (see Fig. 6.6).

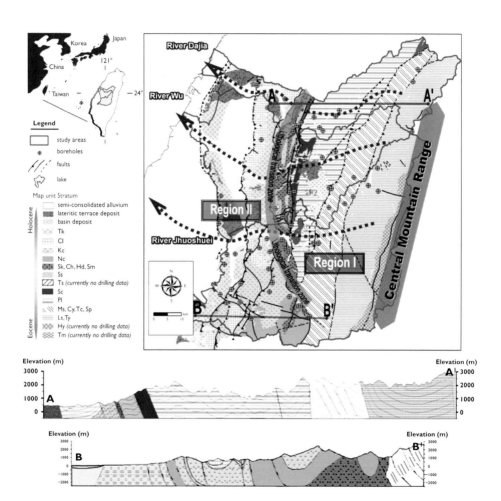

Plate 22 Geologic setting (up) and two geological cross sections (east-west) (down) of the study area. (see *Fig. 7.1*).

Plate 23 Photos taken during packer testing, (A) Equipment calibration, (B) Rig transfer, (C) Data recording and analysis. (*see Fig. 7.3*).

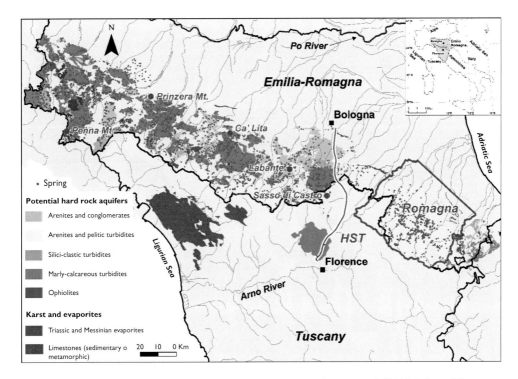

Plate 24 Aquifers of Northern Apennines along with location of test-sites. HST: High Speed Train tunnels path. (*see Fig. 8.1*).

Plate 25 Outcrop of turbiditic hard rock aquifer: Premilcuore member of Marnoso-Arenacea Formation (Upper Bidente valley, Romagna). (*see Fig. 8.2*).

Plate 26 Ophiolitic hard rock aquifer: peridotite block of Ragola Mt. (near Prinzera Mt. in Fig. 1). (*see Fig. 8.3*).

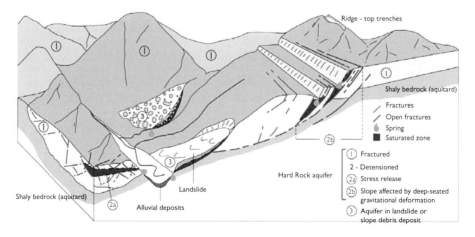

Plate 27 Hydrogeological sketch map of groundwater flow systems (saturated zone) in an ophiolitic block involved by fracturing and gravitational processes. (*see Fig. 8.4*).

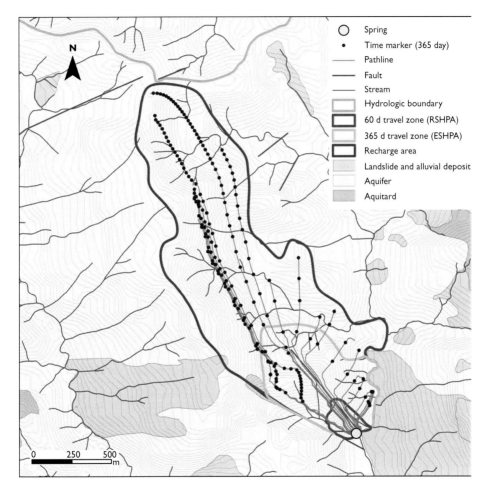

Plate 28 Recharge area and travel time zones for *Brenziga* spring calculated from advective transport simulation. (see *Fig. 8.10*).

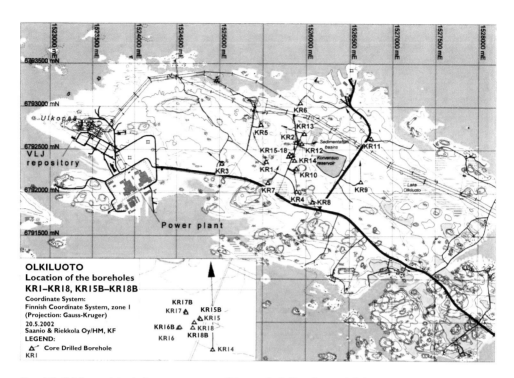

Plate 29 Olkiluoto Island. (Image courtesy of Posiva Oy.) (See Figure 9.1 for a complete explanation).

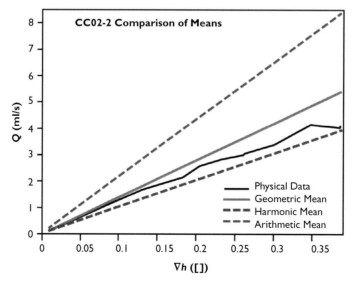

Plate 30 Comparison of means as a function of hydraulic gradient (∇h) for flow tests on sample CC02-2). Hydraulic apertures are between the geometric and harmonic means. The geometric mean is the best approximation at low gradients. (*see Fig. 10.3*).

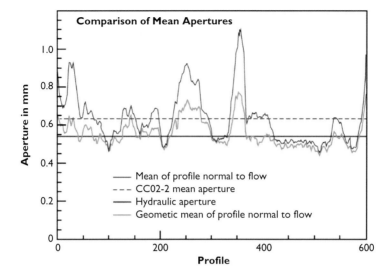

Plate 31 Arithmetic and geometric means of scanlines normal to flow through sample CC02-2 with actual hydraulic aperture and total fracture arithmetic mean aperture. (*see Fig. 10.4*).

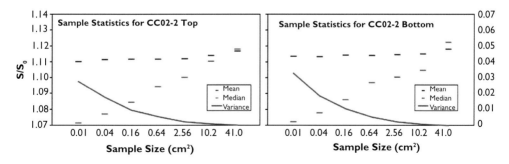

Plate 32 Roughness mean, median, and variance as a function of area scanned for the top and bottom surfaces of sample CC02-2. (*see Fig. 10.5*).

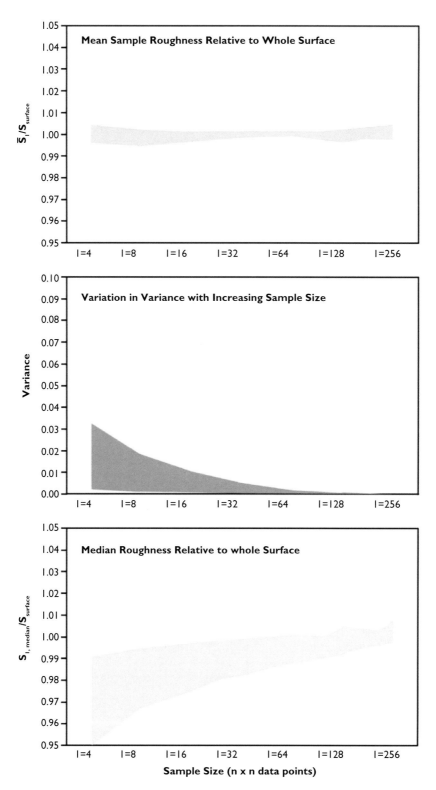

Plate 33 Range of S/S_0 statistics for surfaces (Table 1) as a function of sample size. (a) Arithmetic mean; (b) Variance of sample means; (c) Median value of means. (*see Fig. 10.6*).

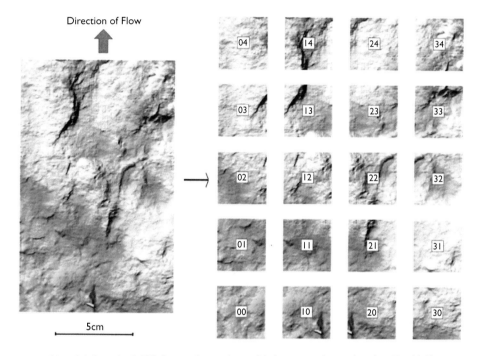

Plate 34 Sample CC02-2 transformed into 20 discrete subsamples. (*see Fig. 10.7*).

(a)

(b)

(c)

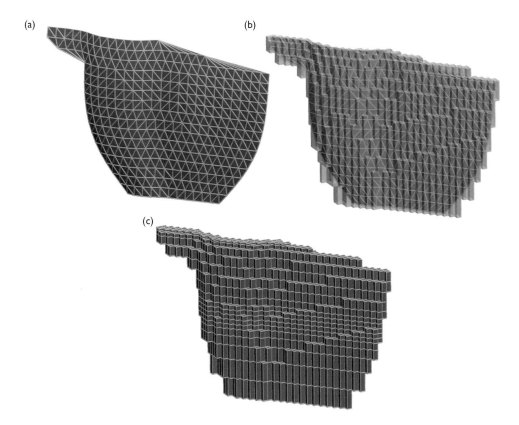

Plate 35 (a) View of a single FZNM fracture zone; (b) View of the single fracture zone and the model grid blocks that are intersected by the fracture zone; and (c) View of the model grid orthogonal faces that best represent the fracture zone. (*see Fig. 12.2*).

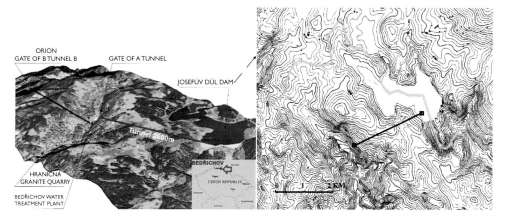

Plate 36 The site location and tunnel position. (left), the digital terrain model (DTM) with the problem domain and the tunnel position (right). (*see Fig. 15.1*).

Plate 37 Left: a part of the tunnel excavated by boring machine with wet and dry strips visible; Middle: a part by drill-and-blast method, shotcrete covered, with more visible inflow; Right: an uncovered shaft to the collection canal (conductivity measurement during the dilution experiment). (see *Fig. 15.2*).

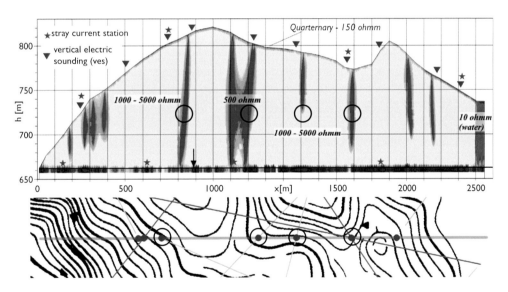

Plate 38 (Upper) Electrical resistivity cross section along the tunnel. (Lower) Tectonic lineaments in the map. Circles denote the corresponding locations from both sources, represented in the model. (see *Fig. 15.5*).

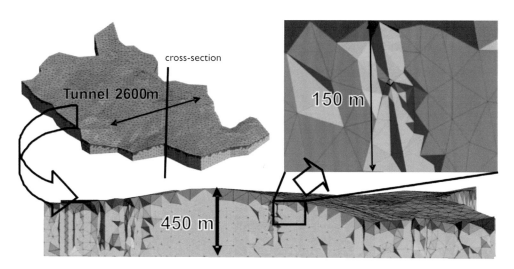

Plate 39 Discretisation of the model with tetrahedra in 3D – (upper left) the full view, (lower) a vertical cross-section, and (upper right) a detail around the tunnel. The colour corresponds to the subdomains of different hydraulic conductivity. (*see Fig. 15.7*).

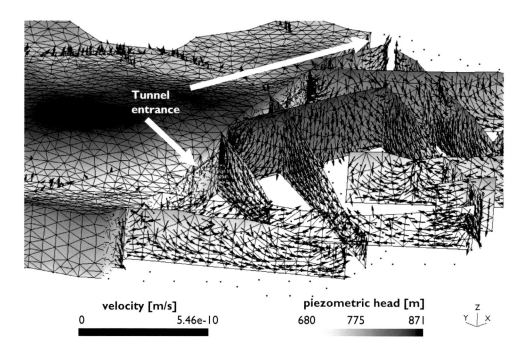

Plate 40 Piezometric head and velocity field visualisation – the section along the tunnel separates a part with 3D domain visible and a part with only 2D fault structures visible. (*see Fig. 15.10*).

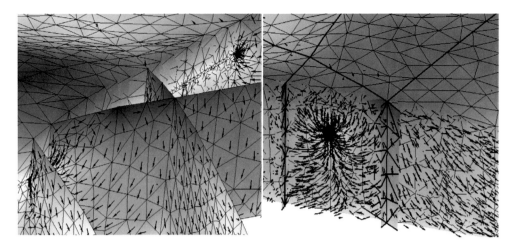

Plate 41 Details of the flow field: in the left the flow along the fault structures, in the right the flow in the 3D continuum separated by the fault structures between the tunnel-controlled and the topography-controlled parts. The scale is the same as in Figure 15.10. (*see Fig. 15.11*).

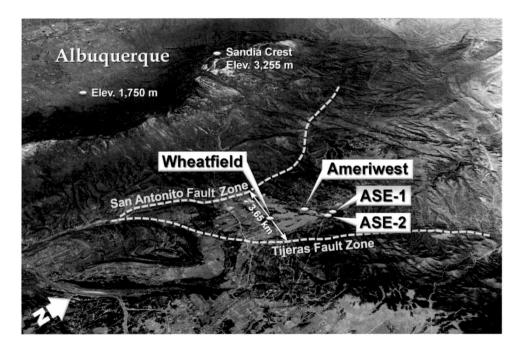

Plate 42 Site location. (*see Fig. 18.1*).

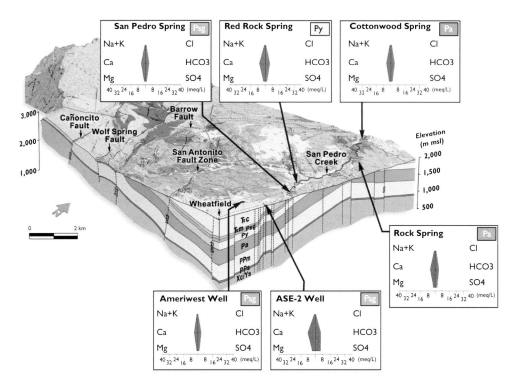

Plate 43 Hydrogeological block diagram. (*see Fig. 18.3*).

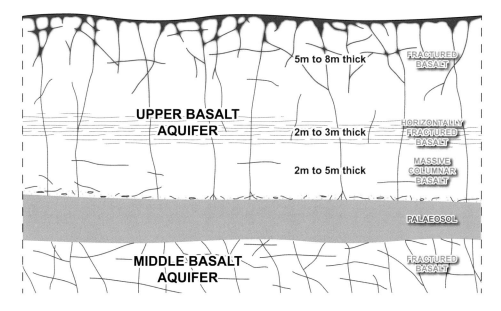

Plate 44 Distribution of main zones within the Upper Basalt Aquifer. (*see Fig. 20.5*).

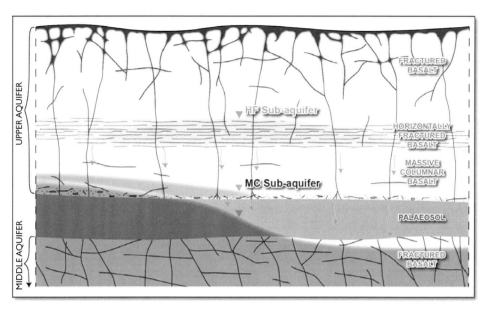

Plate 45 Sub-aquifers within the Upper Basalt Aquifer and distribution of groundwater saturation zones. HF – Horizontal fractured zone, MC – Massive columnar basalt. (*see Fig. 20.6*).

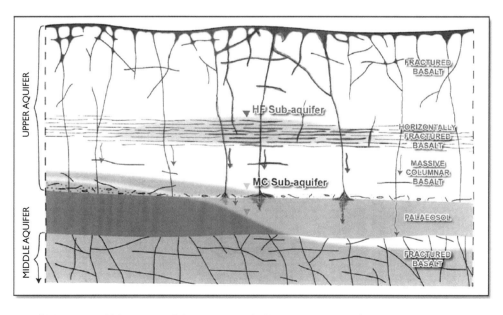

Plate 46 Downward Movement of Contaminants (NAPL indicated in red) HF – Horizontal fractured zone, MC – Massive columnar basalt. (*see Fig. 20.7*).

Series IAH-selected papers

14. Advances in Subsurface Pollution of Porous Media: Indicators, Processes
    and Modelling
    Edited by: Lucila Candela, Iñaki Vadillo and Francisco Javier Elorza
    2008, ISBN Hb: 978-0-415-47690-4

15. Groundwater Governance in the Indo-Gangetic and Yellow River Basins –
    Realities and Challenges
    Edited by: Aditi Mukherji, Karen G. Villholth, Bharat R. Sharma
            and Jinxia Wang
    2009, ISBN Hb: 978-0-415-46580-9

16. Groundwater Response to Changing Climate
    Edited by: Makoto Taniguchi and Ian P. Holman
    2010, ISBN Hb: 978-0-415-54493-1

17. Groundwater Quality Sustainability
    Edited by: Piotr Maloszewski, Stanisław Witczak and Grzegorz Malina
    2013, ISBN Hb: 978-0-415-69841-2

18. Groundwater and Ecosystems
    Edited by: Luís Ribeiro, Tibor Y. Stigter, António Chambel, M. Teresa Condesso
            de Melo, José Paulo Monteiro and Albino Medeiros
    2013, ISBN Hb: 978-1-138-00033-9

19. Assessing and Managing Groundwater in Different Environments
    Edited by: Jude Cobbing, Shafick Adams, Ingrid Dennis and Kornelius Riemann
    2013, ISBN Hb: 978-1-138-00100-8

Printed and bound by CPI Group (UK) Ltd, Croydon, CR0 4YY

24/10/2024

01778309-0003